DISCRETE MATHEMATICS

DISCRETE MATHEMATICS

JOHN A. DOSSEY

ALBERT D. OTTO

LAWRENCE E. SPENCE

CHARLES VANDEN EYNDEN

Illinois State University

HarperCollins*Publishers*

To our wives Anne, Judy, Linda, and Joan.

Cover and chapter opener details: Martin, Kenneth, *Chance and Order 10 (Monastral Blue)*. 1972. By permission of the Tate Gallery and the estate of Kenneth Martin.

Library of Congress Cataloging-in-Publication Data

Discrete mathematics.

   Bibliography: p.
   Includes index.
   1. Mathematics—1961–   . 2. Electronic data
processing—Mathematics. I. Dossey, John A.
QA39.2.D57 1986   510   86-26210
ISBN 0-673-18191-X

3 4 5 6 – RRC – 91

PREFACE

Today an increasing proportion of the applications of mathematics involve discrete rather than continuous models. The main reason for this trend is the integration of the computer into more and more of modern society. This book is intended for a one-semester introductory course in discrete mathematics. It has a strong algorithmic emphasis that serves to unify the material. Algorithms are presented in English so that knowledge of a particular programming language is not required.

The choice of topics is based upon the recommendations of various professional organizations, including those of the MAA's Panel on Discrete Mathematics in the First Two Years and the CUPM's recommendations for the mathematical training of teachers. Although designed for a one-semester course, the book contains more material than can be covered in either one semester or two quarters. Consequently, instructors will have considerable freedom to choose topics tailored to the particular needs and interests of their students.

The sequence of chapters also allows considerable flexibility in teaching a course from this book. The following diagram shows the logical dependence of the chapters. The dashed lines indicate that only the initial sections of Chapter 3 are needed for Chapters 5 and 6. Although this book assumes only the familiarity with logic and proof ordinarily gained in high-school geometry, an appendix is provided for those who prefer a more formal treatment. If the appendix is covered, it may be taught at any time as an independent unit or in combination with Chapter 9.

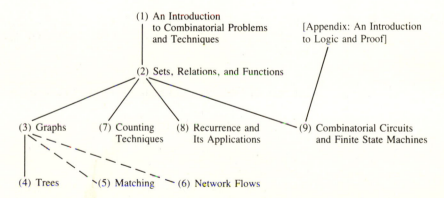

Although a course taught from this book requires few formal mathematical prerequisites, students are assumed to have the mathematical maturity ordinarily obtained by taking at least two years of high school mathematics, including problem-solving and algorithmic skills, and the ability to think abstractly.

Courses of various levels of sophistication can be taught from this book. For example, the topic of computational complexity is of great importance, and so attention is given to the complexity of many algorithms in this text. Yet it is a difficult topic, and the detail with which it is treated should correspond to the intended level of the course and the preparation of students. Our practice has been to include proofs of theorems when they are not too long or technical, and when they give insight into the ideas involved. Proofs of other theorems are left to the exercises.

The exercise sets in this book have also been designed for flexibility. Many straightforward computational and algorithmic exercises are included after each section. These give students hands-on practice with the concepts and algorithms of discrete mathematics and are especially important for students whose mathematical backgrounds are weak. Other exercises extend the material in the text or introduce new concepts not treated there. Blue exercise numbers indicate the more challenging problems. An instructor should choose those exercises appropriate to his or her course and students.

Chapters 1 and 2 are introductory in nature. Chapter 1, which should be covered fairly quickly, gives a sampling of the sort of discrete problems the course treats. Some questions are raised that will not be answered until later in the book. The chapter ends with a discussion of complexity that some instructors may want to omit or delay until students have had more experience with algorithms.

Chapter 2 reviews various basic topics, including sets, relations, functions, and mathematical induction. It can be taught more or less rapidly depending on the mathematical backgrounds of the students and the level of the course. It should be possible for students with good high-school mathematics backgrounds to be able to read much of Chapter 2 on their own.

The remaining chapters are, as the diagram above shows, independent except that Chapter 4 depends on Chapter 3, and the beginning concepts of Chapter 3 are needed in Chapters 5 and 6. A course emphasizing graph theory and its applications would cover most of Chapters 3–6, while a course with less graph theory would omit Chapters 5 and 6 and concentrate on Chapters 7–9. Two sample three-semester-hour courses along these lines are indicated below.

First Course		Second Course	
Chapter	*Hours*	*Chapter*	*Hours*
1 (skip 1.4)	3	1	4
2	6	2	6
3	6	3	6
4	6	4	7
7	8	5	6
8	5	6	4
Appendix	3	7	8
9	4		

Since both the proper place in the mathematics curriculum for discrete mathematics courses and the topics to include are still somewhat unsettled, we have tried to write a book that provides flexibility for the instructor. Our hope is that we have created a text which instructors in varied situations can teach from with both success and enjoyment.

We would like to thank the following mathematics professionals whose reviews guided the preparation of this text: Dorothee Blum, Virginia Commonwealth University; Richard Brualdi, University of Wisconsin, Madison; John L. Bryant, Florida State University; Richard Crittenden, Portland State University; Klaus Fischer, George Mason University; Dennis Grantham, East Texas State University; William R. Hare, Clemson University; Christopher Hee, Eastern Michigan University; Frederick Hoffman, Florida Atlantic University; Julian L. Hook, Florida International University; Carmelita Keyes, Broome Community College; Richard K. Molnar, Macalester College; Catherine Murphy, Purdue University, Calumet; Charles Nelson, University of Florida; Fred Schuurmann, Miami University; Karen Sharp, Charles S. Mott Community College; and Donovan H. Van Osdol, University of New Hampshire.

John A. Dossey
Albert D. Otto
Lawrence E. Spence
Charles Vanden Eynden

CONTENTS

1

AN INTRODUCTION TO COMBINATORIAL PROBLEMS AND TECHNIQUES

Combinatorial Analysis is an area of mathematics concerned with solving problems for which the number of possibilities is finite (though possibly quite large). These problems may be broken into three main categories: determining existence, counting, and optimization. Sometimes it is not clear whether a problem has a solution or not. This is a question of **existence.** In other cases solutions are known to exist, but we want to know how many there are. This is a **counting** problem. Or a solution may be desired that is ''best'' in some sense. This is an **optimization** problem. We will give a simple example of each type.

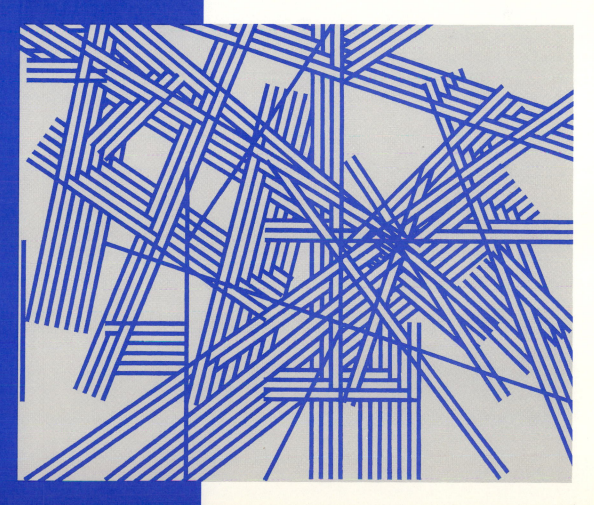

Determining Existence

Four married couples play mixed doubles tennis on two courts each Sunday night. They play for two hours, but switch partners and opponents after each half hour period. Does a schedule exist so that each man plays with and against each woman exactly once, and plays against each other man at least once and at most twice?

Counting

A six-person investment club decides to rotate the positions of president and treasurer each year. How many years can pass before they will have to repeat the same people in the same offices?

Optimization

An employer has three employees, Pat, Quentin, and Robin, who are paid $6, $7, and $8 per hour, respectively. She has 3 jobs to assign. The following table shows how long each employee requires to do each job.

	Pat	Quentin	Robin
Job 1	7.5 hr	6 hr	6.5 hr
Job 2	8 hr	8.5 hr	7 hr
Job 3	5 hr	6.5 hr	5.5 hr

How should she assign one job to each person to get the work done as cheaply as possible?

Often the solution we develop for a combinatorial problem will involve an **algorithm,** that is, an explicit step-by-step procedure for solving the problem. Many algorithms lend themselves well to implementation by a computer, and the importance of combinatorial mathematics has increased because of the wide use of these machines. However, even with a large computer, solving a combinatorial problem by simply running through all possible cases is often impossible. More sophisticated methods of attack are needed. In this chapter we will present more complicated examples of combinatorial problems and some analysis of how they might be solved.

1.1   The Time to Complete a Project

The Problem

A large department store is having a Fourth of July sale (which will actually start July 2), and plans to send out an 8-page advertisement for it. This advertisement must be mailed out at least 10 days before July 2 to be effective, but various tasks must be done and decisions made first. The managers of each department decide which items in stock to put on sale, and buyers decide what merchandise should be brought in from outside for the sale. Then a management committee decides which items to put in the advertisement and sets their sale prices.

The art department prepares pictures of the sale items, and a writer provides copy describing them. Then the final design of the advertisement, integrating words and pictures, is put together.

A mailing list for the advertisement is compiled from several sources. It will depend on the items put on sale. Then mailing labels are printed. After the advertisement itself is printed, labels are attached, and the finished product, sorted by zip code, is taken to the post office.

Of course, all these operations take time. Unfortunately, it is already June 2, so only 30 days are available for the whole operation, including delivery. There is some concern whether the advertisements can be gotten out in time, and so estimates are made for the number of days needed for each task, based on past experience. These times are listed in the table below.

Task	Time in days
choose items (department managers)	3
choose items (buyers)	2
choose and price items for ad	2
prepare art	4
prepare copy	3
design advertisement	2
compile mailing list	3
print labels	1
print advertisement	5
affix labels	2
deliver advertisements	10

If the time needed for all the jobs is added up, we get 37 days, which is more than is available. Some tasks can be done simultaneously, however. For example, the department managers and the buyers can be working on what they want to put on sale at the same time. On the other hand, many tasks cannot even be started until others are completed. For example, the mailing list cannot be compiled until it is decided exactly what items will be advertised.

In order to examine which jobs need be done before which other jobs, we label them A, B, . . ., K and list after each job any job which must immediately precede it.

Task	Preceding tasks
A choose items (department managers)	none
B choose items (buyers)	none
C choose and price items for ad	A, B
D prepare art	C
E prepare copy	C
F design advertisement	D, E
G compile mailing list	C
H print labels	G
I print advertisement	F
J affix labels	H, I
K deliver advertisements	J

For example, the letters A and B are listed after task C because the items to be advertised and their prices cannot be decided until the department managers and buyers decide what they want to put on sale. Likewise, the letter C is listed after task D because the art cannot be prepared until the items to be advertised are decided. Notice that tasks A and B must also precede the preparation of the art, but this information is omitted because it is implied by what is given. That is, since A and B must precede C, and since C must precede D, logically A and B must go before D also, so this need not be said explicitly.

Let us assume that workers are available to start on each task as soon as it is possible to do so. Even so, it is not clear whether the advertisement can be prepared in time, although we have all the relevant information. Here we have a problem of *existence*. Does a schedule exist that will allow the advertisement to be sent out in time for the sale?

Analysis

Sometimes a body of information can be understood more easily if it is presented in graphical form. Let us represent each task by a point, and draw an arrow from one point to another if the task represented by the first point must immediately precede the task represented by the second. For example, tasks A and B must precede task C, and C must precede D, so we start as in Figure 1.1.

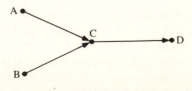

FIGURE 1.1

Continuing in this way produces the diagram of Figure 1.2.

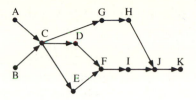

FIGURE 1.2

If we agree that all arrows go from left to right, we can omit the arrowheads, which we will do from now on.

This picture makes the whole project seem somewhat more comprehensible, but we must still take into account the time needed to do each task. Let us introduce these times into our diagram in Figure 1.3 by replacing each point with a circle containing the time in days needed for the corresponding task.

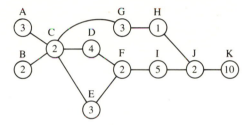

FIGURE 1.3

Now we will determine the smallest number of days after the start of the whole project that each task can be finished. For example, task A can be started at once, so it will be done after 3 days. We will write the number 3 to the right of the corresponding circle to indicate this. Likewise, we write a 2 to the right of circle B.

How we treat task C is the key to the whole algorithm we will develop. This task cannot be started until both A and B are done. This will be after 3 days, since that is the time needed for A. Then task C will take 2 days. Thus, 5 days are needed until C can be completed, and this is the number we write by the circle for C. So far our diagram looks as in Figure 1.4.

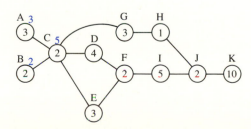

FIGURE 1.4

We carry on in the same fashion. Notice that if more than one line comes into a point from the left, then we add to the time for that point the *maximum* of all the incoming times to determine when it can first be completed. For example, it will take 9 days until D is finished and 8 days until E is finished. Since task F must wait for both of these, it will not be done for

$$(\text{maximum of 8 and 9}) + 2 = 11 \text{ days.}$$

The reader should check the numbers on the completed diagram in Figure 1.5.

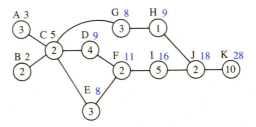

FIGURE 1.5

We see that the advertisement can be produced and delivered in 28 days, in time for the sale!

Critical Path Analysis

The method just described is called **PERT,** which stands for Program Evaluation and Review Technique. The PERT method (in a somewhat more complicated form) was developed in 1958 for the U.S. Navy Polaris submarine and missile project, although similar techniques were invented at about the same time at the E.I. du Pont de Nemours chemical company and in England, France, and Germany. Its usefulness in scheduling and estimating completion times for large projects, involving hundreds of steps and subcontracts, is obvious; and in various forms it has become a standard industrial technique. Any large library will contain dozens of books on the subject (look under PERT, Critical Path Analysis, or Network Analysis).

More information may be gleaned from the diagram we have just created. Let us work backwards, starting from task K, and see what makes the project take all of 28 days. Clearly it takes 28 days to finish K because it is 18 days until J is completed. Tasks H and I lead into J, but it is the 16 days needed to finish I that is important. Of course, task I cannot be completed before F is finished. So far we have traced a path back from K to F as shown in color in Figure 1.6.

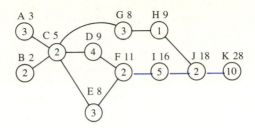

FIGURE 1.6

In the same way we work back from F to D (since the reason it takes 11 days to finish F is because it cannot be started until the 9 days it takes to complete D), then C, and finally A. The path A-C-D-F-I-J-K which is in color in Figure 1.7 is called a **critical path.**

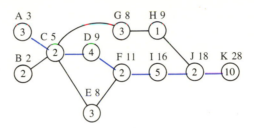

FIGURE 1.7

A critical path is important because the tasks on it are those that determine the total project time. If this time is to be reduced, then some task on a critical path must be done faster. For example, if the mailing list is compiled in 3 days instead of 2, it will still take 28 days to prepare and deliver the advertisement, since compiling the list (task G) is not on a critical path. Shortening the printing time (task I) by a day, however, would reduce the total time to 27 days; I is on the critical path. (Note, however, that changing the time for one task may change the critical path, altering whether other tasks are on it or not.)

A Construction Example

The following table gives the steps necessary in building a house, the number of days needed for each step, and the immediately preceding steps.

Task	Time in days	Preceding steps
A site preparation	4	none
B foundation	6	A
C drains and services	3	A
D framing	10	B
E roof	5	D
F windows	2	E
G plumbing	4	C, E
H electrical work	3	E
I insulation	2	G, H
J shell	6	F
K drywall	5	I, J
L cleanup and paint	3	K
M floors and trim	4	L
N inspection	10	I

We prepare the diagram in Figure 1.8 showing times and precedences.

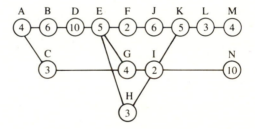

FIGURE 1.8

Working first from left to right, then from right to left, we analyze the total times to complete each task as follows. The critical path is marked with color in Figure 1.9. The only decision to be made in finding it comes in working back from K, where it is the 33 days needed to complete J that is important.

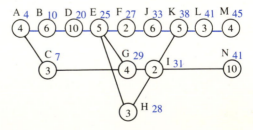

FIGURE 1.9

We see that a total of 45 days are needed to build the house, and the critical path is A-B-D-E-F-J-K-L-M.

The technique of representing a problem by a diagram of points, with lines between some of them, is useful in many other situations, and will be used throughout this book. The formal study of such diagrams will begin in Chapter 3.

EXERCISES 1.1

In Exercises 1–8 use the PERT method to determine the total project time and all the critical paths.

1.
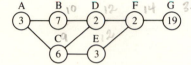

2.
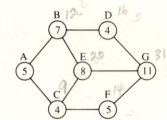

3.
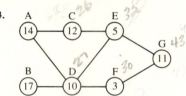

4.
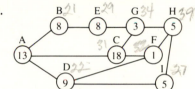

5.
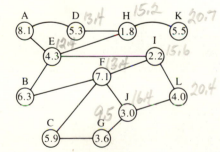

6.
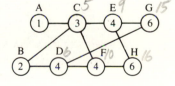

7.
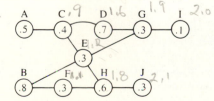

8.
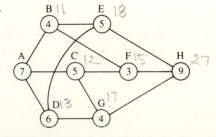

In Exercises 9–16 a table is given telling the time needed for each of a number of tasks and which tasks must immediately precede them. Make a PERT diagram for each problem, and determine the project time and critical path.

9.

Task	Time	Preceding tasks
A	5	none
B	2	A
C	3	B
D	6	A
E	1	B, D
F	8	B, D
G	4	C, E, F

10.

Task	Time	Preceding tasks
A	5	none
B	6	A
C	7	A
D	10	B
E	8	B, C
F	7	C
G	6	D, E, F

11.

Task	Time	Preceding tasks
A	3	none
B	5	none
C	4	A, B
D	2	A, B
E	6	C, D
F	7	C, D
G	8	E, F

12.

Task	Time	Preceding tasks
A	10	none
B	12	none
C	15	none
D	6	A, C
E	3	A, B
F	5	B, C
G	7	D, F
H	6	D, E
I	9	E, F

13.

Task	Time	Preceding tasks
A	3.3	none
B	2.1	none
C	4.6	none
D	7.2	none
E	6.1	none
F	4.1	A, B
G	1.3	B, C
H	2.0	F, G
I	8.5	D, E, G
J	6.2	E, H

14.

Task	Time	Preceding tasks
A	6	none
B	9	A, D
C	10	B, I
D	8	none
E	9	B
F	13	I
G	5	C, E, F
H	9	none
I	6	D, H

15.

Task	Time	Preceding tasks
A	.05	none
B	.09	A
C	.10	A, F
D	.07	B, C
E	.02	none
F	.04	E
G	.11	E
H	.09	F, G
I	.06	D, H

16.

Task	Time	Preceding tasks
A	11	none
B	13	none
C	12	none
D	14	none
E	8	A, C, D
F	6	A, B, D
G	10	A, B, C
H	5	B, C, D
I	9	E, F, H
J	7	F, G, H

17. A small purse manufacturer has a single machine that makes the metal parts of a purse. This takes 2 minutes. Another single machine makes the cloth parts in 3 minutes. Then it takes a worker 4 minutes to sew the cloth and metal parts together. Only one worker has the skill to do this. How long will it take to make 6 purses?

18. What is the answer to the previous problem if the worker can do the sewing in 2 minutes?

19. A survey is to be made of grocery shoppers in Los Angeles, Omaha, and Miami. First a preliminary telephone survey is made in each city to identify consumers in certain economic and ethnic groups willing to cooperate, and also to determine what supermarket characteristics they deem important. This will take 5 days in Los Angeles, 4 days in Miami, and 3 days in Omaha. After the telephone survey in each city, a list of shoppers to be visited in person is prepared for that city. This takes 6 days for Miami and 4 days each for Omaha and Los Angeles. After all three telephone surveys are made, a standard questionnaire is prepared. This takes 3 days. In each city when the list of consumers to be visited in that city has been prepared and the questionnaire is ready, the questionnaire is administered in that city. This takes 5 days in Los Angeles and Miami and 6 days in Omaha. How long will it take until all 3 cities are surveyed?

1.2 A Matching Problem

The Problem

An airline flying out of New York has 7 long flights on its Monday morning schedule: to Los Angeles, Seattle, London, Frankfort, Paris, Madrid, and Dublin. Fortunately, seven capable pilots are available: Alfors, Timmack, Jelinek, Tang, Washington, Rupp, and Ramirez. There is a complication, however. Pilots are allowed to request particular destinations, and these requests are to be honored so far as possible. The pilots requesting each city are listed below.

Los Angeles: Timmack, Jelinek, Rupp

Seattle: Alfors, Timmack, Tang, Washington

London: Timmack, Tang, Washington

Frankfort: Alfors, Tang, Rupp, Ramirez

Paris: Jelinek, Washington, Rupp

Madrid: Jelinek, Ramirez

Dublin: Timmack, Rupp, Ramirez

This information could also be represented by a diagram (Figure 1.10(a)), where we draw a black line between a city and a pilot if the former is on the pilot's request list.

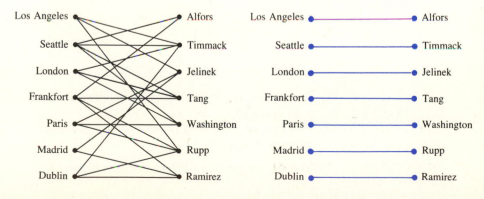

(a)   **(b)**

FIGURE 1.10

The person assigning the flights would like to please all the pilots if this can be done, and if not, would like to accommodate as many as possible. This may be thought of as an *optimization* problem. We desire a matching of pilots with flights such that the number of pilots who get flights they have requested is as large as possible.

Analysis

Let us start with a very crude attack on our matching problem. We could simply list all possible ways of assigning one pilot to each flight, and count for each the number of pilots who are assigned to flights they requested. For example, one matching would be to take the flights and pilots in the order they were listed.

Flight	Pilot	Requested?
Los Angeles	Alfors	no
Seattle	Timmack	yes
London	Jelinek	no
Frankfort	Tang	yes
Paris	Washington	yes
Madrid	Rupp	no
Dublin	Ramirez	yes

This matching is indicated by the colored lines in Figure 1.10(b). Here four of the pilots would get flights they want, but perhaps a different matching would do even better.

If we agree always to list the flights in the same order, say that of our original list, then an assignment will be determined by some arrangement of the seven pilots' names. For example, the arrangement

Timmack, Alfors, Jelinek, Tang, Washington, Rupp, Ramirez

would send Timmack to Los Angeles and Alfors to Seattle, while assigning the same pilots to the other flights as previously. Likewise, the arrangement

Ramirez, Rupp, Washington, Tang, Jelinek, Timmack, Alfors

would send Ramirez to Los Angeles, Rupp to Seattle, etc. The reader should check that this matching will accommodate only 3 pilots' wishes.

Several questions come to mind concerning our plan for solving this problem.

(1) How much work will this be? In particular, how many arrangements will we have to check?

(2) How can we generate all possible arrangements so that we are sure we have not missed any?

The second question is somewhat special, and we will not answer it until

Chapter 7, but the first question is easier. (Note that it is a *counting* problem.) In order to make the count, we will invoke a simple principle that will be useful many times in this book.

The Multiplication Principle   Suppose that a procedure can be divided into a sequence of k steps and that the first step can be performed in n_1 ways, the second step can be performed in n_2 ways, and, in general, the ith step can be performed in n_i ways. Then the number of different ways in which the entire procedure can be performed is $n_1 n_2 \ldots n_k$.

EXAMPLE 1.1   A certain Japanese car is available in 6 colors, with 3 different engines and either a manual or automatic transmission. What is the total number of ways the car can be ordered?

We can apply the multiplication principle with $k = 3$, $n_1 = 6$, $n_2 = 3$, and $n_3 = 2$. The number of ways is $(6)(3)(2) = 36$.   ■

Now we return to the problem of counting the number of ways the 7 flights can be assigned. Let us start with the Los Angeles flight. There are 7 pilots who can be assigned to it. We pick one, and turn to the Seattle flight. Now only 6 pilots are left. Choosing one of these leaves 5 from which to pick for the London flight. We can continue in this manner all the way to the Dublin flight, at which time only one pilot will be left. Thus, the total number of matchings we can devise will be $7 \cdot 6 \cdot 5 \cdot 4 \cdot 3 \cdot 2 \cdot 1$.

The same argument will work when we have any number of flights and the same number of available pilots, producing

$$n(n - 1)(n - 2) \ldots 3 \cdot 2 \cdot 1$$

possible matchings if there are n flights and n pilots.

Permutations

The reader is probably aware that there is a shorter notation for a product of the type we just developed. If n is any nonnegative integer, we define **n factorial,** which is denoted by $n!$, as follows:

$0! = 1$, $1! = 1$, $2! = 1 \cdot 2$, and, in general, $n! = 1 \cdot 2 \cdot \ldots \cdot (n - 1)n$.

Notice that if $n > 1$, then $n!$ is just the product of the integers from 1 to n.

By a **permutation** of a set of objects, we mean any ordering of those objects. For example, the permutations of the letters a, b, and c are:

$$abc, \ acb, \ bac, \ bca, \ cab, \ cba.$$

THEOREM 1.1   There are exactly $n!$ permutations of a set of n objects.

There is a generalization of the idea of a permutation that often arises. Suppose in the flight assignment problem the flights to Madrid and Dublin are cancelled because of bad weather. Now 7 pilots are available for the 5 remaining flights. There are 7 ways to choose a pilot for the Los Angeles flight, then 6 pilots to choose from for the Seattle flight, etc. Since only 5 pilots need be chosen, there are a total of

$$7 \cdot 6 \cdot 5 \cdot 4 \cdot 3$$

possible ways to make the assignments. (Notice that this product has 5 factors.) The same argument works in general.

THEOREM 1.2   The number of ways an ordered list of r objects can be chosen from a group of n objects is

$$n(n - 1) \ldots (n - r + 1) = \frac{n!}{(n - r)!}.$$

Proof.   The first object can be chosen in n ways, the second in $n - 1$ ways, etc. Since it is easy to check that the product on the left has r factors, by the multiplication principle it counts the total number of arrangements. As for the second expression, note that

$$n(n - 1) \ldots (n - r + 1)$$
$$= \frac{n(n - 1) \ldots (n - r + 1)(n - r)(n - r - 1) \ldots 2 \cdot 1}{(n - r)(n - r - 1) \ldots 2 \cdot 1}$$
$$= \frac{n!}{(n - r)!}. \quad \blacksquare$$

EXAMPLE 1.2   The junior class at Taylor High School is to elect a president, vice president, and secretary from among its 30 members. How many different choices are possible?
    We are to choose an ordered list of 3 officers from 30 students. The number of possibilities is

$$30 \cdot 29 \cdot 28 = 24,360. \quad \blacksquare$$

The number of ordered lists of r objects chosen from n objects is denoted by $P(n, r)$. These lists are called **permutations of n objects, taken r at a time.**

For example, we have just seen that $P(30, 3) = 24,360$, and, in general, according to the last theorem,

$$P(n, r) = \frac{n!}{(n - r)!}.$$

The Practicality of Our Solution to the Airline Problem

We were going to run through all the ways of assigning a pilot to each of the seven flights to see which would please the most pilots. We now know the number of possible assignments is $7! = 5040$. This number is large enough to discourage us from trying this method by hand. If a computer were available, however, the method would look more promising. We would need a way to tell the computer how to generate these 5040 permutations, that is, an algorithm. This would amount to an explicit answer to question (2) above.

Of course, 7 flights and 7 pilots are really unrealistically small numbers. For example, at O'Hare field in Chicago an average of more than 900 airplanes take off every day. Let us consider a small airline with 20 flights and 20 pilots to assign to them, and consider the practicality of running through all possible assignments. The number of these is 20!, which we calculate on a hand calculator to be about $2.4 \cdot 10^{18}$. This is a number of 19 digits, and a computer is apparently required. Let us suppose our computer can generate one million assignments per second and check for each of them how many pilots get their requested flights. How long would it take to run through them all?

The answer is not hard to calculate. Doing $2.4 \cdot 10^{18}$ calculations at 1,000,000 per second would take

$$\frac{2.4 \cdot 10^{18}}{1,000,000} = 2.4 \cdot 10^{12} \text{ seconds,}$$

$$\text{or} \quad \frac{2.4 \cdot 10^{12}}{60} = 4 \cdot 10^{10} \text{ minutes,}$$

$$\text{or} \quad \frac{4 \cdot 10^{10}}{60} \approx 6.7 \cdot 10^{8} \text{ hours,}$$

$$\text{or} \quad \frac{6.7 \cdot 10^{8}}{24} \approx 2.8 \cdot 10^{7} \text{ days,}$$

$$\text{or} \quad \frac{2.8 \cdot 10^{7}}{365} \approx 7.6 \cdot 10^{4} \text{ years.}$$

The calculation would take about 76,000 years, just for 20 flights and 20 pilots.

The point of this calculation is that even with a computer you sometimes have to be clever. In Chapter 5 we will explain a much more efficient way to solve our matching problem. This method will allow a person to handle 7 flights and 7 pilots in a few minutes, and a computer to deal with hundreds of flights and pilots in a reasonable time.

EXERCISES 1.2

In Exercises 1–16 calculate the number shown.

1. 5!

2. 6!

3. $\dfrac{8!}{3!}$

4. $\dfrac{7!}{4!}$

5. $\dfrac{8!}{2!6!}$

6. $\dfrac{9!}{3!6!}$

7. $P(7, 4)$

8. $P(8, 4)$

9. $P(10, 7)$

10. $P(11, 9)$

11. $\dfrac{P(9, 4)}{5!}$

12. $\dfrac{P(10, 2)}{3!5!}$

13. $P(6, 6)$

14. $P(7, 7)$

15. $\dfrac{P(8, 3)}{P(3, 3)}$

16. $\dfrac{P(9, 5)}{P(5, 5)}$

17. A baseball manager has decided who his 9 starting hitters are to be, but not the order in which they will bat. How many possibilities are there?

18. A president, vice president, and treasurer are to be chosen from a club with 7 members. In how many ways can this be done?

19. A record company executive must decide the order in which to present 6 songs on one side of an album. How many choices does she have?

20. A Halloween makeup kit contains 3 different moustaches, 2 different sets of eyebrows, 4 different noses, and a set of ears. (It is not necessary to use any moustache, etc.) How many disguises using at least one of these items are possible?

21. A man has 5 sport coats, 4 pairs of slacks, 6 shirts, and one tie. How many combinations of these can he wear, if he must wear at least slacks and a shirt?

22. Different prizes for first place, second place, and third place are to be awarded to 3 of the 12 finalists in a beauty contest. How many ways is this possible?

23. Seven actresses have auditioned for the parts of the three daughters of King Lear: Goneril, Regan, and Cordelia. In how many ways can the roles be filled?

24. A farmer with 7 cows likes to milk them in a different order each morning. How many days can he do this without repeating?

25. A busy summer resort motel has 5 empty rooms and 3 travelers who want rooms. In how many ways can the motel manager assign to each guest a room of his or her own?

26. An Alaskan doctor visits each of 5 isolated settlements by plane once a month. He can use either of two planes, but once he starts out he visits all 5 settlements in some order before returning home. How many possibilities are there?

27. A tennis coach must pick her top 6 varsity and top 6 junior varsity players in order from among 9 varsity and 11 junior varsity players. In how many ways is this possible?

28. A dinner special for 4 at a Chinese restaurant allows one shrimp dish (from 3), one beef dish (from 5), one chicken dish (from 4), and one pork dish (from 4). Also each diner can choose either soup or an egg roll. How many different orders might be sent to the kitchen?

29. How many ways can 6 keys be placed on a circular key ring? Both sides of the ring are the same, and there is no way to tell which is the "first" key on the ring.

30. Show that if $n > 1$, then $P(n, 2) = n^2 - n$.

31. Show that if $n > 0$, then $P(n, n - 1) = n!$.

32. Show that if $0 < r < \dfrac{n}{2}$, then $\dfrac{P(n, 2r)}{P(n, r)} = P(n - r, r)$.

1.3   A Knapsack Problem

The Problem

A U.S. satellite is to be put into orbit around the earth, and 700 kilograms of its payload are allotted to experiments designed by scientists. Researchers from around the country apply for the inclusion of their experiments. Of course, they must specify how much the equipment they want taken into orbit will weigh. A panel of reviewers then decides which proposals are reasonable. These are then rated from 1 to 10 (the highest possible score) on their potential importance to science. The ratings are listed below.

Experiment	Weight in kilograms	Rating	Ratio
1 Cosmic rays	80	6	.0750 ⑦
2 Weightless vines	25	3	.1200 ⑤
3 Binary stars	224	4	.0179 ⑫
4 Mice tumors	65	8	.1231 ④
5 Space dust	127	7	.0551 ⑧
6 Solar power	188	7	.0372 ⑨
7 Relativity	104	8	.0769 ⑥
8 Seed viability	7	5	.7143 ①
9 Sun spots	92	2	.0217 ⑪
10 Speed of light	324	8	.0247 ⑩
11 Cloud patterns	36	6	.1667 ③
12 Yeast	22	4	.1818 ②

It is decided to choose experiments so that the total of all their ratings is as large as possible. Since there is also the limitation that the total weight cannot exceed 700 kilograms, it is not clear how to do this. If we just start down the list, for example, experiments 1, 2, 3, 4, and 5 have a total weight of 521 kilograms. Now we cannot take experiment 6, since its 188 kilograms would put us over the 700 kilogram limit. We could include experiments 7 and 8, however, which would bring us up to 632 kilograms. The following table shows how we might go down the list this way, keeping a running total of the weight and putting in whichever experiments do not put us over 700 kilograms.

Experiment	Weight	Include?	Total weight	Rating
1	80	yes	80	6
2	25	yes	105	3
3	224	yes	329	4
4	65	yes	394	8
5	127	yes	521	7
6	188	no	521	—
7	104	yes	625	8
8	7	yes	632	5
9	92	no	632	—
10	324	no	632	—
11	36	yes	668	6
12	22	yes	690	4

Note that the ratings total of the experiments chosen this way is

$$6 + 3 + 4 + 8 + 7 + 8 + 5 + 6 + 4 = 51.$$

The question is whether we can do better than this. Since we just took the experiments as they came, without paying any attention to their ratings, it seems likely that we can. Perhaps it would be better to start with the experiments with the highest rating and include as many of them as we can, then go on to those with the next highest rating, and so on. If two experiments have the same rating, we would naturally choose the lighter one first. The following table shows how such a tactic would work.

Experiment	Rating	Weight	Include?	Total weight
4	8	65	yes	65
7	8	104	yes	169
10	8	324	yes	493
5	7	127	yes	620
6	7	188	no	620
11	6	36	yes	656
1	6	80	no	656
8	5	7	yes	663
12	4	22	yes	685
3	4	224	no	685
2	3	25	no	685
9	2	92	no	685

Using this method we choose experiments 4, 7, 10, 5, 11, 8, and 12, giving a rating total of

$$8 + 8 + 8 + 7 + 6 + 5 + 4 = 46.$$

This is worse than our previous total!

Another idea would be to start with the experiment of smallest weight (number 8), then include the next lightest (number 12), and so on, continuing until we reach the 700 kilogram limit. The reader should check that this would mean including experiments 1, 2, 4, 5, 7, 8, 9, 11, and 12 for a rating total of 49, still inferior to our first attempt.

Still another idea would be to compute a rating points-per-kilogram ratio for each experiment, and to include, whenever possible, the experiments for which this ratio is highest. We will illustrate this idea with a case where only three experiments are submitted, with the limit of 700 kilograms still in effect.

Experiment	Weight	Rating	Ratio
1	390	8	$8/390 \approx .0205$
2	350	6	$6/350 \approx .0171$
3	340	5	$5/340 \approx .0147$

Using our new scheme we would choose experiment 1, since it has the highest ratio, and then not be able to include either of the other two. The total rating would be 8. But this is not as good as choosing experiments 2 and 3 for a rating total of 11. The ratio method does not assure us of the best selection either.

Analysis

We could play around with this problem, taking experiments out and putting experiments in, and perhaps find a collection of experiments with a higher rating total than 51. Even then, it would be hard for us to be sure we could not do even better somehow. Notice that this is another *optimization* problem. We want to find a selection of experiments from the 12 given whose total weight is less than 700 kilograms and whose rating total is as large as possible.

As in the case of the matching problem of the previous section, we will turn to the tedious method of trying all the possibilities. If we could get a computer to do the calculations, even if there are many experiments, this might be a practical way to attack the problem. Since the experiments are numbered from 1 to 12, we will save time by simply using the numbers. We will introduce some language (with which the reader is probably already familiar) in order to state the problem in a compact way.

We need the idea of a **set.** Although we cannot give a definition of a set in terms of simpler ideas, we think of a set as a collection of objects of some sort such that given any object, we can tell whether that object is in the set or not. If the object x is in the set S we write $x \in S$, and if not we write $x \notin S$.

EXAMPLE 1.3   Let P be the set of all presidents of the United States. Then

$$\text{George Washington} \in P,$$

but

$$\text{Benjamin Franklin} \notin P.$$

If U is the set of all integers from 1 to 12, then

$$5 \in U, \text{ but } 15 \notin U. \quad \blacksquare$$

If a set has only a finite number of objects, one way to define it is simply to list them all between curly braces. For example, the set U of the example could also be defined by

$$U = \{1, 2, 3, 4, 5, 6, 7, 8, 9, 10, 11, 12\}.$$

If the set has more elements than we care to list, we may use three dots to indicate some elements. For example, we could also write

$$U = \{1, 2, 3, \ldots, 11, 12\}.$$

Another way to express a set is to enclose inside curly braces a variable standing for a typical element of the set, followed by a colon, followed by a description of what condition or conditions the variable must satisfy in order to be in the set. For example,

$$U = \{x \colon x \text{ is an integer and } 0 < x < 13\},$$

and

$$P = \{x \colon x \text{ is a president of the U.S.}\}.$$

The latter expression is read "the set of all x such that x is a president of the United States." In these two examples the use of x for the variable is arbitrary; any other letter having no previous meaning could be used just as well.

Let A and B be sets. We say that A is a **subset** of B, and write

$$A \subseteq B,$$

if every element of A is also in B. In this case we also say that A **is contained in** B, and that B **contains** A. An equivalent notation is

$$B \supseteq A.$$

EXAMPLE 1.4   If U is the set defined above and

$$T = \{1, 2, 3, 4, 5, 7, 8, 11, 12\},$$

then $T \subseteq U$. Likewise, if

$$C = \{\text{Lincoln, A. Johnson, Grant}\},$$

then $C \subseteq P$, where as before P is the set of all American presidents. On the other hand, $P \subseteq C$ is false.   ▬

If A is a set, we will denote by $|A|$ the number of elements in A. For example, if C, T, U, and P are as defined above, we have $|C| = 3$, $|T| = 9$, $|U| = 12$, and (in 1985) $|P| = 39$. (Although Ronald Reagan is often listed as the 40th president of the United States, this number is achieved by counting Grover Cleveland twice, because Benjamin Harrison was president between Cleveland's two terms. But an element is either in a set or is not; it cannot be in the set more than once.) The **empty set** is the set which has no elements at all. We denote it by $\varnothing$. Thus, if A is a set, then $A = \varnothing$ if and only if $|A| = 0$.

We say that two sets are **equal** if every element in the first is also in the second, and conversely, every element in the second is also in the first. Thus, $A = B$ if and only if $A \subseteq B$ and $B \subseteq A$.

The Knapsack Problem Revisited

Armed with the language of sets, we return to the question of selecting experiments. The set of all experiments corresponds to the set

$$U = \{1, 2, \ldots, 11, 12\},$$

and each selection corresponds to some subset of U. For example, the choice of experiments 1, 2, 3, 4, 5, 7, 8, 11, and 12 corresponds to the subset

$$T = \{1, 2, 3, 4, 5, 7, 8, 11, 12\}.$$

This happens to be the selection of our first (and, so far, most successful) attempt to solve the problem, with a rating total of 51.

Of course, some subsets of U will be unacceptable because their total weight will exceed 700 kilograms. An example of such a subset is

$$\{3, 6, 10\},$$

with a total weight of 736 kilograms.

We could simply go through all the subsets of U, computing for each its total weight. If this does not exceed 700, then we will add up the ratings of the corresponding experiments. Eventually we will find which subset (or subsets) has the maximal rating total.

As in the last section, two questions arise:

(1) How many subsets are there? (Another counting problem.)
(2) How can we list all the subsets without missing any?

We will start with problem (1), saving problem (2) for Section 1.4. Let us start with some smaller sets to get the idea.

Set	Subsets	Number of subsets
$\{1\}$	$\varnothing, \{1\}$	2
$\{1, 2\}$	$\varnothing, \{1\}, \{2\}, \{1, 2\}$	4
$\{1, 2, 3\}$	$\varnothing, \{1\}, \{2\}, \{1, 2\}, \{3\}, \{1, 3\},$ $\{2, 3\}, \{1, 2, 3\}$	8

We see that a set with 1 element has 2 subsets, a set with 2 elements has 4 subsets, and a set with 3 elements has 8 subsets. This suggests the following theorem, which will be proved in Section 2.6.

THEOREM 1.3   A set with n elements has exactly 2^n subsets.

The set U has 12 elements, and so by the theorem it has exactly $2^{12} = 4096$ subsets. This is more than we would like to run through by hand, but would be easy enough for a computer. In fact, as n gets large the quantity 2^n does not grow as fast as the quantity $n!$ that arose in the previous section. For example, 2^{20} is only about a million. Our hypothetical computer that could check one million subsets per second could run through the possible selections from 20 experiments in about a second, which is considerably less than the 76,000 years we found it would take to check the 20! ways of assigning 20 pilots to 20 flights. Even so, 2^n can get unreasonably large for modest values of n. For example, 2^{50} is about $1.13 \cdot 10^{15}$, and our computer would take about 36 years to run through this number of subsets.

Our problem of choosing experiments is an example of the **knapsack problem.** The name comes from the idea of a hiker who has only so much room in his knapsack, and must choose which items—food, first aid kit, water, tools, etc.—to include. Each item takes up a certain amount of space and has a certain value to the hiker, and the idea is to choose items that fit with the greatest total value.

In contrast to the matching problem of the previous section, there is no efficient way known to solve the knapsack problem. What exactly is meant by an "efficient way" will be made clearer when complexity theory is discussed in the next section.

EXERCISES 1.3

In Exercises 1–14 let A = $\{1, 2\}$, B = $\{2, 3, 4\}$, C = $\{2\}$, D =
$\{x: x$ is an odd positive integer$\}$, and E = $\{3, 4\}$. Tell whether each statement is true or false.

1. $A \subseteq B$

2. $C \subseteq A$

3. $2 \subseteq A$

4. $B \subseteq D$

5. $10^6 \in D$

6. $C \in B$

7. $A \subseteq A$

8. $B = \{C, E\}$

9. $2 \in \{C, E\}$

10. $|\{C, E\}| = 2$

11. $|\{2, 3, 4, 3, 2\}| = 5$

12. $|\{\varnothing, \varnothing\}| = 1$

13. $\varnothing \in A$

14. $\varnothing \subseteq \{C, E\}$

In Exercises 15–18 the given sets represent a selection of satellite experiments from among the 12 given in the text. Determine whether each selection is acceptable (i.e., not over 700 kilograms). If it is, then find the total rating.

15. $\{2, 3, 8, 9, 10, 12\}$

16. $\{2, 3, 9, 10, 11\}$

17. $\{2, 4, 7, 8, 9, 10, 11\}$

18. $\{2, 3, 4, 6, 7, 9\}$

19. Suppose that the rating/kilogram ratio is computed for each of the 12 proposed satellite experiments. Experiments are chosen by including those with the highest ratio that do not push the total weight over 700 kilograms. What set of experiments does this produce, and what is their total rating?

20. List the subsets of $\{1, 2, 3, 4\}$. How many are there?

21. How many subsets does {Sunday, Monday, . . ., Saturday} have?

22. How many subsets does {Dopey, Happy, . . ., Doc} have?

23. How many subsets does {Chico, Harpo, Groucho, Zeppo, Gummo} have?

24. How many subsets does $\{13, 14, . . ., 22\}$ have?

25. Suppose m and n are positive integers with $m < n$. How many elements does $\{m, m + 1, . . ., n\}$ have?

26. How many subsets does $\{2, 4, 8, . . ., 256\}$ have?

27. A draw poker player may discard some of his 5 cards and be dealt new ones. The rules say he cannot discard all 5. How many sets of cards can be discarded?

28. Suppose in the previous problem no more than 3 cards may be discarded. How many choices does a player have?

29. How long would it take a computer that can check one million subsets per second to run through the subsets of a set of 40 elements?

1.4   Algorithms and Their Efficiency

Comparing Algorithms

In previous sections we developed algorithms for solving certain practical combinatorial problems. We also saw that in some cases solving a problem of reasonable size, even using a high-speed computer, can take an unreasonable amount of time. Obviously an algorithm is not practical if its use will cost more than we are prepared to pay, or if it will provide a solution too late to be of value. In this section we will examine in more detail the construction and efficiency of algorithms.

Our point of view is that a digital computer will handle the actual implementation of the algorithms we develop. This means a precise set of instructions (a "program") must be prepared telling the computer what to do. When an algorithm is presented to human beings, it is explained in an informal way, with examples and references to familiar techniques. (Think of how the operation of long division is explained to children in grade school.) Telling a computer how to do something requires a more organized and precise presentation.

Of course, most computer programs are written in some specific higher-level computer language, such as FORTRAN, BASIC, COBOL, or Pascal. In this book we will not write our algorithms in any particular computer language, but rather use English in a form that is sufficiently organized and precise that a program could easily be written from them. Usually this will mean a numbered sequence of steps, with precise instructions on how to proceed from one step to the next.

Of course a big, complicated problem will require a big, complicated solution, no matter how good an algorithm we find for it. For the problem of choosing an optimal set of experiments to place in a satellite, introduced in Section 1.3, choosing from among 12 submitted experiments requires looking at about 4000 subsets, while if there are 20 experiments, then there are about 1,000,000 subsets, 250 times as many. A reasonable measure of the "size" of the problem in this example would be n, the number of experiments. For each type of problem we may be able to identify some number n that measures the amount of information upon which a solution must be based. Admittedly, choosing the quantity to be labeled n is often somewhat arbitrary. The precise choice may not be important, however, for the purpose of comparing two algorithms that do the same job, so long as n represents the same quantity in both algorithms.

We will also try to measure the amount of work done in computing a solution to a problem. Of course this will depend on n, and for a desirable algorithm it will not grow too quickly as n gets larger. We need some unit in which to measure the size of an algorithmic solution. In the satellite problem, for example, we saw that a set of n experiments has 2^n subsets that need to be checked. "Checking" a subset itself involves certain computations, however. The weights of the experiments in the subset must be added to see if the 700 kilogram limit is exceeded. If not, then the ratings of the subset must be added and compared with the previous best total. How much work this will be for a particular subset will also depend on the total number of experiments.

We will take the conventional course of measuring the size of an algorithm by counting the total number of elementary operations it involves, where an elementary operation is defined as the addition, subtraction, multiplication, division, or comparison of two numbers. For example, adding up the k numbers $a_1, a_2, \ldots, a_k$ involves $k - 1$ additions, as we compute

$$a_1 + a_2, (a_1 + a_2) + a_3, \ldots, (a_1 + \ldots + a_{k-1}) + a_k.$$

We will call the total number of elementary operations required the **complexity** of the algorithm.

There are two disadvantages to measuring the complexity of an algorithm this way.

(1) This method essentially tries to measure the time an algorithm will take to implement, assuming that each elementary operation will take just as long. But computers are also limited by their memory. An algorithm may require storage of more data than a given computer can hold. Or memory to which access is slower may have to be used, slowing down the solution. In any case, computer storage itself has a monetary value which our simple counting of elementary operations does not take into account.

(2) It may be that not all operations take the same amount of computer time; for example, division may take longer than addition. Also the time an elementary operation takes may depend on the size of the numbers involved; computations with larger numbers take longer. Just the assignment of a value to a variable also takes computer time, time which we are not taking into account.

In spite of these criticisms that can be made of our proposed method of measuring the complexity of an algorithm, we will use it to maintain simplicity and to avoid knowing the internal workings of particular computers.

Evaluating Polynomials

We will consider some examples of algorithms and their complexity. Let us start with the problem of evaluating x^n, where x is some number and n is a positive integer. When we break this down into elementary operations, we see we need to compute $x^2 = x \cdot x$, then $x^3 = (x^2)x, \ldots$, until we get to x^n. Since computing x^2 takes 1 multiplication, computing x^3 takes 2 multiplications, etc., a total of $n - 1$ multiplications is necessary. An algorithm for this process might be as follows.

Algorithm for Evaluating x^n

Given a real number x and a positive integer n, this algorithm computes $P = x^n$.

Step 1 (initialization). Let $P = x$ and $k = 1$.

Step 2 (next power). If $k < n$, then replace P by Px, replace k by $k + 1$, and go to step 2.

Step 3 (output). $P = x^n$.

Here it is understood that after each step we go on to the next one unless some other step is specified. Thus, step 2 is repeated for $k = 1, 2, \ldots, n - 1$, requiring $n - 1$ multiplications. Only when $k = n$ do we proceed to step 3, where the algorithm stops. As a matter of fact we see that not only are $n - 1$ multiplications involved in computing x^n, but also n comparisons and $n - 1$ additions (in step 2), for a total of $3n - 2$ elementary operations.

Since our method of estimating the number of operations in an algorithm involves various inaccuracies anyway, usually we are not interested in an exact count. Knowing that computing x^n takes about $3n$ operations, or even a number of operations that is less than some constant multiple of n, may satisfy us. We are mainly interested in avoiding, when possible, algorithms whose number of operations grows very quickly as n gets large.

Later in this book we will often write algorithms in a less formal way that may make impractical an exact count of the elementary operations involved. For example, in the last algorithm instead of incrementing k by 1 at each step and then comparing its new value to that of n, we might simply have said something like "for $k = 1$ to $n - 1$ replace P by Px." High level computer languages usually allow loops to be defined by some such language. Although the exact number of operations required for each value of k is obscured, we can still see that the complexity of the algorithm does not exceed an expression of the form Cn, where C is some constant. (We could actually take $C = 3$ for the algorithm just presented.) Often knowing the precise value of C is not important to us. We will indicate why this is true later in this section, after we have more examples of algorithms.

By a **polynomial of degree n in x** we mean an expression of the form

$$P(x) = a_n x^n + a_{n-1} x^{n-1} + \ldots + a_1 x + a_0,$$

where $a_n, a_{n-1}, \ldots, a_0$ are constants and $a_n \neq 0$. Thus, a polynomial in x is a sum of terms, each of which is either a constant or else a constant times a positive integral power of x. We will consider two algorithms for computing the value of a polynomial. The first one will start with a_0, then add $a_1 x$ to that, then add $a_2 x^2$ to that, etc.

Polynomial Evaluation Algorithm

This algorithm computes $P(x) = a_n x^n + \ldots + a_0$, given the positive integer n and real numbers $x, a_0, a_1, \ldots, a_n$.

Step 1 (initialization). Let $S = a_0$ and $k = 1$.

Step 2 (add next term). If $k \leq n$, then replace S by $S + a_k x^k$ and k by $k + 1$ and go to step 2.

Step 3 (output). $P(x) = S$.

In this algorithm we will execute step 2 a total of $n + 1$ times, with $k = 1, 2, \ldots, n + 1$. For a particular value of $k \leq n$ this will entail 1 comparison (to check that $k \leq n$), 2 additions (in computing the new values of S and k), and one multiplication (multiplying a_k by x^k), a total of 4 operations. But this assumes we know what x^k is. We just saw that this number takes $3k - 2$ operations to compute. Thus, for a given value of $k \leq n$ we will use $4 + (3k - 2) = 3k + 2$ operations.

Letting $k = 1, 2, \ldots, n$ accounts for a total of

$$5 + 8 + 11 + \ldots + (3n + 2)$$

operations.

The last expression is an arithmetic progression with n terms, first term 5, and common difference 3, and its value is $\frac{1}{2}(3n^2 + 7n)$. (A proof of this will be given in Section 2.5, along with the definition of an arithmetic progression.) The extra comparison when $k = n + 1$ gives a total of $\frac{1}{2}(3n^2 + 7n) + 1$ operations.

Notice that here the complexity of our algorithm is itself a polynomial in n, namely $1.5n^2 + 3.5n + 1$. Given a polynomial in n of degree k, say $a_k n^k + a_{k-1} n^{k-1} + \ldots + a_0$, the term $a_k n^k$ will exceed the sum of all the other terms in absolute value if n is sufficiently large. Thus, if the complexity of an algorithm is a polynomial in n, we are interested mainly in the degree of that polynomial. Even the coefficient of the highest power of n appearing is of secondary importance. That the complexity of the algorithm just presented is a polynomial with n^2 as its highest power of n (instead of n or n^3, for example) is more interesting to us than the fact that the coefficient of n^2 is 1.5.

We will say that an algorithm has **order at most $f(n)$**, where $f(n)$ is some expression in n, in case the complexity of the algorithm does not exceed $Cf(n)$ for some constant C. Recall that the polynomial evaluation algorithm has complexity $1.5n^2 + 3.5n + 1$. It is not hard to see that $3.5n + 1 \leq 4.5n^2$ for all positive integers n. Then

$$1.5n^2 + 3.5n + 1 \leq 1.5n^2 + 4.5n^2 = 6n^2$$

for all positive integers n. Thus, (taking $C = 6$) we see that the polynomial evaluation algorithm has order at most n^2. A similar proof shows that in general an algorithm whose complexity is no more than a polynomial of degree k has order no more than n^k.

The algorithms we have presented so far are simple enough that we can compute their complexities exactly. Often, however, the exact number of operations necessary may not depend only on n. An algorithm to sort n numbers into numerical order, for example, may entail more or fewer steps depending on how the numbers are arranged initially. Here we might count the number of operations in the worst possible case. The actual number of operations will then be less than or equal to this number.

Later we may present more complicated algorithms in an informal way. An exact analysis of their complexity would entail a more detailed description revealing each elementary operation, akin to an actual program in some computer language. In this case the statement that the algorithm has order no more than $f(n)$ is to be interpreted as saying that a computer implementation exists for which the number of elementary operations does not exceed $Cf(n)$ for some constant C.

Now we will present a more efficient algorithm for polynomial evaluation. It was first published in 1819 by W. G. Horner, an English schoolmaster. The idea behind it is illustrated by the following computation with $n = 3$:

$$a_3x^3 + a_2x^2 + a_1x + a_0 = x(a_3x^2 + a_2x + a_1) + a_0$$
$$= x(x(a_3x + a_2) + a_1) + a_0 = x(x(x(a_3) + a_2) + a_1) + a_0.$$

Horner's Polynomial Evaluation Algorithm

This algorithm computes $P(x) = a_nx^n + \ldots + a_0$, given the positive integer n and real numbers $x, a_0, \ldots, a_n$.

Step 1 (initialization). Let $S = a_n$ and $k = 1$.

Step 2 (compute next expression). If $k \leq n$, replace S by $xS + a_{n-k}$ and k by $k + 1$.

Step 3 (output). $P(x) = S$.

The following table shows how this algorithm works for $n = 3$, $P(x) = 5x^3 - 2x^2 + 3x + 4$ (so $a_3 = 5$, $a_2 = -2$, $a_1 = 3$, and $a_0 = 4$), and $x = 2$.

	S	k
Start:	$a_3 = 5$	1
	$xS + a_2 = 2(5) + (-2) = 8$	2
	$xS + a_1 = 2(8) + 3 = 19$	3
	$xS + a_0 = 2(19) + 4 = 42$	4

We run through step 2 for $k = 1, 2, \ldots, n + 1$, a total of $n + 1$ times, and each time except the last requires just 1 comparison, 1 multiplication, and 2 additions. Thus, this algorithm evaluates a polynomial of degree n using just $4n + 1$ operations, as opposed to $\frac{1}{2}(3n^2 + 7n) + 1$ for our first version. If n were 10, the second algorithm would take 41 operations while the first would take 186; and for larger values of n the difference would be even more marked.

In broader terms Horner's polynomial evaluation algorithm is superior to the previous polynomial evaluation algorithm because it has order no more than n, while we could only say that the first algorithm had order no more than n^2.

A Subset-Generating Algorithm

Now we will consider an algorithm for generating subsets of a set, as our solution of the satellite problem requires. If a set S consists of the n elements $x_1, x_2, \ldots, x_n$, a compact way of representing a subset of S is as a string of 0's and 1's, where the kth entry in the string is 1 if $x_k \in S$ and 0 otherwise. If $n = 3$, for example, the $8 = 2^3$ strings and the subsets to which they correspond are shown below.

$$
\begin{array}{ll}
000 & \varnothing \\
100 & \{x_1\} \\
010 & \{x_2\} \\
110 & \{x_1, x_2\} \\
001 & \{x_3\} \\
101 & \{x_1, x_3\} \\
011 & \{x_2, x_3\} \\
111 & \{x_1, x_2, x_3\}
\end{array}
$$

By examining this list we see how these digits might be generated in the order given. We look for the left-most 0, change it to a 1, and then change any digits still further to the left to 0's. If we let

$$a_1 a_2 \ldots a_n$$

be a given string with n 0's and 1's, the following algorithm generates the next string.

Next Subset Algorithm

Given a positive integer n and the string $a_1 \ldots a_n$ of 0's and 1's corresponding to a subset of a set with n elements, this algorithm computes the string corresponding to the next subset.

Step 1 (initialization). Let $k = 1$.

Step 2 (look for left-most 0). If $k \leq n$ and $a_k = 1$, replace k by $k + 1$ and go to step 2.

Step 3 (string of all 1's?). If $k = n + 1$, stop. (All digits of the string are 1.)

Step 4 (change left-most 0 to 1). Replace a_k by 1. Let $j = 1$.

Step 5 (change further left 1's to 0). If $j < k$, replace a_j by 0, replace j by $j + 1$, and go to step 5.

Step 6 (output). Now $a_1 \ldots a_n$ corresponds to the next subset.

In this algorithm step 2 finds the left-most string digit a_k that is 0. (If all digits are 1 we stop at step 3.) Then a_k is changed to 1 in step 4, and the digits to its left changed to 0's in step 5. The actual number of arithmetic operations required will depend on the string $a_1 \ldots a_n$ we start with, although since step 2 and step 5 each can be repeated at most n times, the number of operations will be no more than some constant multiple of n.

Let us consider how this algorithm might be applied to the satellite problem. We will restrict our attention to deciding whether the subset we have generated has a total weight of less than 700 kilograms. Let W_i be the weight of the ith experiment. Then we need to compute

$$a_1 W_1 + \ldots + a_n W_n$$

and see if this exceeds 700 or not. If we are including experiment i when a_i is 1 and excluding it when a_i is 0, then this sum gives the total weight of the included experiments. The reader should check that the sum may be computed using n multiplications and $n - 1$ additions, for a total of $2n - 1$ operations, not counting any comparisons or additions of an index.

Since if there are n experiments then there are 2^n subsets of experiments, and since generating and checking each subset takes a multiple of n operations, the complexity of this method of finding the best choice of experiments is $Cn \cdot 2^n$, where C is some constant. This is an expression that gets large quite quickly as n increases. The following table shows how long a computer that executes one million operations per second would take to do $n \cdot 2^n$ operations for various values of n. For comparison purposes we also show how long $1000n^2$ operations would take.

n	10	20	30	40	50
$n2^n$ operations	0.01 sec	21 sec	9 hr	1.4 years	1785 years
$1000n^2$ operations	0.1 sec	0.4 sec	0.9 sec	1.6 sec	2.5 sec

This table indicates that simply increasing computer speed may not make an algorithm practical, even for modest values of n. If, for example, our computer were capable of one *billion* operations per second instead of one million (making it 1000 times faster), performing $n2^n$ operations for $n = 50$ would still require 1.785 years.

In general an algorithm is considered ''good'' if its complexity is no more than some polynomial in n, although of course in practice a non-polynomial complexity may be acceptable if only small values of n arise.

Comparison of complexities involves knowledge of how fast various expressions depending on n get large as n increases. In general, expressions with n as an exponent grow faster than any polynomial in n; and $n!$ grows even faster. On the other hand, $\log_2 n$, which is explained in Section 2.4, although it increases without bound as n does, increases more slowly than any positive power of n. The mathematical comparison of these expressions entails analytic techniques not appropriate for this course, but we offer the following table to give an idea of how fast various expressions grow. The time given is that for a computer executing one million operations per second to run through the given number of operations.

n	10	20	30	40	50	60
expression						
$\log_2 n$	.0000033 S	.0000043 S	.0000049 S	.0000053 S	.0000056 S	.0000059 S
$n^{1/2}$	.0000032 S	.0000045 S	.0000055 S	.0000063 S	.0000071 S	.0000077 S
n	.00001 S	.00002 S	.00003 S	.00004 S	.00005 S	.00006 S
n^2	.0001 S	.0004 S	.0009 S	.0016 S	.0025 S	.0036 S
$n^2 + 10n$	.0002 S	.0006 S	.0012 S	.0020 S	.003 S	.0042 S
n^{10}	2.8 H	119 D	19 Y	333 Y	3097 Y	19174 Y
2^n	.001 S	1 S	18 M	13 D	36 Y	36559 Y
$n!$	3.6 S	77147 Y	$8.4 \cdot 10^{18}\,Y$	$2.6 \cdot 10^{34}\,Y$	$9.6 \cdot 10^{50}\,Y$	$2.6 \cdot 10^{68}\,Y$

EXERCISES 1.4

In Exercises 1–6 tell whether the given expression is a polynomial in x or not, and if so give its degree.

1. $5x^2 - 3x + \frac{1}{2}$

2. 16

3. $x^3 - \dfrac{1}{x^2}$

4. $2^x + 3x$

5. $\dfrac{1}{2x^2 + 7x + 1}$

6. $2x + 3x^{1/2} + 4$

In Exercises 7–10 compute the various values S takes on when the polynomial evaluation algorithm is used to compute P(x). Then do the same thing using Horner's polynomial evaluation algorithm.

7. $P(x) = 5x + 3, x = 2$

8. $P(x) = 3x^2 + 2x - 1, x = 5$

9. $P(x) = -x^3 + 2x^2 + 5x - 7, x = 2$

10. $P(x) = 2x^3 + 5x^2 - 4, x = 3$

In Exercises 11–14 tell what next string will be produced by the next subset algorithm.

11. 110101

12. 110111

13. 001101

14. 001001

In Exercises 15–18 make a table listing the values of k, j, and $a_1, \ldots, a_n$ after each step when the next subset algorithm is applied to the given string.

15. 101

16. 111

17. 1101

18. 1110

In Exercises 19–22 estimate how long a computer doing one million operations per second would take to do 3^n and $100n^3$ operations.

19. $n = 20$

20. $n = 30$

21. $n = 40$

22. $n = 50$

In Exercises 23 and 24 tell how many arithmetic operations the given algorithm will use. (It will depend on n.)

23. Algorithm for evaluating $n!$

Step 1 Let $k = 0$ and $P = 1$.

Step 2 If $k < n$, then replace k by $k + 1$, replace P by kP, and go to step 2.

Step 3 $P = n!$.

24. Algorithm for computing the sum of an arithmetic progression with first term a and common difference d

Step 1 Let $S = a$, $k = 1$, and $t = a$.

Step 2 If $k < n$, then replace t by $t + d$, S by $S + t$, and k by $k + 1$, and go to step 2.

Step 3 The sum is S.

Suggested Readings

1. Graham, Ronald L. ''The Combinatorial Mathematics of Scheduling.'' *Scientific American* (March 1978): 124–132.
2. Lawler, Eugene L. *Combinatorial Optimization: Networks and Matroids*. New York: Holt, Rinehart, and Winston, 1976.
3. Lewis, Harry R., and Christos H. Papadimitriou. ''The Efficiency of Algorithms.'' *Scientific American* (March 1978): 96–109.
4. Lockyer, K. G. *An Introduction to Critical Path Analysis*. 3rd ed. London: Pitman and Sons, 1969.
5. Niven, Ivan. *Mathematics of Choice or How to Count Without Counting*. New York: L. W. Singer, 1965.
6. Pipenger, Nicholas. ''Complexity Theory.'' *Scientific American* (June 1978): 114–124.

2

SETS, RELATIONS AND FUNCTIONS

As we saw in Chapter 1, discrete mathematics is concerned with solving problems in which the number of possibilities is finite. Often, as in the analysis of the knapsack problem in Section 1.3, the discussion of a problem requires consideration of all the possibilities for a solution. Such an approach can be made easier by the use of sets. In other situations we may need to consider relationships between the elements of sets. Such relationships can frequently be expressed using the mathematical ideas of relations and functions. In this chapter we will study these basic concepts as well as the principle of mathematical induction, an important method of proof in discrete mathematics.

2.1   Set Operations

In Section 1.3 we presented some of the basic ideas about sets. In this section we will discuss several ways in which sets can be combined to produce new sets.

Suppose that in the example discussed in Section 1.3 it is decided that the space shuttle will carry experiments on two successive trips. If $S = \{1, 5, 6, 8\}$ is the set of experiments carried on the first trip and $T = \{2, 4, 5, 8, 9\}$ is the set of experiments carried on the second trip, then $\{1, 2, 4, 5, 6, 8, 9\}$ is the set of experiments carried on the first or second trip or both. This set is called the union of S and T.

More generally, by the **union** of sets A and B we mean the set consisting of all the elements that are in A or in B. Note that, as always in mathematics, the word "or" in this definition is used in the inclusive sense. Thus, an element x is in the union of sets A and B in each of the following cases:

(**1**) $x \in A$ and $x \notin B$,

(**2**) $x \notin A$ and $x \in B$, or

(**3**) $x \in A$ and $x \in B$.

The union of sets A and B is denoted $A \cup B$. Thus,

$$A \cup B = \{x: x \in A \text{ or } x \in B\}.$$

Another set of interest in the space shuttle example is the set $\{5, 8\}$ of experiments carried on both trips. This set is called the intersection of S and T. In general, the **intersection** of sets A and B is the set consisting of all the elements that are in both A and B. This set is denoted $A \cap B$. So

$$A \cap B = \{x: x \in A \text{ and } x \in B\}.$$

If the intersection of the two sets is the empty set, then these sets are said to be **disjoint.**

EXAMPLE 2.1   If $A = \{1, 2, 4\}$, $B = \{2, 4, 6, 8\}$, and $C = \{3, 6\}$, then

$$A \cup B = \{1, 2, 4, 6, 8\} \quad \text{and} \quad A \cap B = \{2, 4\},$$
$$A \cup C = \{1, 2, 3, 4, 6\} \quad \text{and} \quad A \cap C = \varnothing, \text{ and}$$
$$B \cup C = \{2, 3, 4, 6, 8\} \quad \text{and} \quad B \cap C = \{6\}.$$

So A and C are disjoint sets.   ■

The set $\{1, 6\}$ of elements carried on the first space shuttle trip but not on the second trip is called the difference of S and T. More specifically, the **difference** of sets A and B, denoted $A - B$, is the set consisting of the elements in A that are not in B. Thus,

$$A - B = \{x: x \in A \text{ and } x \notin B\}.$$

Note that, as the following example shows, the sets $A - B$ and $B - A$ are not usually equal.

EXAMPLE 2.2   If A and B are as in Example 2.1, then $A - B = \{1\}$ and $B - A = \{6, 8\}$. ■

In many situations all of the sets under consideration are subsets of a set U. For example, in our discussion of the space shuttle example in Section 1.3, all of the sets were subsets of the set of experiments

$$U = \{1, 2, 3, 4, 5, 6, 7, 8, 9, 10, 11, 12\}.$$

Such a set containing all of the elements of interest in a particular situation is called a **universal set.** Since there are many different sets that could be used as a universal set, the particular universal set being considered must always be described explicitly. Given a universal set U and a subset A of U, the set $U - A$ is called the **complement** of A and is denoted $\overline{A}$.

EXAMPLE 2.3   If $A = \{1, 2, 4\}$, $B = \{2, 4, 6, 8\}$, and $C = \{3, 6\}$ are the sets in Example 2.1 and

$$U = \{1, 2, 3, 4, 5, 6, 7, 8\}$$

is the universal set, then

$$\overline{A} = \{3, 5, 6, 7, 8\},$$
$$\overline{B} = \{1, 3, 5, 7\}, \text{ and}$$
$$\overline{C} = \{1, 2, 4, 5, 7, 8\}. \quad ■$$

The theorem below lists some elementary properties of set operations. These properties follow immediately from the definitions given above.

THEOREM 2.1   Let U be a universal set. For any subsets A, B, and C of U, the following are true.

(a) $A \cup B = B \cup A$ and   (commutative laws)
$\quad A \cap B = B \cap A$

(b) $(A \cup B) \cup C = A \cup (B \cup C)$ and   (associative laws)
$\quad (A \cap B) \cap C = A \cap (B \cap C)$

(c) $A \cup (B \cap C) = (A \cup B) \cap (A \cup C)$ and   (distributive laws)
$\quad A \cap (B \cup C) = (A \cap B) \cup (A \cap C)$

(d) $\overline{\overline{A}} = A$

(e) $A \cup \overline{A} = U$

(f) $A \cap \overline{A} = \varnothing$

(g) $A \subseteq A \cup B$ and $B \subseteq A \cup B$

(h) $A \cap B \subseteq A$ and $A \cap B \subseteq B$

(i) $A - B = A \cap \overline{B}$

Relationships among sets can be pictured in **Venn diagrams,** which are named after the English logician John Venn (1834–1923). In a Venn diagram, the universal set is represented by a rectangular region and subsets of the universal set are usually represented by circular disks drawn within the rectangular region. Sets that are not known to be disjoint should be represented by overlapping circles as in Figure 2.1.

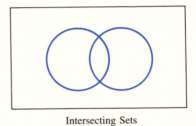

Intersecting Sets

FIGURE 2.1

Figure 2.2 contains Venn diagrams for the four set operations defined above. In each diagram the colored region depicts the set being represented.

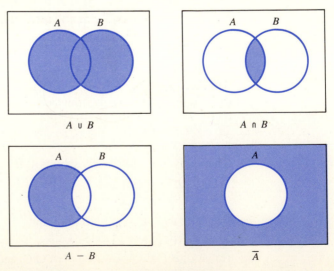

FIGURE 2.2

Venn diagrams depicting more complicated sets can be constructed by combining the basic diagrams found in Figure 2.2. For example, Figure 2.3 shows how to construct a Venn diagram for $\overline{(\overline{A} \cup B)}$.

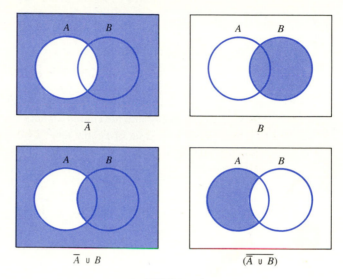

FIGURE 2.3

The theorem below enables us to determine the complement of a union or intersection of sets. This result will be needed in Section 7.6 to help us find the number of elements in the union of several sets.

THEOREM 2.2   *De Morgan laws* For any subsets A and B of a universal set U, the following are true.

(a) $\overline{(A \cup B)} = \overline{A} \cap \overline{B}$

(b) $\overline{(A \cap B)} = \overline{A} \cup \overline{B}$

Proof.   To prove that $\overline{(A \cup B)} = \overline{A} \cap \overline{B}$, we will show that each of the sets $\overline{(A \cup B)}$ and $\overline{A} \cap \overline{B}$ is a subset of the other.

First suppose that $x \in \overline{(A \cup B)}$. Then $x \notin A \cup B$. But since this is true, $x \notin A$ and $x \notin B$. So $x \in \overline{A}$ and $x \in \overline{B}$. It follows that $x \in \overline{A} \cap \overline{B}$. Thus, $\overline{(A \cup B)} \subseteq \overline{A} \cap \overline{B}$.

Now suppose that $x \in \overline{A} \cap \overline{B}$. Then $x \in \overline{A}$ and $x \in \overline{B}$. Hence, $x \notin A$ and $x \notin B$. It follows that $x \notin A \cup B$. So $x \in \overline{(A \cup B)}$. Thus, $\overline{A} \cap \overline{B} \subseteq \overline{(A \cup B)}$.

Because we have $\overline{(A \cup B)} \subseteq \overline{A} \cap \overline{B}$ and $\overline{A} \cap \overline{B} \subseteq \overline{(A \cup B)}$, it follows that $\overline{(A \cup B)} = \overline{A} \cap \overline{B}$. This proves part (a).

The proof of part (b) is similar and will be left as an exercise.   ∎

EXAMPLE 2.4   We can compute $\overline{(\overline{A} \cup B)}$ using Theorems 2.2(a), 2.1(d), and 2.1(i).

$$\overline{(\overline{A} \cup B)} = \overline{\overline{A}} \cap \overline{B} = A \cap \overline{B} = A - B$$

This equality is illustrated by Figure 2.3.   ■

EXAMPLE 2.5   According to the United States customs laws, a person is not allowed to bring liquor into the United States duty-free if he is not over 21 or if he has brought duty-free liquor into this country in the previous thirty days. Who is allowed to bring duty-free liquor into the United States?

    Let A denote the set of people aged 21 or over, and let B denote the set of people who have brought duty-free liquor into the U.S. in the previous thirty days. Then the persons who are not allowed to bring duty-free liquor into the United States are those in the set $\overline{A} \cup B$. This means that those who are eligible to bring duty-free liquor into this country are those in the set $\overline{(\overline{A} \cup B)}$. Example 2.4 shows that the persons who can bring duty-free liquor into the United States are those in the set $A - B$.   ■

    When listing the elements of a set, the order in which the elements are written is immaterial. Thus, for example, $\{1, 2, 3\} = \{2, 3, 1\} = \{3, 1, 2\}$. Often, however, we need to be able to distinguish the order in which two elements are written. In an **ordered pair** of elements a and b, denoted (a, b), the order in which the entries is written is taken into account. Thus, $(1, 2) \neq (2, 1)$, and $(a, b) = (c, d)$ if and only if $a = c$ and $b = d$.

    The final set operation that we will consider is the Cartesian product, which arises in connection with relations (to be studied in Section 2.2). Given sets A and B, the **Cartesian product** of A and B is the set consisting of all the ordered pairs (a, b), where $a \in A$ and $b \in B$. The Cartesian product of A and B is denoted $A \times B$. Thus,

$$A \times B = \{(a, b): a \in A \text{ and } b \in B\}.$$

The Cartesian product is often encountered in discussions of the Euclidean plane; for if R denotes the set of all real numbers, then $R \times R$ is the set of all ordered pairs of real numbers, which can be pictured as the Euclidean plane.

EXAMPLE 2.6   Let $A = \{1, 2, 3\}$ and $B = \{3, 4\}$. Then

$$A \times B = \{(1, 3), (1, 4), (2, 3), (2, 4), (3, 3), (3, 4)\} \text{ and}$$
$$B \times A = \{(3, 1), (3, 2), (3, 3), (4, 1), (4, 2), (4, 3)\}. \quad ■$$

As Example 2.6 shows, usually $A \times B \neq B \times A$.

EXAMPLE 2.7   A public opinion poll is to be taken to see how each of the leading Democratic presidential candidates in 1988 would fare against each of the leading Republican presidential candidates. Suppose that the set of Democratic candidates to be considered is

$$D = \{\text{Cuomo, Hart, Kennedy}\}$$

and the set of Republican candidates to be considered is

$$R = \{\text{Baker, Bush, Kemp}\}.$$

How many pairings of a Democratic and a Republican candidate are there?

The set of all possible pairs of a Democratic and Republican candidate is $D \times R$. The elements of this set are the pairs

(Cuomo, Baker), (Cuomo, Bush), (Cuomo, Kemp),

(Hart, Baker), (Hart, Bush), (Hart, Kemp),

(Kennedy, Baker), (Kennedy, Bush), and (Kennedy, Kemp).

There are 9 different pairings of a Democratic and a Republican candidate. ∎

EXERCISES 2.1

In Exercises 1–4 evaluate $A \cup B$, $A \cap B$, $A - B$, $\overline{A}$, and $\overline{B}$ for each of the given sets A and B. In each case assume that the universal set is $U = \{1, 2, 3, \ldots, 9\}$.

1. $A = \{2, 3, 5, 7, 8\}$ and $B = \{1, 3, 4, 5, 6, 9\}$ **2.** $A = \{1, 4, 6, 9\}$ and $B = \{1, 2, 4, 5, 6, 7, 9\}$

3. $A = \{1, 2, 4, 8, 9\}$ and $B = \{3, 7\}$ **4.** $A = \{3, 4, 6, 7, 8, 9\}$ and $B = \{2, 5, 7, 9\}$

In Exercises 5–8 compute $A \times B$ for each of the given sets A and B.

5. $A = \{1, 2, 3, 4\}$ and $B = \{7, 8\}$ **6.** $A = \{3, 4, 5\}$ and $B = \{1, 2, 3\}$

7. $A = \{a, e\}$ and $B = \{x, y, z\}$ **8.** $A = \{p, q, r, s\}$ and $B = \{a, c, e\}$

Draw Venn diagrams representing the sets in Exercises 9–12.

9. $(\overline{A \cap \overline{B}})$ **10.** $\overline{A} - \overline{B}$ **11.** $\overline{A} \cap (B \cup C)$ **12.** $A \cup (B - C)$

13. Give an example of sets for which $A \cup C = B \cup C$, but $A \neq B$.

14. Give an example of sets for which $A \cap C = B \cap C$, but $A \neq B$.

15. Give an example of sets for which $A - C = B - C$, but $A \neq B$.

16. Give an example of sets A, B, and C for which $(A - B) - C \neq A - (B - C)$.

Use Theorems 2.1 and 2.2 as in Example 2.4 to simplify the sets in Exercises 17–24.

17. $A \cap (B - A)$ **18.** $(A - B) \cup (A \cap B)$ **19.** $(A - B) \cap (A \cup B)$ **20.** $\overline{A} \cap (A \cap B)$

21. $\overline{A} \cap (A \cup B)$ **22.** $(\overline{A - B}) \cap A$ **23.** $A \cap (\overline{A \cap B})$ **24.** $A \cup (\overline{A \cup B})$

25. If A is a set containing m elements and B is a set containing n elements, how many elements are there in $A \times B$?

26. Under what conditions is $A - B = B - A$?   **27.** Under what conditions is $A \cup B = A$?

28. Under what conditions is $A \cap B = A$?   **29.** Prove parts (c) and (i) of Theorem 2.1.

30. Prove part (b) of Theorem 2.2 by using an argument similar to that in the proof of part (a).

31. Note that if $A = B$, then $\overline{A} = \overline{B}$. Use this fact to prove part (b) of Theorem 2.2 from part (a).

32. Let A and B be subsets of a universal set U. Prove that if $A \subseteq B$, then $\overline{B} \subseteq \overline{A}$.

Prove the set equalities in Exercises 33–38.

33. $(A - B) \cup (B - A) = (A \cup B) - (A \cap B)$ **34.** $(A - B) \cup (A - C) = A - (B \cap C)$

35. $(A - B) - C = (A - C) - (B - C)$ **36.** $(A \times C) \cap (B \times D) = (A \cap B) \times (C \cap D)$

37. $(A - B) \cap (A - C) = A - (B \cup C)$ **38.** $(A - C) \cap (B - C) = (A \cap B) - C$

39. Give an example where $(A \times C) \cup (B \times D) \neq (A \cup B) \times (C \cup D)$.

40. Prove that $(A \times C) \cup (B \times D) \subseteq (A \cup B) \times (C \cup D)$.

2.2   Equivalence Relations

In Section 1.2 we considered a problem involving the matching of pilots to flights having different destinations. Recall that the seven destinations and the pilots who requested them are as shown below.

> Los Angeles: Timmack, Jelinek, Rupp
>
> Seattle: Alfors, Timmack, Tang, Washington
>
> London: Timmack, Tang, Washington
>
> Frankfurt: Alfors, Tang, Rupp, Ramirez
>
> Paris: Jelinek, Washington, Rupp
>
> Madrid: Jelinek, Ramirez
>
> Dublin: Timmack, Rupp, Ramirez

This list establishes a relation between the set of destinations and the set of pilots, where a pilot is related to one of the seven destinations whenever that destination was requested by the pilot.

From this list we can construct a set of ordered pairs in which the first entry of each ordered pair is a destination and the second entry is a pilot who requested that destination. For example, the pairs (Los Angeles, Timmack), (Los Angeles, Jelinek), and (Los Angeles, Rupp) correspond to the three pilots who requested the flight to Los Angeles. Let

$$S = \{(\text{Los Angeles, Timmack}), (\text{Los Angeles, Jelinek}), \ldots ,$$

$$(\text{Dublin, Ramirez})\}$$

denote the set of all 22 ordered pairs of destinations and the pilots who requested them. This set contains exactly the same information as the original list of pilots and their requested destinations. Thus, the relation between the destinations and the pilots who requested them can be completely described by a set of ordered pairs. Notice that S is a subset of $A \times B$, where A is the set of destinations and B is the set of pilots.

Generalizing from the example above, we define a **relation from a set A to a set B** to be any subset of the Cartesian product $A \times B$. If R is a relation from set A to set B and (x, y) is an element of R, we will say that **x is related to y by R** and write $x R y$ instead of $(x, y) \in R$.

EXAMPLE 2.8   Suppose that among three college professors Lopez speaks Dutch and French; Parr speaks German and Russian; and Zak speaks Dutch. Let

$$A = \{Lopez, Parr, Zak\}$$

denote this set of professors and

$$B = \{Dutch, French, German, Russian\}$$

denote the set of foreign languages they speak. Then

$$R = \{(Lopez, Dutch), (Lopez, French), (Parr, German), (Parr, Russian),$$
$$(Zak, Dutch)\}$$

is a relation from A to B in which x is related to y by R whenever Professor x speaks language y. So, for instance, Lopez R French and Parr R German are both true, but Zak R Russian is false. ■

Often we need to consider a relation between the elements of some set. In Section 1.1, for instance, we considered the problem of determining how long it would take to produce and deliver the advertisements for a department store sale. In analyzing this problem, we listed all the tasks that needed to be done and represented each task by a letter. Then we determined which tasks immediately preceded each task. This listing established a relation on the set of tasks

$$S = \{A, B, C, D, E, F, G, H, I, J, K\}$$

in which task X is related to task Y if X immediately precedes Y. The resulting relation

$$\{(A, C), (B, C), (C, D), (C, E), (D, F), (E, F), (C, G), (G, H), (F, I),$$
$$(H, J), (I, J), (J, K)\}$$

is a relation from set S to itself, that is, a subset of $S \times S$.

A relation from a set S to itself is called a **relation on S**.

EXAMPLE 2.9   Let $S = \{1, 2, 3, 4\}$. Define a relation R on S by letting $x\,R\,y$ mean $x < y$. Then 1 is related to 4, but 4 is not related to 2. Likewise $2\,R\,3$ is true, but $4\,R\,2$ is false.   ■

A relation R on a set S may have any of the following special properties.

(1) If for each x in S, $x\,R\,x$ is true, then R is called **reflexive.**
(2) If $y\,R\,x$ is true whenever $x\,R\,y$ is true, then R is called **symmetric.**
(3) If $x\,R\,z$ is true whenever $x\,R\,y$ and $y\,R\,z$ are both true, then R is called **transitive.**

The relation R in Example 2.9 is not reflexive since $1\,R\,1$ is false. Likewise, it is not symmetric because $4\,R\,1$ is false but $1\,R\,4$ is true. However, R is transitive because if x is less than y and y is less than z, then x is less than z.

EXAMPLE 2.10   Let S be the set of positive integers, and define $x\,R\,y$ to mean that x divides y (that is, $\dfrac{y}{x}$ is an integer). Thus, $3\,R\,6$ and $7\,R\,35$ are true, but $8\,R\,4$ and $6\,R\,9$ are false.

Then R is a relation on S. Furthermore, R is reflexive since every positive integer divides itself, and R is transitive since if x divides y and y divides z, then x divides z. (To see why this is so, note that if $\dfrac{y}{x}$ and $\dfrac{z}{y}$ are integers, so is $\dfrac{z}{x} = \dfrac{y}{x} \cdot \dfrac{z}{y}$.)

However, R is not symmetric because $2\,R\,8$ is true, but $8\,R\,2$ is false.   ■

EXAMPLE 2.11   Let S denote the set of all nonempty subsets of $\{1, 2, 3, 4, 5\}$ and define $A\,R\,B$ to mean that $A \cap B \neq \varnothing$. Then R is clearly reflexive and symmetric. But R is not transitive since $\{1, 2\}\,R\,\{2, 3\}$ and $\{2, 3\}\,R\,\{3, 4\}$ are true, but $\{1, 2\}\,R\,\{3, 4\}$ is false.   ■

A relation that is reflexive, symmetric, and transitive is called an **equivalence relation.** The most familiar example of an equivalence relation is the relation of equality. Two more examples follow.

EXAMPLE 2.12   On the set of students attending a particular university, define one student to be related to another whenever their surnames begin with the same letter. This relation is easily seen to be an equivalence relation on the set of students at this university.   ■

EXAMPLE 2.13   An integer greater than 1 is called **prime** if its only positive integer divisors are itself and 1. The first few prime numbers are 2, 3, 5, 7, 11, 13, 17, 19, and 23. We will see in Theorem 2.17 that every integer greater than 1 is either prime or a product of primes. For example, 67 is prime and $65 = 5 \cdot 13$ is a product of primes. On the set S of integers greater than 1, define $x\,R\,y$ to mean that x has the same number of *distinct* prime divisors as y. Thus, for example, $12\,R\,55$ since $12 = 2 \cdot 2 \cdot 3$ and $55 = 5 \cdot 11$ both have two distinct prime divisors. Then R is an equivalence relation on S. ■

EXAMPLE 2.14   Let S denote the set of all people in the United States. Define a relation R on S by letting $x\,R\,y$ mean that x has the same mother or father as y. Then R is easily seen to be reflexive and symmetric. But R is not transitive, for x and y may have the same mother and y and z may have the same father, but x and z may have no parent in common. Hence, R is not an equivalence relation on S. ■

If R is an equivalence relation on a set S and $x \in S$, the set of elements of S that are related to x is called the **equivalence class** containing x and is denoted $[x]$. Thus,

$$[x] = \{y \in S : y\,R\,x\}.$$

Note that by the reflexive property of R, $x \in [x]$ for each element x in S. In the equivalence relation described in Example 2.12 there are 26 possible equivalence classes; namely the set of students whose surnames begin with A, the set of students whose surnames begin with B, and so forth. In this example different equivalence classes are disjoint sets. This fact is true in general, as the following theorem shows.

THEOREM 2.3   Let R be an equivalence relation on a set S. Then:

    (a) If x and y are elements of S, then x is related to y by R if and only if $[x] = [y]$.

    (b) Two equivalence classes of R are either equal or disjoint.

Proof.   (a) Let x and y be elements of S such that $x\,R\,y$. We will prove that $[x] = [y]$ by showing that $[x] \subseteq [y]$ and $[y] \subseteq [x]$.

If $z \in [x]$, then z is related to x, that is, $z\,R\,x$. But if $z\,R\,x$ and $x\,R\,y$, then $z\,R\,y$ by the transitive property of R. So $z \in [y]$. This proves that $[x] \subseteq [y]$.

If $z \in [y]$, then $z\,R\,y$ as above. By assumption $x\,R\,y$, and so $y\,R\,x$ is true by the symmetric property of R. But then $z\,R\,y$ and $y\,R\,x$ imply $z\,R\,x$ by the transitive property. Hence, $z \in [x]$, proving that $[y] \subseteq [x]$. Since we have both $[x] \subseteq [y]$ and $[y] \subseteq [x]$, it follows that $[x] = [y]$.

Conversely, suppose that $[x] = [y]$. Now $x \in [x]$ by the reflexive property, and so $x \in [y]$ because $[x] = [y]$. But if $x \in [y]$, then $x \, R \, y$. This completes the proof of part (a).

(b) Let $[u]$ and $[v]$ be any two equivalence classes of R. If $[u]$ and $[v]$ are not disjoint, then they contain a common element w. Since $w \in [u]$, part (a) shows that $[w] = [u]$. Likewise, $[w] = [v]$. It follows that $[u] = [v]$. Hence, $[u]$ and $[v]$ are either disjoint or equal. ■

Because of part (b) of Theorem 2.3, the equivalence classes of an equivalence relation R on set S divide S into disjoint subsets. This family of subsets has the following properties:

(1) No subset is empty.
(2) Each element of S belongs to some subset.
(3) Two distinct subsets are disjoint.

Such a family of subsets of S is called a **partition** of S.

EXAMPLE 2.15   Let $A = \{1, 3, 4\}$, $B = \{2, 6\}$, and $C = \{5\}$. Then $\mathcal{P} = \{A, B, C\}$ is a partition of $S = \{1, 2, 3, 4, 5, 6\}$. See Figure 2.4. ■

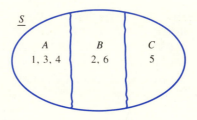

FIGURE 2.4

We have seen that every equivalence relation on S gives rise to a partition of S by taking the family of subsets in the partition to be the equivalence classes of the equivalence relation. Conversely, if $\mathcal{P}$ is a partition of S, we can define a relation R on S by letting $x \, R \, y$ mean that x and y lie in the same member of $\mathcal{P}$. Using the partition in Example 2.15, for instance, we obtain the relation

$$\{(1, 1), (1, 3), (1, 4), (3, 1), (3, 3), (3, 4), (4, 1), (4, 3), (4, 4), (2, 2), (2, 6),$$
$$(6, 2), (6, 6), (5, 5)\}.$$

Then clearly R is an equivalence relation on S, and the equivalence classes of R are precisely the members of $\mathcal{P}$. We will state these facts formally as our next theorem.

THEOREM 2.4

(a) An equivalence relation R gives rise to a partition $\mathscr{P}$ in which the members of $\mathscr{P}$ are the equivalence classes of R.

(b) Conversely, a partition $\mathscr{P}$ induces an equivalence relation R in which two elements are related by R whenever they lie in the same member of $\mathscr{P}$. Moreover, the equivalence classes of this relation are the members of $\mathscr{P}$.

Although the definitions of an equivalence relation and a partition appear to be quite different, as a result of Theorem 2.4 we see that these two concepts are actually just different ways of describing the same situation.

EXERCISES 2.2

In Exercises 1–12 determine which of the reflexive, symmetric, and transitive properties are satisfied by the given relation R defined on set S.

1. $S = \{1, 2, 3\}$ and $R = \{(1, 1), (1, 2), (2, 1), (2, 2)\}$.

2. $S = \{1, 2, 3\}$ and $R = \{(1, 1), (1, 3), (2, 2), (2, 3), (3, 1), (3, 2), (3, 3)\}$.

3. S is the set of all Illinois residents and $x\,R\,y$ means that x has the same mother as y.

4. S is the set of all citizens of the United States and $x\,R\,y$ means that x has the same weight as y.

5. S is the set of all students at Illinois State University and $x\,R\,y$ means that the height of x differs from the height of y by no more than one inch.

6. S is the set of all teenagers and $x\,R\,y$ means that x has a grandfather in common with y.

7. S is the set of all graduates of Michigan State University and $x\,R\,y$ means that x graduated in the same year as y.

8. S is the set of all residents of Los Angeles and $x\,R\,y$ means that x is a brother of y.

9. S is the set of all real numbers and $x\,R\,y$ means that $x^2 = y^2$.

10. S is the set of positive integers and $x\,R\,y$ means that x divides y or y divides x.

11. S is the set of all subsets of $\{1, 2, 3, 4\}$ and $X\,R\,Y$ means $X \subseteq Y$.

12. S is the set of ordered pairs of real numbers and $(x_1, x_2)\,R\,(y_1, y_2)$ means that $x_1 = y_1$ and $x_2 < y_2$.

In Exercises 13–18 show that the given relation R is an equivalence relation on set S. Then describe the distinct equivalence classes of R.

13. Let S be the set of integers and define $x\,R\,y$ to mean that $x - y$ is even.

14. Let S be the set of residents of New York City and define $x\,R\,y$ to mean that the surname of x contains the same number of letters as the surname of y.

15. Let S be the set of integers greater than 1 and define $x\,R\,y$ to mean that the largest prime divisor of x equals the largest prime divisor of y.

X7 1

$S = \{\phi, \{1\}, \{2\}, \{3\}, \ldots \}$

16. Let S be the set of all subsets of $\{1, 2, 3, 4, 5\}$ and define $X\,R\,Y$ to mean that $X \cap \{1, 3, 5\} = Y \cap \{1, 3, 5\}$.

17. Let S be the set of ordered pairs of real numbers and define $(x_1, x_2)\,R\,(y_1, y_2)$ to mean that $x_1^2 + x_2^2 = y_1^2 + y_2^2$.

18. Let S be the set of ordered pairs of positive integers and define $(x_1, x_2)\,R\,(y_1, y_2)$ to mean that $x_1 + y_2 = y_1 + x_2$.

19. Write the equivalence relation on $\{1, 2, 3, 4, 5\}$ that is induced by the partition with $\{1, 5\}, \{2, 4\}$, and $\{3\}$ as its partitioning subsets.

20. Write the equivalence relation on $\{1, 2, 3, 4, 5, 6\}$ that is induced by the partition with $\{1, 3, 6\}, \{2, 5\}$, and $\{4\}$ as its partitioning subsets.

21. Let R be an equivalence relation on a set S. Prove that if x and y are any elements in S, $x\,R\,y$ is false if and only if $[x] \cap [y] = \varnothing$.

22. Let R be an equivalence relation on a set S, and let x and y be elements of S. Prove that if $a \in [x]$, $b \in [y]$, and $[x] \neq [y]$, then $a\,R\,b$ is false.

23. What is wrong with the following argument that attempts to show that if R is a relation on set S that is both symmetric and transitive, then R is also reflexive?
    Since $x\,R\,y$ implies $y\,R\,x$ by the symmetric property, $x\,R\,y$ and $y\,R\,x$ imply $x\,R\,x$ by the transitive property. Thus, $x\,R\,x$ is true for each $x \in S$, and so R is reflexive.

24. Let R_1 and R_2 be equivalence relations on sets S_1 and S_2, respectively. Define a relation R on $S_1 \times S_2$ by letting $(x_1, x_2)\,R\,(y_1, y_2)$ mean that $x_1\,R_1\,y_1$ and $x_2\,R_2\,y_2$. Prove that R is an equivalence relation on $S_1 \times S_2$ and describe the equivalence classes of R.

25. Determine the number of relations on a set S containing n elements.

26. Call a relation R "circular" if $x\,R\,y$ and $y\,R\,z$ imply $z\,R\,x$. Prove that R is an equivalence relation if and only if R is both reflexive and circular.

27. Let S be a set containing n elements, where n is a positive integer. How many ways are there to partition S into two subsets?

28. How many partitions are there of a set containing three elements?

29. How many partitions are there of a set containing four elements?

30. Prove Theorem 2.4.

31. Let S be any nonempty set and f any function with domain S. Define $s_1\,R\,s_2$ to mean that $f(s_1) = f(s_2)$. Prove that R is an equivalence relation on S.

32. State and prove a converse to Exercise 31.

33. Let $p_m(n)$ be the number of partitions of a set of n elements into m subsets. Show that for $1 \leq m \leq n$, $p_m(n + 1) = mp_m(n) + p_{m-1}(n)$.

2.3   Congruence

In this section we will consider an important equivalence relation on the set of integers. This relation will lead to the study of number systems containing only finitely many elements. Such number systems arise naturally in the study of computer arithmetic.

We will begin by recalling some familiar ideas from arithmetic. If n and m are integers and $m > 0$, the process of long division enables us to express n in the form $n = qm + r$, where $0 \le r < m$. The numbers q and r are called the **quotient** and **remainder,** respectively, in the division of n by m. Thus, for instance, in the division of 34 by 9, the quotient is 3 and the remainder is 7 since $34 = 3 \cdot 9 + 7$ and $0 \le 7 < 9$. If the remainder in the division of n by m is 0, then we say that n **is divisible by** m (or that m **divides** n). Thus, to say that n is divisible by m means that $\dfrac{n}{m}$ is an integer.

Now let m be an integer greater than 1. If p and q are integers, we say that p **is congruent to** q **modulo** m if $p - q$ is divisible by m. If p is congruent to q modulo m, we write $p \equiv q \pmod{m}$; otherwise we write $p \not\equiv q \pmod{m}$. We call this relation on the set of integers **congruence modulo** m.

EXAMPLE 2.16

Clearly, $3 \equiv 24 \pmod 7$ because $3 - 24 = -21$ is divisible by 7. And similarly, $98 \equiv 43 \pmod{11}$ because $98 - 43 = 55$ is divisible by 11. But $42 \not\equiv 5 \pmod 8$ since $42 - 5 = 37$ is not divisible by 8, and $4 \not\equiv 29 \pmod 6$ since $4 - 29 = -25$ is not divisible by 6. ■

The most common situation in which congruence occurs is in connection with the telling of time. Standard clocks and watches keep track of time modulo 12. Thus, we say that 15 hours after 7 o'clock is 10 o'clock because $7 + 15 \equiv 10 \pmod{12}$. Transportation schedules (such as train schedules) usually keep track of time modulo 24 because there are 24 hours per day.

EXAMPLE 2.17

Congruences often occur in applications involving error-correcting codes. In this example we will describe an application of such a code in the publishing industry.

Since 1972 a book published anywhere in the world has carried a 10-digit numerical code number called an International Standard Book Number (ISBN). For instance, the ISBN for *Finite Mathematics* by Lawrence E. Spence is 0–06–046369–4. By providing a standard numerical identifier for books, these numbers have allowed publishers and bookstores to computerize their inventories and billing procedures more easily than if each book had to be referred to by author, title, and edition.

An ISBN consists of four parts: a group code, a publisher code, an identifying number assigned by the publisher, and a check digit. In the ISBN 0–06–046369–4, the group code (0) denotes that the book was published in an English-speaking country (either Australia, Canada, New Zealand, South Africa, the United Kingdom, or the United States). The next group of digits (06) identifies the publisher, Harper and Row, Publishers; and the third group of digits (046369) designates this particular book among all those published by Harper and Row. The final digit of the ISBN (4) is the check digit, which is used to detect errors in copying or transmitting the ISBN. By using the check digit, publishers are often able to detect an incorrect ISBN so as to prevent the costly shipping charges that would result from filling an incorrect order.

The check digit in the ISBN has eleven possible values, 0, 1, 2, 3, 4, 5, 6, 7, 8, 9, or X. (A check digit of X represents the number 10.) This digit is determined in the following way: multiply the first nine digits of the ISBN by 10, 9, 8, 7, 6, 5, 4, 3, and 2 respectively, and add these nine products to obtain a number y. The check digit d is then chosen so that $y + d \equiv 0 \pmod{11}$. For example, the check digit in the ISBN for Spence's *Finite Mathematics* is 4 since

$$10(0) + 9(0) + 8(6) + 7(0) + 6(4) + 5(6) + 4(3) + 3(6) + 2(9) =$$
$$0 + 0 + 48 + 0 + 24 + 30 + 12 + 18 + 18 = 150$$

and $150 + 4 = 154 \equiv 0 \pmod{11}$.

Likewise, the ISBN for *Algebra Lineal* by Stephen H. Friedberg, Arnold J. Insel, and Lawrence E. Spence (published in Spanish by Publicaciones Cultural, S.A.), is 968–439–197–8. Here the check digit is 8 because

$$10(9) + 9(6) + 8(8) + 7(4) + 6(3) + 5(9) + 4(1) + 3(9) + 2(7) =$$
$$90 + 54 + 64 + 28 + 18 + 45 + 4 + 27 + 14 = 344$$

and $344 + 8 = 352 \equiv 0 \pmod{11}$. ∎

It is easily seen that p is congruent to q modulo m precisely when $p = km + q$ for some integer k. Hence, p is congruent to q modulo m if and only if p and q have the same remainder when divided by m. (See Exercise 49.) From this fact the theorem below follows immediately.

THEOREM 2.5   Congruence modulo m is an equivalence relation.

The equivalence classes for congruence modulo m are called **congruence classes modulo m.** The set of all the congruence classes modulo m will be denoted by Z_m. It follows from Theorem 2.3 that any two congruence classes modulo m are either equal or disjoint. Moreover, in Z_m, $[p] = [q]$ if and only if $p \equiv q \pmod{m}$.

EXAMPLE 2.18   In Z_3 the distinct congruence classes are

$$[0] = \{\ldots, -6, -3, 0, 3, 6, 9, \ldots\},$$
$$[1] = \{\ldots, -5, -2, 1, 4, 7, 10, \ldots\}, \text{ and}$$
$$[2] = \{\ldots, -4, -1, 2, 5, 8, 11, \ldots\}.$$

Notice that each of the congruence classes in Z_3 has many possible representations. For instance, $[0] = [3] = [9] = [-12]$ and $[2] = [-4] = [11] = [32]$. ▪

In general, there are m distinct congruence classes in Z_m, namely $[0]$, $[1]$, $[2]$, $\ldots$, $[m - 1]$. These classes correspond to the m possible remainders when dividing by m.

We would like to define addition and multiplication in Z_m. There is a natural way to do this using the addition and multiplication of integers; simply define

$$[p] + [q] = [p + q] \text{ and } [p][q] = [pq].$$

However, in order for this definition to make sense we must be sure that our definition does not depend on the way in which congruence classes are represented. In other words, we must be certain that these definitions depend only on the congruence classes themselves. For example, in Z_3 we have $[0] = [9]$ and $[2] = [11]$; so we must be certain that the sums $[0] + [2]$ and $[9] + [11]$ give the same answer. The following result gives us that assurance.

THEOREM 2.6   If $p \equiv x \pmod{m}$ and $q \equiv y \pmod{m}$, then:

(a) $p + q \equiv x + y \pmod{m}$ and

(b) $pq \equiv xy \pmod{m}$.

Proof.   If $p \equiv x \pmod{m}$ and $q \equiv y \pmod{m}$, then there are integers r and s such that $p = rm + x$ and $q = sm + y$.

(a) Thus, $p + q = (rm + x) + (sm + y) = (r + s)m + (x + y)$. So,

$$(p + q) - (x + y) = (r + s)m,$$

so that $(p + q) - (x + y)$ is divisible by m. This proves (a).

(b) Likewise, $pq = (rm + x)(sm + y) = (rms + ry + xs)m + xy$, so that

$$pq - xy = (rms + ry + xs)m.$$

Hence, $pq - xy$ is divisible by m, proving (b). ▪

Notice that part (b) of Theorem 2.6 implies that if $p \equiv x \pmod{m}$, then $p^n \equiv x^n \pmod{m}$ for all positive integers n. Moreover, the definition of multiplication in Z_m shows that $[x]^n = [x^n]$.

EXAMPLE 2.19   In Z_6 we have

$$[5] + [3] = [5 + 3] = [8] = [2]$$

since $8 \equiv 2 \pmod 6$. Also,

$$[5][3] = [5 \cdot 3] = [15] = [3]$$

because $15 \equiv 3 \pmod 6$. And

$$[8]^4 = [2]^4 = [2^4] = [16] = [4]$$

since $8 \equiv 2 \pmod 6$ and $16 \equiv 4 \pmod 6$. ■

EXAMPLE 2.20   In Z_8 we have

$$[4] + [7] = [4 + 7] = [11] = [3]$$

since $11 \equiv 3 \pmod 8$. Also

$$[4][7] = [4 \cdot 7] = [28] = [4]$$

because $28 \equiv 4 \pmod 8$. And

$$[7]^9 = [-1]^9 = [(-1)^9] = [-1] = [7]$$

since $7 \equiv -1 \pmod 8$. ■

EXAMPLE 2.21   A scientific recording instrument uses one foot of paper per hour. If a new roll of paper 100 feet long is installed at 11 A.M., at what time of day will the instrument run out of paper?

To answer this question, we will number the hours of a day with midnight being hour 0, 1 A.M. being hour 1, etc. Using arithmetic in Z_{24}, we see that the paper will run out at time $[11] + [100] = [111] = [15]$. Since hour 15 corresponds to 3 P.M., we see that the paper will run out at 3 P.M. ■

EXAMPLE 2.22   In the Apple Pascal programming language, integer variables must have values between $-32{,}768$ and $32{,}767$ inclusive. This range permits integer variables to have $65{,}536 = 2^{16}$ different values. Moreover, all integer arithmetic is done modulo 65,536, with answers given in the range above. Thus,

$$60{,}000 + 10{,}000 = 4464$$

because $70{,}000 \equiv 4464 \pmod{65{,}536}$. Likewise,

$$23{,}000 + 13{,}000 = -29{,}536$$

since $36{,}000 \equiv 36{,}000 - 65{,}536 \equiv -29{,}536 \pmod{65{,}536}$. Similarly, we see that $400 \cdot 500 = 3392$ and $123 \cdot 487 = -5635$. ∎

EXAMPLE 2.23

On a Sharp model EL–506S calculator, the value of 2^{30} is given as $1{,}073{,}741{,}820$. If this value is correct, then the last digit of $2^{28} = 2^{30}/4$ must be 5. But clearly no power of 2 can be odd, so the last digit of 2^{30} must be wrong. What is the correct last digit of 2^{30}?

It is easy to see that two positive integers have the same last digit if and only if they are congruent modulo 10. But in Z_{10}

$$[2^{30}] = [2^5]^6 = [32]^6 = [2]^6 = [2^6] = [64] = [4].$$

Hence, the last digit of 2^{30} is 4. Actually $2^{30} = 1{,}073{,}741{,}824$. ∎

EXERCISES 2.3

In Exercises 1–8 find the quotient and remainder in the division of n by m.

1. $n = 67$ and $m = 9$

2. $n = 39$ and $m = 13$

3. $n = 25$ and $m = 42$

4. $n = 103$ and $m = 8$

5. $n = 54$ and $m = 6$

6. $n = 75$ and $m = 23$

7. $n = 89$ and $m = 10$

8. $n = 57$ and $m = 11$

In Exercises 9–16 determine if $p \equiv q \pmod{m}$.

9. $p = 29, q = -34$, and $m = 7$

10. $p = 47, q = 8$, and $m = 11$

11. $p = 96, q = 35$, and $m = 10$

12. $p = 21, q = 53$, and $m = 8$

13. $p = 39, q = -46$, and $m = 2$

14. $p = 75, q = -1$, and $m = 19$

15. $p = 91, q = 37$, and $m = 9$

16. $p = 83, q = -23$, and $m = 6$

In Exercises 17–36 perform the indicated calculations in Z_m. Write your answer in the form [r] with $0 \le r < m$.

17. $[8] + [6]$ in Z_{12}

18. $[9] + [11]$ in Z_{15}

19. $[5] + [10]$ in Z_{11}

20. $[9] + [8]$ in Z_{13}

21. $[23] + [15]$ in Z_8

22. $[12] + [25]$ in Z_7

23. $[16] + [9]$ in Z_6

24. $[43] + [31]$ in Z_{22}

25. $[8][7]$ in Z_6

26. $[9][3]$ in Z_4

27. $[4][11]$ in Z_9

28. $[3][20]$ in Z_{11}

29. $[5][12]$ in Z_8

30. $[8][11]$ in Z_5

31. $[9][6]$ in Z_{10}

32. $[16][3]$ in Z_7

33. $[9]^7$ in Z_7

34. $[11]^8$ in Z_5

35. $[11]^9$ in Z_{13}

36. $[13]^6$ in Z_{15}

37. A newspaper teletypewriter that is in constant operation uses four feet of paper per hour. If a new roll of paper 200 feet long is installed at 6 P.M., at what time of day will the machine run out of paper?

38. A hospital monitoring device uses two feet of paper per hour. If it is attached to a patient at 8 A.M. with a supply of paper 150 feet long, at what time of day will the device run out of paper?

39. Use Example 2.17 to determine the correct check digit for the ISBN that has 3–540–90518 as its first nine digits. $200 + x = 0 \bmod 11$ $x = 9$

40. Use Example 2.17 to determine the correct check digit for the ISBN that has 0–553–10310 as its first nine digits.

41. Use Example 2.22 to determine the result of the operations $26{,}793 + 28{,}519$ and $418 \cdot 697$ if performed on integer variables in the Apple Pascal programming language.

42. Use Example 2.22 to determine the result of the operations $4{,}082 + 30{,}975$ and $863 \cdot 729$ if performed on integer variables in the Apple Pascal programming language.

43. Let A denote the equivalence class containing 4 in Z_6 and B denote the equivalence class containing 4 in Z_8. Is $A = B$?

44. In Z_8 which of the following congruence classes are equal: $[2]$, $[7]$, $[10]$, $[16]$, $[39]$, $[45]$, $[-1]$, $[-3]$, $[-6]$, $[-17]$, and $[-23]$?

45. Let R be the equivalence relation defined in Example 2.13. Give an example to show it is possible that $p \mathrel{R} x$ and $q \mathrel{R} y$ are both true, yet $(p + q) \mathrel{R} (x + y)$ and $pq \mathrel{R} xy$ are both false. Thus, the definitions $[p] + [q] = [p + q]$ and $[p][q] = [pq]$ do not define meaningful operations on the equivalence classes of R.

46. Given an example to show that in Z_m it is possible that $[x] \neq [0]$ and $[y] \neq [0]$ but $[x][y] = [0]$.

47. Let m and n be positive integers such that m divides n. Define a relation R on Z_n by $[x] \mathrel{R} [y]$ in case $x \equiv y \pmod{m}$. Prove that R is an equivalence relation on Z_n. What can be said if m does not divide n?

48. **(a)** Show that if a, b, and c are integers and $x \equiv y \pmod{m}$, then $ax^2 + bx + c \equiv ay^2 + by + c \pmod{m}$.

   (b) Show that the result in part (a) may be false if a, b, and c are not all integers, even if $ax^2 + bx + c$ and $ay^2 + by + c$ are both integers.

49. **(a)** Prove that $p \equiv q \pmod{m}$ if and only if $p = km + q$ for some integer k.

   (b) Prove that $p \equiv q \pmod{m}$ if and only if p and q have the same remainder when divided by m.

2.4   Functions

In the matching problem described in Section 1.2 we are seeking an assignment of flights and pilots such that as many pilots as possible are assigned to flights that they requested. We saw in Section 2.2 that the list of destinations requested by the pilots gives rise to a relation between the set of destinations and the set of pilots. An assignment of flights and pilots can therefore be thought of as a special type of relation between the set of destinations and the set of pilots in which exactly one pilot is assigned to each destination. We will now study this special type of relation.

   If X and Y are sets, a **function** f **from** X **to** Y is a relation from X to Y having the property that for each element x in X there is *exactly one* element y in Y such that $x \, f \, y$. Note that because a relation from X to Y is simply a subset of $X \times Y$, a function is a subset S of $X \times Y$ such that for each $x \in X$ there is a unique $y \in Y$ with (x, y) in S.

EXAMPLE 2.24   Let $X = \{1, 2, 3, 4\}$ and $Y = \{5, 6, 7, 8, 9\}$. Then

$$f = \{(1, 5), (2, 8), (3, 7), (4, 5)\}$$

is a function from X to Y because, for each $x \in X$, there is exactly one $y \in Y$ with (x, y) in f. Note that in this case not every element of Y occurs as the second entry of an ordered pair in f (6 and 9 do not occur in any ordered pair in f), and some element of Y (namely 5) occurs as the second entry of several ordered pairs in f.

On the other hand,

$$g = \{(1, 5), (1, 6), (2, 7), (3, 8), (4, 9)\}$$

is not a function from X to Y because there is more than one $y \in Y$ (namely 5 and 6) such that $(1, y)$ belongs to g. And

$$h = \{(1, 5), (2, 6), (4, 7)\}$$

is not a function from X to Y because there is no element of Y associated with some element (namely 3) of X. However, h is a function from $\{1, 2, 4\}$ to Y.   ▬

We denote that f is a function from set X to set Y by writing $f: X \rightarrow Y$. The sets X and Y are called the **domain** and **codomain** of the function, respectively. The unique element of Y such that $x \, f \, y$ is called the **image of x under f** and is written $f(x)$, read "f of x." For the function f defined in Example 2.24, for instance, $f(1) = 5, f(2) = 8, f(3) = 7$, and $f(4) = 5$. Thus, writing $y = f(x)$ is another way of expressing that (x, y) belongs to f.

It is often useful to regard a function $f: X \rightarrow Y$ as a pairing of each element x in X with a unique element $f(x)$ in Y. (See Figure 2.5.) In fact, functions are often defined by giving a formula that expresses $f(x)$ in terms of x; for example, $f(x) = 7x^2 - 5x + 4$.

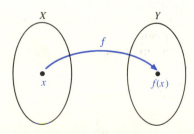

FIGURE 2.5

Note that in order for a set X to be the domain of a function g, it is necessary that $g(x)$ be defined for all x in X. Thus, $g(x) = \sqrt{x}$ cannot have the set of all real numbers as its domain and codomain, for $g(x)$ is not a real number if $x < 0$. Likewise, $g(x) = \dfrac{1}{x}$ cannot have the set of all nonnegative real numbers as its domain because $g(x)$ is not defined if $x = 0$.

EXAMPLE 2.25   Let $X = \{-1, 0, 1, 2\}$ and $Y = \{-4, -2, 0, 2\}$. The function $f: X \rightarrow Y$ defined by $f(x) = x^2 - x$ behaves as follows.

The image of -1 under f is the element $(-1)^2 - (-1) = 2$ in Y.
The image of 0 under f is the element $(0)^2 - (0) = 0$ in Y.
The image of 1 under f is the element $(1)^2 - (1) = 0$ in Y.
The image of 2 under f is the element $(2)^2 - (2) = 2$ in Y.

Thus, $f(-1) = 2, f(0) = 0, f(1) = 0$, and $f(2) = 2$. See Figure 2.6. ■

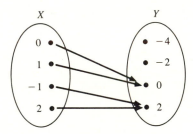

FIGURE 2.6

EXAMPLE 2.26   Let X denote the set of all real numbers and Y denote the set of all nonnegative real numbers. The function $g: X \rightarrow Y$ defined by $g(x) = |x|$ assigns to each element of X its absolute value $|x|$. (Recall that the absolute value of x is defined to be x if $x \geq 0$ and is defined to be $-x$ if $x < 0$.) The domain of g is X and the codomain is Y. ■

EXAMPLE 2.27   Let X be the set of all real numbers between 0 and 100 inclusive, and let Y be the set of all real numbers between 32 and 212 inclusive. The function $F: X \rightarrow Y$ that assigns to each Celsius temperature c its corresponding Fahrenheit temperature $F(c)$ is defined by $F(c) = \frac{9}{5}c + 32$.

Unlike the preceding examples it is not immediately clear that the image under F of each element in X is an element of Y. To see that this is so, we must show that $32 \leq F(c) \leq 212$ if $0 \leq c \leq 100$. But if

$$0 \leq c \leq 100,$$

then

$$0 \leq \frac{9}{5}c \leq \frac{9}{5} \cdot 100 = 180.$$

So

$$32 \leq \frac{9}{5}c + 32 \leq 212.$$

Hence, $F(c)$ is an element of Y. Thus, F is a function with domain X and codomain Y. ■

EXAMPLE 2.28

Let Z denote the set of integers. The function $G: Z \rightarrow Z$ that assigns to each integer m the number $2m$ is defined by $G(m) = 2m$. The domain and the codomain of G are both equal to Z. ■

EXAMPLE 2.29

Let Z denote the set of integers and Z_{12} the set of congruence classes modulo 12. The function $h: Z \rightarrow Z_{12}$ defined by $h(x) = [x]$ is the function that assigns to each integer its congruence class in Z_{12}. Here the domain of h is Z and the codomain is Z_{12}. ■

EXAMPLE 2.30

In Apple Pascal there is a built-in function named MOD that behaves as follows: If N and M are positive integers, the value of the expression N MOD M is the remainder in the division of N by M. Therefore, we can regard MOD as a function with $\{(N, M): N \text{ and } M \text{ are positive integers}\}$ as its domain and the set of nonnegative integers as its codomain. ■

EXAMPLE 2.31

Let X denote the set of all subsets of $U = \{1, 2, 3, 4, 5\}$, and let $Y = \{y: y \text{ is a nonnegative integer less than 20}\}$. If S is an element of X (i.e., if S is a subset of U), define $H(S)$ to be the number of elements in S. Then $H: X \rightarrow Y$ is a function with domain X and codomain Y. ■

We have already noted that it is possible for a function to assign the same element of the codomain to different elements in the domain. The function

$g(x) = |x|$ in Example 2.26, for example, assigns to both -4 and 4 in the domain the element 4 in the codomain. If this does not occur, that is, if no two distinct elements of the domain are assigned the same element in the codomain, then the function is said to be **one-to-one.** Thus, to show that a function $f: X \rightarrow Y$ is one-to-one, we must show that $f(x_1) = f(x_2)$ implies $x_1 = x_2$.

It is also possible that one or more elements of the codomain are not paired by a function to any element in the domain. The function f in Example 2.25, for instance, does not pair the elements -4 and -2 in the codomain with any elements in the domain. Thus, in this case only the elements 0 and 2 in the codomain are paired by f with elements in the domain. The subset of the codomain consisting of the elements that are paired with elements of the domain is called the **range** of the function. In Example 2.25 the range of f is $\{0, 2\}$. If the range and codomain of a function are equal, then the function is called **onto.** Hence, to show that a function $f: X \rightarrow Y$ is onto, we must show that if $y \in Y$, then there is an $x \in X$ such that $y = f(x)$.

A function that is both one-to-one and onto is called a **one-to-one correspondence.** Note that if $f: X \rightarrow Y$ is a one-to-one correspondence, then for each $y \in Y$ there is *exactly one* $x \in X$ such that $y = f(x)$.

EXAMPLE 2.32   The function f in Example 2.25 is neither one-to-one nor onto. It is not one-to-one because f assigns the same element of the codomain (namely 0) to both 0 and 1, that is, because 0 and 1 are distinct elements of the domain for which $f(0) = f(1)$. And f is not onto because, as noted above, the elements -4 and -2 in the codomain of f are not elements in the range of f.   ■

EXAMPLE 2.33   Let X be the set of real numbers. We will show that the function $f: X \rightarrow X$ defined by $f(x) = 2x - 3$ is both one-to-one and onto and, hence, is a one-to-one correspondence.

In order to show that f is one-to-one, we must show that if $f(x_1) = f(x_2)$, then $x_1 = x_2$. Let $f(x_1) = f(x_2)$. Then

$$2x_1 - 3 = 2x_2 - 3$$

$$2x_1 = 2x_2$$

$$x_1 = x_2.$$

Hence, f is one-to-one.

In order to show that f is onto, we must show that if y is an element of the codomain of f, then there is an element of the domain x such that $y = f(x)$. Since the domain and codomain of f are both the set of real numbers, we need to show that for any real number y, there is a real number x such that $y = f(x)$. Take $x =$

$\frac{1}{2}(y + 3)$. (This value was found by solving $y = 2x - 3$ for x.) Then

$$f(x) = f\left(\tfrac{1}{2}(y + 3)\right)$$
$$= 2\left[\tfrac{1}{2}(y + 3)\right] - 3$$
$$= (y + 3) - 3$$
$$= y.$$

Thus, f is onto and therefore is a one-to-one correspondence. ■

EXAMPLE 2.34   The function $G: Z \rightarrow Z$ in Example 2.28 is one-to-one. For if $G(x_1) = G(x_2)$, then $2x_1 = 2x_2$; so $x_1 = x_2$. But G is not onto because there is no element x in the domain Z for which $G(x) = 5$. In fact it is easy to see that the range of G is the set of all even integers, and so the range and codomain of G are not equal. ■

EXAMPLE 2.35   The function $h: Z \rightarrow Z_{12}$ in Example 2.29 is easily seen to be onto. But h is not one-to-one since $1 \neq 13$ but $h(1) = [1] = [13] = h(13)$ in Z_{12}. ■

Before continuing with our discussion of functions, we will describe an application of the preceding material that arises in computer science. An important use of computers is to process data, often in enormous quantities. Consider, for instance, a data set that contains information about the owners of a company's products. Such a data set consists of a large number of records, each of which contains information about a particular customer. If this data set is very large, it may be impractical to maintain the data set in alphabetical order because of the need to update it whenever a new customer is added. But if the records are unsorted, we will need some way to locate a particular record in the data set without searching the records one-by-one.

Let us suppose that we want to be able to locate the record in a large unsorted data set that contains a particular piece of identifying information (a customer's name, for example). One way to find the desired record is to create a function that has as its domain the set of all customer names and has as its values the addresses of the records in our data set. Such a function is called a **hashing function.** Ideally we would like a hashing function to be a one-to-one correspondence between the set of customer names and the set of addresses. But we are almost never able to find an easily computed hashing function that is a one-to-one correspondence, and so we are usually content with finding a hashing function that is one-to-one. There are two basic techniques for constructing hashing functions, one based on division (modular arithmetic) and the other based on multiplication. The interested reader should consult [5] in the suggested readings at the end of this chapter.

EXAMPLE 2.36   Suppose that we have a data set containing the names of a company's ten customers: Smith, Cohen, Moore, Young, Romano, Armstrong, Garcia, O'Brien, Walters, and Kennedy. (Of course, with such a small data set, there is no real need for a hashing function because it is very easy to locate each customer's record. But with ten thousand customers the use of a hashing function may save considerable time if a particular record must be found.)

Although we will not attempt to explain how to construct such a function, we will describe a hashing function h that works for these customer names. First we will assign a numerical value to each customer name according to the position in the alphabet of the name's first letter. So Smith is assigned the value 19, Cohen is assigned the value 3, etc. Then to each customer name h will assign the remainder when the name's numerical value is divided by 13. So h(Smith) $= 6$ because the remainder is 6 when 19 is divided by 13 and h(Cohen) $= 3$ because the remainder is 3 when 3 is divided by 13. Likewise, h(Moore) $= 0$, h(Young) $= 12$, h(Romano) $= 5$, h(Armstrong) $= 1$, h(Garcia) $= 7$, h(O'Brien) $= 2$, h(Walters) $= 9$, and h(Kennedy) $= 11$. Thus, the hashing function h would lead to the following ordering of the records:

(0) Moore	**(5)** Romano	**(10)**
(1) Armstrong	**(6)** Smith	**(11)** Kennedy
(2) O'Brien	**(7)** Garcia	**(12)** Young
(3) Cohen	**(8)**	
(4)	**(9)** Walters	

It is easy to see that h is a one-to-one function. (This fact depends on the particular set of customer names in our data set.) Notice, however, that although h is a one-to-one function from the set of 10 names to the set of 13 possible remainders when dividing by 13, it is not a one-to-one correspondence. So there are $13 - 10 = 3$ unused records, namely records 4, 8, and 10. (These can be regarded as blank records that do not contain a customer's name.) Thus, we have increased the speed with which we can locate a record at the expense of additional computer memory required to hold the 3 unneeded records. Unfortunately there is no simple means for improving this function; if we try to reduce the number of unneeded records by dividing by a positive integer smaller than 13, the resulting function is no longer one-to-one. ■

Since functions are sets of ordered pairs, the definition of equality for functions follows from the definition of equality for sets. That is, $f: X \rightarrow Y$ and $g: V \rightarrow W$ are equal if

$$\{(x, f(x)): x \in X\} = \{(v, g(v)): v \in V\}.$$

It follows that $f = g$ if and only if $X = V$ and $f(x) = g(x)$ for all x in X. It is possible for functions that appear different to be equal, as the following example shows.

EXAMPLE 2.37

Let $X = \{-1, 0, 1, 2\}$ and $Y = \{-4, -2, 0, 2\}$. The functions $f: X \rightarrow Y$ and $g: X \rightarrow Y$ defined by

$$f(x) = x^2 - x \quad \text{and} \quad g(x) = 2\left|x - \tfrac{1}{2}\right| - 1$$

are equal since they have the same domains and

$$f(-1) = 2 = g(-1), \qquad f(0) = 0 = g(0),$$
$$f(1) = 0 = g(1), \qquad \text{and} \quad f(2) = 2 = g(2). \quad \blacksquare$$

If f is a function from X to Y and g is a function from Y to Z, then it is possible to combine them to obtain a function gf from X to Z. The function gf is called the **composition** of g and f and is defined by taking the image of x under gf to be $g(f(x))$. Thus, $gf(x) = g(f(x))$ for all $x \in X$. The composition of g and f is therefore obtained by first applying f to x to obtain $f(x)$, an element of Y, and then applying g to $f(x)$ to obtain $g(f(x))$, an element of Z. (See Figure 2.7.) Note that in evaluating $gf(x)$ we first apply f and then apply g. If $X = Z$, it is also possible to define the function fg; here we first apply g and then apply f. In general, however, the functions gf and fg are not equal.

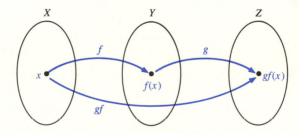

FIGURE 2.7

EXAMPLE 2.38

Let X denote the set of all subsets of $\{1, 2, 3, 4, 5\}$, $Y = \{y: y$ is a nonnegative integer less than 20$\}$, and Z be the set of nonnegative integers. If S is an element of X, define $f(S)$ to be the number of elements in set S, and if $y \in Y$ define $g(y) = 2y$. Then for $S = \{1, 3, 4\}$ we have $gf(S) = g(f(S)) = g(3) = 6$. More generally $gf(S) = g(f(S)) = 2 \cdot f(S)$; thus, gf assigns to S the integer that is twice the number of elements in S. It follows that gf is a function with domain X and codomain Z. Note that in this case the function fg is not defined because $g(y)$ does not lie in X, the domain of f. $\blacksquare$

EXAMPLE 2.39

Let each of X, Y, and Z be the set of real numbers. Define $f: X \rightarrow Y$ and $g: Y \rightarrow Z$ by $f(x) = |x|$ for all $x \in X$ and $g(y) = 3y + 2$ for all $y \in Y$. Then $gf: X \rightarrow Z$ is the function such that

$$gf(x) = g(f(x)) = g(|x|) = 3|x| + 2.$$

In this case we can also define the function fg, but

$$fg(x) = f(g(x)) = f(3x + 2) = |3x + 2|.$$

So $gf \neq fg$ because $gf(-1) = 5 \neq 1 = fg(-1)$. ■

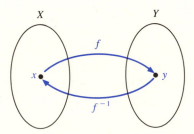

FIGURE 2.8

Suppose that $f: X \rightarrow Y$ is a one-to-one correspondence. Then for each $y \in Y$ there is exactly one $x \in X$ such that $y = f(x)$. Hence, we may define a function with domain Y and codomain X by associating to each $y \in Y$ the unique $x \in X$ such that $y = f(x)$. This function is denoted by f^{-1} and is called the **inverse** of function f. (See Figure 2.8.) The theorem below lists some properties that follow immediately from the definition of an inverse function.

THEOREM 2.7

Let $f: X \rightarrow Y$ be a one-to-one correspondence. Then
(a) $f^{-1}: Y \rightarrow X$ is a one-to-one correspondence.
(b) The inverse function of f^{-1} is f.
(c) For all $x \in X$, $f^{-1}f(x) = x$; and for all $y \in Y$, $ff^{-1}(y) = y$.

EXAMPLE 2.40

Theorem 2.7(c) can be used to compute the inverse of a given function. Suppose, for instance, that S is the set of real numbers,

$$X = \{x \in S: -1 < x \leq 3\},$$
$$Y = \{y \in S: 6 < x \leq 14\},$$

and $f: X \rightarrow Y$ is defined by $f(x) = 2x + 8$. It can be shown that f is a one-to-one correspondence and, hence, has an inverse.

If $y = f(x)$, then by Theorem 2.7(c) $f^{-1}(y) = f^{-1}f(x) = x$. Thus, if we solve the equation $y = f(x)$ for x, we will obtain $f^{-1}(y)$. This calculation can be done as follows.

$$y = 2x + 8$$

$$y - 8 = 2x$$

$$\tfrac{1}{2}(y - 8) = x$$

Hence, $f^{-1}(y) = \tfrac{1}{2}(y - 8)$, and so $f^{-1}(x) = \tfrac{1}{2}(x - 8)$. ■

We will conclude this section by discussing an important inverse function that frequently arises in discussions about the complexity of algorithms. Recall that for any positive integer n, 2^n denotes the product of n factors of 2. Also

$$2^0 = 1, \text{ and } 2^{-n} = \frac{1}{2^n}.$$

It is possible to extend the definition of an exponent to include any real number in such a way that all of the familiar exponent properties hold. When this is done, the equation $f(x) = 2^x$ defines a function with the set of real numbers as its domain and the set of positive real numbers as its range. We call f the **exponential function with base 2.** The behavior of this function is shown in Figure 2.9.

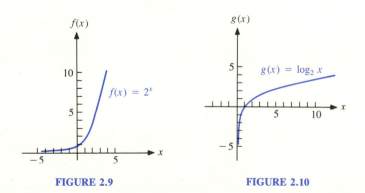

FIGURE 2.9   **FIGURE 2.10**

It can be seen in Figure 2.9 that the exponential function with base 2 is a one-to-one correspondence because each element of the range is associated with exactly one element of the domain. Hence, this function has an inverse g called the **logarithmic function with base 2.** We denote this inverse function by $g(x) = \log_2 x$. Note that the definition of an inverse function implies that

$$y = \log_2 x \text{ if and only if } x = 2^y.$$

Thus, $\log_2 x$ is the exponent y such that $x = 2^y$. In particular, $\log_2 2^n = n$. So $\log_2 4 = \log_2 2^2 = 2$, $\log_2 8 = \log_2 2^3 = 3$, $\log_2 16 = \log_2 2^4 = 4$, $\log_2 1/2 = \log_2 2^{-1} = -1$, and so forth. Although $\log_2 x$ increases as x increases, the rate of growth of $\log_2 x$ is quite slow. For example,

$$\log_2 1000 < \log_2 1024 = \log_2 2^{10} = 10,$$

and similarly $\log_2 1,000,000 < 20$. The behavior of the function $g(x) = \log_2 x$ is shown in Figure 2.10.

Scientific calculators usually contain a key marked LOG. This key can be used to find values of the logarithmic function with base 2, for

$$\log_2 x = \frac{\log x}{\log 2}.$$

EXAMPLE 2.41   A swarm of killer bees was recently released in South America. Suppose that the bees originally occupy a region with area of one square mile and the region occupied by the bees doubles in area each year. How long will it take for the bees to cover the entire surface of the earth, which is 197 million square miles?

Since the area of the region occupied by the bees doubles every year, after n years the bees will cover 2^n square miles. We must determine x such that $2^x = 197,000,000$. But then

$$x = \log_2 197,000,000 = \frac{\log 197,000,000}{\log 2} \approx 27.55.$$

Hence, the bees will cover the entire surface of the earth in about 27.55 years. ■

EXERCISES 2.4

In Exercises 1–4 determine which of the given relations R are functions with domain X.

1. $X = \{1, 3, 5, 7, 8\}$ and $R = \{(1, 7), (3, 5), (5, 3), (7, 7), (8, 5)\}$

2. $X = \{0, 1, 2, 3\}$ and $R = \{(0, 0), (1, 1), (1, -1), (2, 2), (3, -3)\}$

3. $X = \{-2, -1, 0, 1\}$ and $R = \{(-2, 6), (0, 3), (1, -1)\}$

4. $X = \{1, 3, 5\}$ and $R = \{(1, 9), (3, 9), (5, 9)\}$

In Exercises 5–12 determine if the given g is a function with domain X and some codomain Y.

5. X is the set of residents of Iowa and, for $x \in X$, $g(x)$ is the mother of x.

6. X is the set of computer programs run on a particular computer and, for $x \in X$, $g(x)$ is the programming language in which x is written.

7. X is the set of students at Illinois State University and, for $x \in X$, $g(x)$ is the oldest brother of x.

8. X is the set of Presidents of the United States and, for $x \in X$, $g(x)$ is the year that x was first sworn into the office of President.

9. X is the set of real numbers and, for $x \in X$, $g(x) = \log_2 x$.

10. X is the set of real numbers and, for $x \in X$, $g(x) = x^2 + 3$.

11. X is the set of real numbers and, for $x \in X$, $g(x) = x2^x$.

12. X is the set of real numbers and, for $x \in X$, $g(x) = \dfrac{x}{|x|}$.

In Exercises 13–20 find the value of $f(a)$.

13. $f(x) = 5x - 7, a = 3$ 14. $f(x) = 4, a = 8$ 15. $f(x) = 2^x, a = -2$ 16. $f(x) = 3|x| - 2, a = -5$

17. $f(x) = \sqrt{x - 5}, a = 9$ 18. $f(x) = \dfrac{4}{x}, a = \dfrac{1}{2}$ 19. $f(x) = -x^2, a = -3$ 20. $f(x) = 2x^2 - x - 3, a = -2$

Evaluate the numbers in Exercises 21–28 using the fact that $\log_2 2^n = n$.

21. $\log_2 8$

22. $\log_2 \dfrac{1}{2}$

23. $\log_2 1$

24. $\log_2 64$

25. $\log_2 \dfrac{1}{16}$

26. $\log_2 \dfrac{1}{4}$

27. $\log_2 \dfrac{1}{32}$

28. $\log_2 1024$

Approximate the numbers in Exercises 29–36 using a calculator.

29. $\log_2 37$

30. $\log_2 1.72$

31. $\log_2 0.86$

32. $\log_2 100$

33. $\log_2 1.54$

34. $\log_2 9.31$

35. $\log_2 1000$

36. $\log_2 0.17$

Determine the functions gf and fg in Exercises 37–44.

37. $f(x) = 4x + 7$ and $g(x) = 2x - 3$

38. $f(x) = x^2 + 1$ and $g(x) = \sqrt{x}$

39. $f(x) = 2^x$ and $g(x) = 5x + 7$

40. $f(x) = 3x$ and $g(x) = \dfrac{1}{x}$

41. $f(x) = |x|$ and $g(x) = x \log_2 x$

42. $f(x) = 2^x$ and $g(x) = 5x - x^2$

43. $f(x) = x^2 - 2x$ and $g(x) = x + 1$

44. $f(x) = \dfrac{3x + 1}{2 - x}$ and $g(x) = x - 1$

In Exercises 45–52, Z denotes the set of integers. Determine if each function g is one-to-one or onto.

45. $g: Z \to Z$ is defined by $g(x) = 3x$

46. $g: Z \to Z$ is defined by $g(x) = x - 2$

47. $g: Z \to Z$ is defined by $g(x) = 3 - x$

48. $g: Z \to Z$ is defined by $g(x) = x^2$

49. $g: Z \to Z$ is defined by $g(x) = \begin{cases} \dfrac{1}{2}(x + 1) & \text{if } x \text{ is odd} \\ \dfrac{1}{2}x & \text{if } x \text{ is even} \end{cases}$

50. $g: Z \to Z$ is defined by $g(x) = 3x - 5$

51. $g: Z \to Z$ is defined by $g(x) = |x|$

52. $g: Z \to Z$ is defined by $g(x) = \begin{cases} x - 1 & \text{if } x > 0 \\ x & \text{if } x \leq 0 \end{cases}$

In Exercises 53–60, X denotes the set of real numbers. Compute the inverse of each function f: X → X if it exists.

53. $f(x) = 5x$ **54.** $f(x) = 3x - 2$ **55.** $f(x) = -x$ **56.** $f(x) = x^2 + 1$

57. $f(x) = \sqrt[3]{x}$ **58.** $f(x) = \dfrac{-1}{|x| + 1}$ **59.** $f(x) = 3 \cdot 2^{x+1}$ **60.** $f(x) = x^3 - 1$

61. Find a subset Y of the set of real numbers X such that $g: X \to Y$ defined by $g(x) = 3 \cdot 2^{x+1}$ is a one-to-one correspondence. Then compute g^{-1}.

62. Find a subset Y of the set of real numbers X such that $g: Y \to Y$ defined by $g(x) = \dfrac{-1}{x}$ is a one-to-one correspondence. Then compute g^{-1}.

63. If X has m elements and Y has n elements, how many functions are there with domain X and codomain Y?

64. If X has m elements and Y has n elements, how many one-to-one functions are there with domain X and codomain Y?

65. Prove that if $f: X \to Y$ and $g: Y \to Z$ are both one-to-one, then $gf: X \to Z$ is also one-to-one.

66. Prove that if $f: X \to Y$ and $g: Y \to Z$ are both onto, then $gf: X \to Z$ is also onto.

67. Let $f: X \to Y$ and $g: Y \to Z$ be functions such that $gf: X \to Z$ is onto. Prove that g must be onto, and give an example to show that f need not be onto.

68. Let $f: X \to Y$ and $g: Y \to Z$ be functions such that $gf: X \to Z$ is one-to-one. Prove that f must be one-to-one, and give an example to show that g need not be one-to-one.

69. Let $f: X \to Y$ and $g: Y \to Z$ be one-to-one correspondences. Prove that gf is a one-to-one correspondence, and that $(gf)^{-1} = f^{-1}g^{-1}$.

70. Let $f: W \to X$, $g: X \to Y$, and $h: Y \to Z$ be functions. Prove that $h(gf) = (hg)f$.

2.5   Mathematical Induction

In Section 1.4 we claimed that for any positive integer n

$$5 + 8 + 11 + \ldots + (3n + 2) = \tfrac{1}{2}(3n^2 + 7n).$$

Since there are infinitely many positive integers, we cannot justify this assertion by verifying that this equation holds for each individual value of n. Fortunately there is a formal scheme for proving statements are true for all positive integers; this scheme is called the principle of mathematical induction.

The Principle of Mathematical Induction   Let $S(n)$ be a statement involving the integer n. Suppose that for some fixed integer n_0

(1) $S(n_0)$ is true (that is, the statement is true if $n = n_0$) and

(2) whenever k is an integer such that $k \geq n_0$ and $S(k)$ is true, then $S(k + 1)$ is true.

Then $S(n)$ is true for all integers $n \geq n_0$.

Thus, a proof by mathematical induction consists of two parts. Part (1) establishes a base for the induction by proving that some statement $S(n_0)$ is true. And part (2), called the **inductive step,** proves that if any statement $S(k)$ is true, then so is the next statement $S(k + 1)$. The induction principle is a basic property of the integers and so we will give no proof of it. But we will give several examples of its use. In our examples n_0, the base for the induction, will usually be either 0 or 1.

EXAMPLE 2.42   Mathematical induction is often used to verify algorithms. To illustrate this use, we will verify the polynomial evaluation algorithm stated in Section 1.4. Recall that this algorithm evaluates a polynomial

$$P(x) = a_m x^m + a_{m-1} x^{m-1} + \ldots + a_1 x + a_0$$

by the following steps.

Step 1 Let $S = a_0$ and $k = 1$.

Step 2 If $k \leq m$, then replace S by $S + a_k x^k$ and k by $k + 1$ and go to step 2.

Step 3 $P(x) = S$.

Let $S(n)$ be the statement: After completing step 2 of the algorithm n times, $S = a_n x^n + a_{n-1} x^{n-1} + \ldots + a_1 x + a_0$. We will prove that $S(n)$ holds for all integers $n \geq 0$.

(1) For $n = 0$ the value of S is the value a_0 given in step 1. But this equality is the statement $S(0)$; so $S(0)$ is true.

(2) To prove the inductive step, we assume that $S(k)$ is true for some positive integer k and show that $S(k + 1)$ is also true. Now $S(k)$ is the equation $S = a_k x^k + a_{k-1} x^{k-1} + \ldots + a_1 x + a_0$, and this is the value that S has after step 2 has been completed k times. Thus, after completing step 2 one more time ($k + 1$ times in all), the value of S is

$$S + a_{k+1} x^{k+1}$$
$$= (a_k x^k + a_{k-1} x^{k-1} + \ldots + a_1 x + a_0) + a_{k+1} x^{k+1}$$
$$= a_{k+1} x^{k+1} + a_k x^k + a_{k-1} x^{k-1} + \ldots + a_1 x + a_0.$$

Thus, $S(k + 1)$ is true, completing the proof of the inductive step.

Since both (1) and (2) are true, the principle of mathematical induction guarantees that $S(n)$ is true for all integers $n \geq 0$. In particular, the statement $S(m)$ is true. But $S(m)$ is the statement that $P(x) = S$.

The proof above shows that the value of S after completing step 2 of the algorithm k times is

$$S = a_k x^k + a_{k-1} x^{k-1} + \ldots + a_1 x + a_0.$$

Since this relationship holds for all iterations of step 2, it is called a **loop invariant.**

The next theorem proves a result used in Section 1.4.

THEOREM 2.8   For any positive integer n, $5 + 8 + 11 + \ldots + (3n + 2) = \frac{1}{2}(3n^2 + 7n)$.

Proof.   We will prove this result by induction on n with $S(n)$ being the statement:

$5 + 8 + 11 + \ldots + (3n + 2) = \frac{1}{2}(3n^2 + 7n)$. Since $S(n)$ is to be proved for all

positive integers n, we will take the base of the induction to be $n_0 = 1$.

(1) For $n = 1$ the left side of $S(n)$ is 5, and the right side is

$$\frac{3(1)^2 + 7(1)}{2} = \frac{3 + 7}{2} = \frac{10}{2} = 5.$$

Hence, $S(1)$ is true.

(2) To prove the inductive step, we assume that $S(k)$ is true for some positive integer k and show that $S(k + 1)$ is also true. Now $S(k)$ is the equation

$$5 + 8 + 11 + \ldots + (3k + 2) = \frac{(3k^2 + 7k)}{2}.$$

To prove that $S(k + 1)$ is true we must show that

$$5 + 8 + 11 + \ldots + (3k + 2) + [3(k + 1) + 2]$$
$$= \frac{1}{2}\left[3(k + 1)^2 + 7(k + 1)\right].$$

But by using $S(k)$, we can evaluate the left side of the equation to be proved as follows.

$$5 + 8 + 11 + \ldots + (3k + 2) + [3(k + 1) + 2]$$
$$= \frac{1}{2}\left(3k^2 + 7k\right) + [3(k + 1) + 2]$$
$$= \left(\tfrac{3}{2}k^2 + \tfrac{7}{2}k\right) + (3k + 3 + 2)$$
$$= \tfrac{3}{2}k^2 + \tfrac{13}{2}k + 5$$
$$= \frac{1}{2}\left(3k^2 + 13k + 10\right)$$

On the other hand, the right side of the equation to be proved is

$$\frac{1}{2}\left[3(k + 1)^2 + 7(k + 1)\right] = \frac{1}{2}\left[3(k^2 + 2k + 1) + 7(k + 1)\right]$$
$$= \frac{1}{2}\left(3k^2 + 6k + 3 + 7k + 7\right)$$
$$= \frac{1}{2}\left(3k^2 + 13k + 10\right).$$

Because the left and right sides are equal in the equation to be proved, $S(k + 1)$ is true.

Since both (1) and (2) are true, the principle of mathematical induction guarantees that $S(n)$ is true for all integers $n \geq 1$, that is, for all positive integers n. ■

The formula proved in Theorem 2.8 is actually a special case of a more general result. Before this result can be stated, some new terminology is needed.

By an **arithmetic progression** we mean a list of numbers such that each number other than the first number can be gotten by adding some fixed number d, called the **common difference**, to the previous one. An example is

$$5, 8, 11, 14, 17, \ldots$$

for which the common difference is 3. We call the numbers in the list the **terms** of the arithmetic progression. Thus, in the example above, the first term is 5, the second term is 8, and so forth.

We can now state the generalization of Theorem 2.8. The proof of this result will be left as an exercise.

THEOREM 2.9   *The kth Term and Sum of an Arithmetic Progression*   If an arithmetic progression with n terms has first term a and common difference d, then

(a) The kth term of the progression is $a + (k - 1)d$.

(b) The sum of all its terms is $n\left[a + \dfrac{(n - 1)d}{2} \right]$.

EXAMPLE 2.43   Using Theorem 2.9(a) with $a = 5$, $d = 3$, and $k = 20$, we see that the twentieth term of the arithmetic progression $5, 8, 11, \ldots$ is

$$a + (k - 1)d = 5 + (20 - 1)(3) = 5 + 19(3) = 62.$$

The sum of the first n terms of this progression is

$$5 + 8 + 11 + \ldots + (3n + 2) = n\left[a + \frac{(n - 1)d}{2} \right]$$

$$= n\left[5 + \frac{(n - 1) \cdot 3}{2} \right] = 5n + (n^2 - n) \cdot \frac{3}{2}$$

$$= \tfrac{1}{2}(3n^2 + 7n).$$

This calculation confirms the formula proved in Theorem 2.8. ■

In our subsequent induction proofs we will follow the usual practice of not stating explicitly what the statement $S(n)$ is. Nevertheless, in every induction proof the reader should formulate this statement carefully.

THEOREM 2.10   For any positive integer n, $3 + 6 + 12 + \ldots + 3(2^{n-1}) = 3(2^n - 1)$.

Proof.   Again we will prove this formula by induction on n with 1 as the base for the induction.

(1) For $n = 1$ the left side of the equation is 3 and the right side is

$$3(2^1 - 1) = 3(2 - 1) = 3(1) = 3.$$

Thus, the equation is true if $n = 1$.

(2) Assume that the equation is true for some positive integer k; that is, assume that

$$3 + 6 + 12 + \ldots + 3(2^{k-1}) = 3(2^k - 1).$$

To prove that the equation is true for $n = k + 1$, we must prove that

$$3 + 6 + 12 + \ldots + 3(2^{k-1}) + 3(2^k) = 3(2^{k+1} - 1).$$

Now

$$3 + 6 + 12 + \ldots + 3(2^{k-1}) + 3(2^k)$$
$$= 3(2^k - 1) + 3(2^k) = 3(2^k - 1 + 2^k)$$
$$= 3(2^k + 2^k - 1) = 3(2 \cdot 2^k - 1)$$
$$= 3(2^{k+1} - 1).$$

Hence, the equation is true for $n = k + 1$.

Since (1) and (2) are both true, the equation is true for all positive integers n by the principle of mathematical induction.   ■

Again the formula that we have just proved is a special case of a more general result that will be stated in Theorem 2.11. But as before, some new terminology is needed first.

By a **geometric progression** we mean a list of numbers such that each number in the list after the first can be gotten by multiplying the previous number by some fixed number r, called the **common ratio.** An example is

$$3, -6, 12, -24, 48, -96, \ldots$$

for which the common ratio is -2.

THEOREM 2.11   *The kth Term and Sum of a Geometric Progression*   If a geometric progression with n terms has first term a and common ratio r, then

(a) The kth term of the progression is ar^{k-1}.

(b) The sum of all its terms is na if $r = 1$, and $\dfrac{a(r^n - 1)}{r - 1}$ otherwise.

EXAMPLE 2.44

Using Theorem 2.11(a) with $a = 3$, $r = -2$, and $k = 30$, we find that the thirtieth term of the geometric progression with first term 3 and common ratio -2 is $ar^{k-1} = 3(-2)^{30-1} = 3(-2)^{29} = -1,610,612,736$. Likewise, by Theorem 2.11(b) the sum of the first 30 terms of this progression is

$$3 - 6 + 12 - 24 + \ldots + 3(-2)^{29} = \frac{a(r^n - 1)}{r - 1}$$

$$= \frac{3[(-2)^{30} - 1]}{-2 - 1}$$

$$= -1[(-2)^{30} - 1]$$

$$= -1[1,073,741,824 - 1]$$

$$= -1,073,741,823. \quad \blacksquare$$

Closely related to the induction principle are what are known as **recursive definitions**. To define an expression recursively for integers $n \geq n_0$, we must give its value for n_0 and a method of computing its value for $k + 1$ whenever we know its value for $n_0, n_0 + 1, \ldots, k$. An example is the quantity $n!$ which was defined in Section 1.2. A recursive definition of $n!$ is the following:

$$0! = 1, \text{ and if } n > 0 \text{ then } n! = n(n - 1)!.$$

By repeatedly using this definition, we can compute $n!$ for any nonnegative integer n. For example,

$$4! = (4)3! = (4)(3)2! = (4)(3)(2)1! = (4)(3)(2)(1)0! = (4)(3)(2)(1)1 = 24.$$

THEOREM 2.12

If $n \geq 4$, then $n! > 2^n$.

Proof. We will apply the principle of mathematical induction with $n_0 = 4$.
(1) If $n = 4$, then $n! = 24$ and $2^n = 16$, so the statement holds.
(2) Suppose $k! > 2^k$ for some $k \geq 4$. Then

$$(k + 1)! = (k + 1)k! \geq (4 + 1)k! > 2k! > 2(2^k) = 2^{k+1}.$$

This is the required inequality for $k + 1$.
Thus, by the induction principle the statement holds for all $n \geq 4$. $\quad \blacksquare$

Another example of a recursive definition is that of the **Fibonacci numbers** $F_1, F_2, \ldots$, which are defined by

$$F_1 = 1, F_2 = 1, \text{ and if } n > 2 \text{ then } F_n = F_{n-1} + F_{n-2}.$$

For example, $F_3 = F_2 + F_1 = 1 + 1 = 2$, $F_4 = F_3 + F_2 = 2 + 1 = 3$, and $F_5 = F_4 + F_3 = 3 + 2 = 5$. Note that since F_n depends on the two previous Fibonacci numbers, it is necessary to define both F_1 and F_2 at the start in order to have a meaningful definition.

In some circumstances a slightly different form of the principle of mathematical induction is needed.

The Strong Principle of Mathematical Induction   Let $S(n)$ be a statement involving the integer n. Suppose that for some fixed integer n_0
(1) $S(n_0)$ is true, and
(2) whenever k is an integer such that $k \geq n_0$ and all of $S(n_0), S(n_0 + 1), \ldots,$ $S(k)$ are true, then $S(k + 1)$ is true.
Then $S(n)$ is true for all integers $n \geq n_0$.

The only difference between the strong principle of induction and the previous form is in (2), where now we are allowed to assume not only that $S(k)$, but also $S(n_0), S(n_0 + 1), \ldots, S(k - 1)$, are true. Thus, logically the strong principle should be easier to apply, since we are allowed to assume more. It is more complicated than the previous form, however, and usually is not needed. As an example of its use we will prove a theorem involving the Fibonacci numbers.

THEOREM 2.13   For any positive integer n, $F_n \leq 2^n$.

Proof.   Since

$$F_1 = 1 \leq 2 = 2^1 \text{ and}$$
$$F_2 = 1 \leq 4 = 2^2,$$

the statement is true for $n = 1$ and $n = 2$. (We must verify the statement for both $n = 1$ and $n = 2$ because we need to assume that $k \geq 2$ in the inductive step in order to use the recursive definition of the Fibonacci numbers.)

Now suppose that for some positive integer $k \geq 2$ the statement holds for $n = 1, n = 2, \ldots, n = k$. Then

$$F_{k+1} = F_k + F_{k-1} \leq 2^k + 2^{k-1}$$
$$\leq 2^k + 2^k = 2 \cdot 2^k = 2^{k+1}.$$

So the statement is true for $n = k + 1$ if it holds for $n = 1, n = 2, \ldots, n = k$.

Thus, by the strong principle of mathematical induction, the statement is true for all positive integers n. ▪

EXERCISES 2.5

In Exercises 1–8 tell if each list of numbers is an arithmetic progression and give the common difference, a geometric progression and give the common ratio, or neither.

1. $1, 3, 5, 7, 9$ **2.** $1, -3, 5, -7, 9$ **3.** $1, -3, 9, -27$ **4.** $1, -\frac{1}{2}, -2, -\frac{7}{2}$

5. $1, \frac{1}{2}, \frac{1}{4}, \frac{1}{8}$ **6.** $16, 4, 1, \frac{1}{4}$ **7.** $\frac{1}{2}, \frac{2}{3}, \frac{3}{4}, \frac{4}{5}$ **8.** $\frac{9}{2}, -3, 2$

Use Theorems 2.9 and 2.11 to find the indicated term of each of the arithmetic or geometric progressions in Exercises 9–16.

9. the 15th term of $1, 4, 7, 10, \ldots$

10. the 39th term of $3, -3, 3, -3, \ldots$

11. the 12th term of $24{,}576, 12{,}288, 6{,}144, 3{,}072, \ldots$

12. the 48th term of $500, 495, 490, 485, \ldots$

13. the 10th term of $2, -6, 18, -54, \ldots$

14. the 53rd term of $6, 13, 20, 27, \ldots$

15. the 25th term of $150, 148, 146, 144, \ldots$

16. the 11th term of $128, -64, 32, -16, \ldots$

Use Theorems 2.9 and 2.11 to compute the sum of each of the progressions in Exercises 17–24.

$$n\left[\frac{a_1 + (n-1)d}{2}\right]$$

17. $2 + 4 + 6 + \ldots + 20$

18. $2 + 4 + 8 + 16 + \ldots + 1024$

$$\frac{a_1(r^n - 1)}{r - 1}$$

19. $3 + 1 + 1/3 + \ldots + 1/81$

20. $120 + 110 + 100 + \ldots + 30$

21. $0.47 + 0.047 + 0.0047 + 0.00047$

22. $1 + 3 + 5 + \ldots + 99$

23. $82 + 79 + 76 + \ldots + 40$

24. $1 - 3/2 + 9/4 - 27/8 + 81/16$

25. Apply one of the formulas of this section to show that if n is a positive integer, then
$$1 + 2 + 3 + \ldots + n = \frac{1}{2}n(n + 1).$$

26. Apply one of the formulas of this section to show that if n is a nonnegative integer, then $1 + \dfrac{1}{2} + \dfrac{1}{4} + \ldots + \dfrac{1}{2^n} = 2 - \dfrac{1}{2^n}.$

27. Compute the Fibonacci numbers F_1 through F_{10}.

28. According to a baseball pitcher's contract, he will get a bonus of $1000 for the first game he wins, $2000 for the second, $3000 for the third, etc. If he wins 20 games, what is his total bonus?

29. Membership in the Evergreen Racquet Club cost $50 in 1970 and has increased by $5 per year. What was the membership cost in 1985?

30. A family membership to a community swimming pool cost $10 in 1975 and has increased by $1 per year. How much will a family have paid in membership fees during the period 1975–1985 if it held membership to the pool throughout this period?

31. A college graduate accepted a job as a computer programmer at a starting salary of $24,000. If she receives yearly salary increases of 5%, what will her salary be in 10 years?

32. A clothing store paid utility costs of $4000 in 1980. If the utility costs increase at a rate of 3% per year, how much will the store have paid for utility costs in the period from 1980–1990?

In Exercises 33–36 determine what is wrong with the given induction arguments.

33. We will prove that 5 divides $5n + 3$ for all positive integers n.

   Assume that for some positive integer k, 5 divides $5k + 3$. Then there is a positive integer p such that $5k + 3 = 5p$. Now

   $$5(k + 1) + 3 = (5k + 5) + 3 = (5k + 3) + 5 = 5p + 5 = 5(p + 1).$$

   Since 5 divides $5(p + 1)$, it follows that 5 divides $5(k + 1) + 3$, which is the statement that we want to prove.

   Hence, by the principle of mathematical induction, 5 divides $5n + 3$ for all positive integers n.

34. We will prove that in any set of n persons, all people have the same age.

   Clearly all people in a set of 1 person have the same age, so the statement is true if $n = 1$.

   Now suppose that in any set of k people all persons have the same age. Let $S = \{x_1, x_2, \ldots, x_{k+1}\}$ be a set of $k + 1$ people. Then by the induction hypothesis all people in each of the sets $\{x_1, x_2, \ldots, x_k\}$ and $\{x_2, x_3, \ldots, x_{k+1}\}$ have the same age. But then $x_1, x_2, \ldots, x_k$ all have the same age and $x_2, x_3, \ldots, x_{k+1}$ all have the same age. It follows that $x_1, x_2, \ldots, x_{k+1}$ all have the same age. This proves the inductive step.

   The principle of mathematical induction therefore shows that for any positive integer n, all people in any set of n persons have the same age.

35. We will prove that for any positive integer n, if the maximum of two positive integers is n, then the integers are equal.

   If the maximum of any two positive integers is 1, then both of the integers must be 1. Hence the two integers are equal. This proves the result for $n = 1$.

   Assume that if the maximum of any two positive integers is k, then the integers are equal. Let x and y be two positive integers for which the maximum is $k + 1$. Then the maximum of $x - 1$ and $y - 1$ is k. So by the induction hypothesis, $x - 1 = y - 1$. But then $x = y$, proving the inductive step.

   It follows by the principle of mathematical induction that for any positive integer n, if the maximum of two positive integers is n, then the integers are equal. Hence, any two positive integers are equal.

36. Let a be a nonzero real number. We will prove that for any nonnegative integer n, $a^n = 1$.

   Since $a^0 = 1$ by definition, the statement is true for $n = 0$.

   Assume that for some integer k, $a^m = 1$ for $0 \le m \le k$. Then

   $$a^{k+1} = \frac{a^k a^k}{a^{k-1}} = \frac{1 \cdot 1}{1} = 1.$$

   The strong principle of induction therefore implies that $a^n = 1$ for every nonnegative integer n.

37. Give a recursive definition of a polynomial based on the polynomial evaluation algorithm in Section 1.4.

38. Give a recursive definition of a polynomial based on Horner's polynomial evaluation algorithm in Section 1.4.

In Exercises 39–50 prove each of the given statements by mathematical induction.

39. $1 + 4 + 9 + \ldots + n^2 = \frac{1}{6}n(n + 1)(2n + 1)$ for all positive integers n.

40. $1 + 8 + 27 + \ldots + n^3 = \frac{1}{4}n^2(n + 1)^2$ for all positive integers n.

41. $1(1!) + 2(2!) + \ldots + n(n!) = (n + 1)! - 1$ for all positive integers n.

42. $\dfrac{1}{1 \cdot 2} + \dfrac{1}{2 \cdot 3} + \ldots + \dfrac{1}{n(n + 1)} = \dfrac{n}{n + 1}$ for all positive integers n.

43. $n! > 3^n$ for all integers $n \geq 7$.

44. $(2n)! < (n!)^2 4^{n-1}$ for all integers $n \geq 5$.

45. $F_n \leq 2F_{n-1}$ for all integers $n \geq 2$.

46. $F_1 + F_2 + \ldots + F_n = F_{n+2} - 1$ for all positive integers n.

47. $F_2 + F_4 + \ldots + F_{2n} = F_{2n+1} - 1$ for all positive integers n.

48. $\left(1 - \dfrac{1}{2}\right)\left(1 - \dfrac{1}{3}\right) \ldots \left(1 - \dfrac{1}{n + 1}\right) = \dfrac{1}{n + 1}$ for all positive integers n.

49. $2^n + 3^n \equiv 5^n \pmod{6}$ for all positive integers n.

50. $16^n \equiv 1 - 10n \pmod{25}$ for all positive integers n.

51. Prove Theorem 2.9.

52. Prove Theorem 2.11.

53. Mr. and Mrs. Lewis hosted a party for n married couples. As the guests arrived, some people shook hands. Later Mr. Lewis asked everyone else (including his wife) how many hands each had shaken. To his surprise, he found that no two people gave him the same answer. If no one shook his or her own hand, no spouses shook hands, and no two persons shook hands more than once, how many hands did Mrs. Lewis shake? Prove your answer by mathematical induction.

2.6   Applications

In this section we will apply the two versions of the principle of mathematical induction stated in Section 2.5 to establish some facts that are needed elsewhere in this book. Our first two results give the maximum number of comparisons that are needed to search and sort lists of numbers; these facts will be used in our discussion of searching and sorting in Chapter 8.

EXAMPLE 2.45   There is a common children's game in which one child thinks of an integer and another tries to discover what it is. After each guess the person trying to determine the unknown integer is told if the last guess was too high or too low. Suppose, for instance, that we must identify an unknown integer between 1 and 64. One way to find the integer would be to guess the integers from 1 through 64 in order, but this method may require as many as 64 guesses to determine the unknown number. A much better way is to guess an integer close to the middle of the possible values, thereby dividing the number of possibilities in half with each guess. For example, the following sequence of guesses will discover that the unknown integer is 37.

Attempt	Guess	Result	Conclusion
1	32	low	integer is between 33 and 64
2	48	high	integer is between 33 and 47
3	40	high	integer is between 33 and 39
4	36	low	integer is between 37 and 39
5	38	high	integer is between 37 and 37
6	37	correct	

If the strategy described above is used, it is not difficult to see that any unknown integer between 1 and 64 can be found with no more than 7 guesses. This simple game is related to the problem of searching a list of numbers by computer to see if a particular target value is in the list. Of course this situation differs from the number-guessing game in that we do not know in advance what numbers are in the list being searched. But when the list of numbers is sorted in nondecreasing order, the most efficient searching technique is essentially the same as that used in the number-guessing game: repeatedly compare the target value to a number in the list that is close to the middle of the range of values in which the target must occur. The theorem below describes the efficiency of this searching strategy.

THEOREM 2.14   For any nonnegative integer n, at most $n + 1$ comparisons are required to determine if a particular number is present in a list of 2^n numbers that are sorted in nondecreasing order.

Proof.   The proof will be by induction on n. For $n = 0$, we need to show that at most $n + 1 = 1$ comparison is required to see if a particular number m is in a list containing $2^0 = 1$ number. Since the list contains only one number, clearly only one comparison is needed to determine if this number is m. This establishes the result when $n = 0$.

Now assume that the result is true for some nonnegative integer k; that is, assume that at most $k + 1$ comparisons are needed to determine if a particular number is present in a sorted list of 2^k numbers. Suppose that we have a list of 2^{k+1} numbers in nondecreasing order. We must show that it is possible to determine if a particular number m is present in this list using at most $(k + 1) + 1 = k + 2$ comparisons. To do so, we will compare m to the number p in position 2^k of the list.

Case 1: $m = p$

In this case we need only one comparison to find that m is in the list. Since $1 \leq k + 2$, the result is true in this case.

Case 2: $m < p$

Since the list is in nondecreasing order, for m to be present in the list it must lie in positions 1 through 2^k. But the numbers in positions 1 through 2^k are a list of 2^k numbers in nondecreasing order. Hence, by the induction hypothesis we can determine if m is present in this list by using at most $k + 1$ comparisons. So in this case, at most $1 + (k + 1) = k + 2$ comparisons are needed to determine if m is present in the original list.

Case 3: $m > p$

Since the list is in nondecreasing order, for m to be present in the list it must lie in positions $2^k + 1$ through 2^{k+1}. Again the induction hypothesis tells us that we can determine if m is present in this sorted list of 2^k numbers with at most $k + 1$ comparisons. Hence, in this case also at most $k + 2$ comparisons are needed to determine if m is present in the original list.

Thus, in each of the three cases, we can determine if m is present in the list of 2^{k+1} sorted numbers with at most $k + 2$ comparisons. This proves the inductive step and, therefore, proves the theorem for all nonnegative integers n. ∎

Although Theorem 2.14 is stated for lists of numbers in nondecreasing order, it is easy to see that the same conclusion is true for lists that are sorted in nonincreasing order. Moreover, the same conclusion is true for lists of words that are in alphabetical order. The next theorem is similar to Theorem 2.14; it gives an upper bound on the number of comparisons needed to merge two sorted lists of numbers into one sorted list. Before stating this result, we will illustrate the merging process to be used in proving Theorem 2.15.

EXAMPLE 2.46

Consider the two lists of numbers in nondecreasing order:

$$2, 5, 7, 9 \quad \text{and} \quad 3, 4, 7.$$

Suppose that we want to merge them into a single list

$$2, 3, 4, 5, 7, 7, 9$$

in nondecreasing order. To combine the lists efficiently, first compare the numbers at the beginning of each list (2 and 3) and take the smaller one (2) as the first number in the combined list. (If the first number in one list is the same as the first number in the other list, choose either of the equal numbers.) Then delete this smaller number from the list that contains it to obtain the lists

$$5, 7, 9 \quad \text{and} \quad 3, 4, 7.$$

Second, compare the beginning numbers in each of these new lists (5 and 3) and take the smaller one (3) as the second number in the combined list. Delete this number from the list that contains it, and continue the process above until all of the original numbers have been merged into a single list. ∎

THEOREM 2.15   Let A and B be two lists containing numbers sorted in nondecreasing order. Suppose that for some positive integer n, there is a combined total of n numbers in the two lists. Then A and B can be merged into a single list of n numbers in nondecreasing order using at most $n - 1$ comparisons.

Proof.   The proof will be by induction on n. If $n = 1$, then either A or B must be an empty list (and the other must contain 1 number). But then the list C obtained by adjoining list B to the end of list A will be in nondecreasing order, and C is obtained by making $0 = n - 1$ comparisons. This proves the theorem when $n = 1$.

Now suppose that the conclusion of the theorem holds for some positive integer k, and let A and B be sorted lists containing a total of $k + 1$ numbers. We must show that A and B can be merged into a sorted list C using at most k comparisons. Compare a and b, the first elements of A and B, respectively.
Case 1: $a \le b$

Let A' be the list obtained by deleting a from A. Then A' and B are sorted lists containing a total of k elements. So by the induction hypothesis, A' and B can be merged into a single sorted list C' using at most $k - 1$ comparisons. Form the list C by adjoining a to C' as the first element. Then C is in nondecreasing order because C' is in nondecreasing order and a precedes all the other numbers in A and B. Moreover, C was formed using 1 comparison to find that $a \le b$ and at most $k - 1$ comparisons to form list C'; so C was formed using at most k comparisons.
Case 2: $a > b$

Delete b from B to form list B'. Then use the induction hypothesis as in case 1 to sort A and B' into a list C' using at most $k - 1$ comparisons. The list C is then obtained by adjoining b to C' as the first element. As in case 1, C is in nondecreasing order and was formed using at most k comparisons.

Thus, in either case we can merge A and B into a sorted list using at most k comparisons. This completes the proof of the inductive step, and so the conclusion is established for all positive integers n. ◼

Our next two results involve the number of subsets of a set. These results arise in connection with the knapsack problem described in Section 1.3 and with counting techniques to be discussed in Chapter 7. Recall that we stated in Section 1.3 that the set $\{1, 2, \ldots, n\}$ has precisely 2^n subsets. If this result is true in general, then increasing n by one doubles the number of subsets. An example will indicate why this is true. Let us take $n = 2$ and consider the subsets of $\{1, 2\}$. They are

$$\varnothing, \{1\}, \{2\}, \text{ and } \{1, 2\}.$$

Now consider the subsets of $\{1, 2, 3\}$. Of course, the four sets we have just listed are also subsets of this larger set; but there are other subsets, namely those containing 3. In fact any subset of $\{1, 2, 3\}$ that is not a subset of $\{1, 2\}$ must contain the element 3. If we removed the 3 we would have a subset of $\{1, 2\}$ again. Thus, the new subsets are just

$$\{3\}, \{1, 3\}, \{2, 3\}, \text{ and } \{1, 2, 3\},$$

formed by including 3 in each of the previous four sets. The total number of subsets has doubled, as our formula indicates. This argument is the basis for a proof of Theorem 1.3.

THEOREM 1.3   If n is any positive integer, then a set with n elements has exactly 2^n subsets.

Proof.   We will prove this result by induction on n.
To establish a base for the induction, we must show that any set having just 1 element has $2^1 = 2$ subsets. But when the set has one element, say a, then its subsets are just the two sets $\varnothing$ and $\{a\}$.
To prove the inductive step, we assume that for some positive integer k, any set having k elements has exactly 2^k subsets. We must prove that any set with $k + 1$ elements has 2^{k+1} subsets. Let S be a set with $k + 1$ elements, say $a_1, a_2,$ $\ldots, a_k, a_{k+1}$, and define a set R by

$$R = \{a_1, a_2, \ldots, a_k\}.$$

Then R has k elements and so has 2^k subsets by our assumption. But each subset of S is either a subset of R or else a set formed by inserting a_{k+1} into a subset of R. Thus, S has exactly

$$2^k + 2^k = 2(2^k) = 2^{k+1}$$

subsets, which is what we are trying to show.
It follows from the principle of mathematical induction that the result is true for all positive integers n. ■

EXAMPLE 2.47   For many years Wendy's Old Fashioned Hamburger Restaurants advertised that they served hamburgers in 256 different ways. This claim can be justified by using Theorem 1.3, because hamburgers can be ordered at Wendy's with any combination of 8 toppings (cheese, ketchup, lettuce, mayonnaise, mustard, onions, pickles, and tomatoes). Since any selection of toppings can be regarded as a subset of the set of 8 toppings, the number of different toppings is the same as the number of subsets, which is $2^8 = 256$. ■

We can say even more about the number of subsets of a set containing n elements. The result below tells us how many of its 2^n subsets contain a specified number of elements.

THEOREM 2.16   Let S be a set containing n elements, where n is a nonnegative integer. If r is an integer such that $0 \le r \le n$, then the number of subsets of S containing exactly r elements is $\dfrac{n!}{r!(n-r)!}$.

Proof.   The proof will be by induction on n, starting with $n = 0$.

If $n = 0$, then S is the empty set and r must be 0 also. But there is exactly 1 subset of $\varnothing$ with 0 elements, namely $\varnothing$ itself. And

$$\frac{n!}{r!(n-r)!} = \frac{0!}{0!0!} = 1$$

since $0! = 1$ by definition. Thus, the conclusion holds for $n = 0$.

Now suppose that the result holds for some integer $k \ge 0$. Let S be a set containing $k + 1$ elements, say $S = \{a_1, a_2, \ldots, a_k, a_{k+1}\}$, and suppose that $0 \le r \le k + 1$. We must count the subsets of S having exactly r elements. Let R be such a subset.

Case 1: $a_{k+1} \notin R$

Then R is a subset of $\{a_1, a_2, \ldots, a_k\}$ having r elements. By the induction hypothesis there are $\dfrac{k!}{r!(k-r)!}$ such subsets.

Case 2: $a_{k+1} \in R$

In this case if we remove a_{k+1} from R, we have a subset of $\{a_1, a_2, \ldots, a_k\}$ containing $r - 1$ elements. By the induction hypothesis there are $\dfrac{k!}{(r-1)![k-(r-1)]!}$ sets like this.

Putting the two cases together we see that S has a total of

$$\frac{k!}{r!(k-r)!} + \frac{k!}{(r-1)!(k-r+1)!}$$

subsets with r elements. But this number equals

$$\frac{k!(k-r+1)}{r!(k-r)!(k-r+1)} + \frac{k!r}{r(r-1)!(k-r+1)!}$$

$$= \frac{k!(k-r+1)}{r!(k-r+1)!} + \frac{k!r}{r!(k-r+1)!}$$

$$= \frac{k!(k-r+1+r)}{r!(k-r+1)!} = \frac{(k+1)!}{r!(k+1-r)!}$$

Since this is what our formula produces when $k + 1$ is substituted for n, the inductive step is proved.

Thus, by the induction principle the theorem is true for all positive integers n. ▪

EXAMPLE 2.48   How many two-person committees can be chosen from a set of 5 people?

This is equivalent to asking how many subsets of $\{1, 2, 3, 4, 5\}$ have exactly 2 elements. Taking $n = 5$ and $r = 2$ in Theorem 2.16 gives the answer

$$\frac{5!}{2!(5 - 2)!} = \frac{5!}{2!3!} = 10.$$

The actual subsets are $\{1, 2\}$, $\{1, 3\}$, $\{1, 4\}$, $\{1, 5\}$, $\{2, 3\}$, $\{2, 4\}$, $\{2, 5\}$, $\{3, 4\}$, $\{3, 5\}$, and $\{4, 5\}$.   ■

The last result in this section proves a very basic result about the positive integers. This fact was referred to in Example 2.13.

THEOREM 2.17   Every integer greater than 1 is either prime or a product of primes.

Proof.   Let n be an integer greater than 1. The proof will be by induction on n, using the strong form of the principle of mathematical induction. Since 2 is a prime number, the statement is true for $n = 2$.

Assume that for some integer $k > 1$, the statement is true for $n = 2, 3, \ldots, k$. We must prove that $k + 1$ is either prime or a product of primes. If $k + 1$ is prime, then there is nothing to prove; so suppose that $k + 1$ is not prime. Then there is a positive integer p other than 1 and $k + 1$ that divides $k + 1$. So $\frac{k + 1}{p} = q$ is an integer. Clearly $q \neq 1$ (for otherwise $p = k + 1$) and $q \neq k + 1$ (for otherwise $p = 1$). Hence, both p and q are integers between 2 and k, inclusive. So the induction hypothesis can be applied to both p and q. It follows that each of p and q is either prime or a product of primes. But then $k + 1 = pq$ is a product of primes. This proves the inductive step, and, therefore, completes the proof of the theorem.   ■

With the development of larger and faster computers, it is possible to discover huge prime numbers. In 1978 for instance, Laura Nickel and Curt Noll, two teenagers from Hayward, California, used 440 hours of computer time to find the prime number $2^{21701} - 1$. At that time this 6533–digit number was the largest known prime number. But finding whether a particular positive integer is prime or a product of primes remains a very difficult problem. Note that although Theorem 2.17 tells us that positive integers greater than 1 are either prime or products of primes, it does not help determine which is the case. In particular, Theorem 2.17 is of no help in actually finding the prime factors of a particular positive integer.

Indeed, the difficulty of finding the prime factors of large numbers is the basis for an important method of cryptography (encoding of data or messages) called the RSA method. (The name comes from the initials of its discoverers, R. L. Rivest, A. Shamir, and L. Adleman.) For more information on the RSA method, see [7] at the end of this chapter.

EXERCISES 2.6

1. How many subsets of the set {1, 3, 4, 6, 7, 9} are there?

2. How many nonempty subsets of the set {a, e, i, o, u} are there?

3. At Avanti's a pizza can be ordered with any combination of the following ingredients: green pepper, ham, hamburger, mushrooms, onion, pepperoni, and sausage. How many different pizzas can be ordered?

4. If a test consists of 12 questions to be answered true or false, in how many ways can all 12 questions be answered?

5. A certain automobile can be ordered with any combination of the following options: air conditioning, automatic transmission, bucket seats, cruise control, power windows, rear window defogger, sun roof, and tape deck. In how many ways can this car be ordered?

6. Jennifer's grandmother has told her she can take as many of her 7 differently colored glass rings as she wants. How many choices are there?

7. How many subsets of {1, 3, 4, 5, 6, 8, 9} contain exactly 5 elements?

8. How many subsets of {a, e, i, o, u, y} contain exactly 4 elements?

9. A basketball coach must choose a 5-person starting team from a roster of 12 players. In how many ways is this possible?

10. A beginning rock group must choose 2 songs to record from among the 9 they know. How many choices are possible?

11. A person ordering a complete dinner at a restaurant may choose three vegetables from among six offered. In how many ways can this be done?

12. A hearts player must pass 3 cards from his 13–card hand. How many choices of cards to pass does he have?

13. Three persons will be elected from among ten candidates running for a city council. How many sets of winning candidates are possible?

14. A sociologist intends to select 4 persons from a list of 9 people for interviewing. How many sets of persons to interview can be chosen?

15. How many 13–card bridge hands can be dealt from a 52–card deck? Leave your answer in factorial notation.

16. A racketeer is allowed to bring no more than 3 of the 7 lawyers representing him to a Senate hearing. How many choices does he have?

Prove each of the statements in Exercises 17–32 by mathematical induction.

17. If $x > -1$, then $(1 + x)^n \geq 1 + nx$ for all positive integers n.

18. $\dfrac{1}{1^2} + \dfrac{1}{2^2} + \ldots + \dfrac{1}{n^2} < 2 - \dfrac{1}{n}$ for all integers $n \geq 2$.

19. $\dfrac{(2n)!}{2^n}$ is an integer for all positive integers n.

20. $\dfrac{(n + 1)(n + 2) \ldots (2n)}{2^n}$ is an integer for all positive integers n.

21. $F_n \leq \left(\dfrac{7}{4}\right)^n$ for all positive integers n.   **22.** $F_n \geq \left(\dfrac{5}{4}\right)^n$ for all integers $n \geq 3$.

23. For all positive integers n, 3 divides $2^{2n} - 1$.   **24.** For all positive integers n, 6 divides $n^3 + 5n$.

25. $\dfrac{(4n)!}{8^n}$ is an integer for all nonnegative integers n.   **26.** $\dfrac{(4n - 2)!}{8^n}$ is an integer for all integers $n \geq 5$.

27. $(1 + 2 + \ldots + n)^2 = 1^3 + 2^3 + \ldots + n^3$ for all positive integers n.

28. $1^2 - 2^2 + \ldots + (-1)^{n+1}n^2 = \dfrac{(-1)^{n+1}n(n + 1)}{2}$ for all positive integers n.

29. For all integers $n \geq 2$, $\dfrac{1}{\sqrt{1}} + \dfrac{1}{\sqrt{2}} + \ldots + \dfrac{1}{\sqrt{n}} > \sqrt{n}$.

30. For all positive integers n, $\dfrac{1 \cdot 3 \cdot 5 \cdot \ldots \cdot (2n - 1)}{2 \cdot 4 \cdot 6 \cdot \ldots \cdot (2n)} \geq \dfrac{1}{2n}$.

31. Let n be a positive integer and $A_1, A_2, \ldots, A_n$ be subsets of a universal set U. Prove by mathematical induction that
$$\overline{(A_1 \cup A_2 \cup \ldots \cup A_n)} = \overline{A_1} \cap \overline{A_2} \cap \ldots \cap \overline{A_n}.$$

32. Let n be a positive integer and $A_1, A_2, \ldots, A_n$ be subsets of a universal set U. Prove by mathematical induction that
$$\overline{(A_1 \cap A_2 \cap \ldots \cap A_n)} = \overline{A_1} \cup \overline{A_2} \cup \ldots \cup \overline{A_n}.$$

33. If n is an integer larger than three, determine the number of diagonals in a regular n-sided polygon. Then prove your answer is correct using mathematical induction.

34. Suppose that for some positive integer n there are n lines in the Euclidean plane such that no two are parallel and no three meet at the same point. Determine the number of regions into which the plane is divided by these n lines, and prove that your answer is correct by mathematical induction.

35. Prove by mathematical induction that any list of 2^n numbers can be sorted into nondecreasing order using at most $n \cdot 2^n$ comparisons.

36. The **well-ordering principle** states that each nonempty set of positive integers contains a smallest element.

(a) Assume the well-ordering principle and use it to prove the principle of mathematical induction.

(b) Assume the principle of mathematical induction holds and use it to prove the well-ordering principle.

Suggested Readings

1. Buck, R. C. "Mathematical Induction and Recursive Definitions." *American Mathematical Monthly,* vol. 70, no. 2 (February 1963): 128–135.
2. Halmos, Paul R. *Naive Set Theory*. New York: Van Nostrand, 1960.
3. Hayden, S. and J. Kennison. *Zermelo-Fraenkel Set Theory*. Columbus, Ohio: Charles Merrill, 1968.
4. Henken, L. "On Mathematical Induction." *American Mathematical Monthly,* vol. 67, no. 4 (April 1960): 323–337.
5. Knuth, Donald E. *The Art of Computer Programming, vol 3: Searching and Sorting*. Reading, Mass.: Addison-Wesley, 1973.
6. Tuchinsky, Phillip M. "International Standard Book Numbers." *The UMAP Journal,* vol. 6, no. 1 (1985): 41–53.
7. Vanden Eynden, Charles. *Elementary Number Theory*. New York: Random House/Birkhäuser, 1987.

3

GRAPHS

Even though graphs have been studied for a long time, the increased use of computer technology has generated a new interest in them. Not only have applications of graphs been found in computer science but in many other areas such as business and science. As a consequence, the study of graphs has become important to many.

3.1   **Graphs and Their Representations**

It is quite common to represent situations involving objects and their relationships by drawing a diagram of points, with segments connecting those points that are related. Let us consider some specific examples of this idea.

EXAMPLE 3.1   Consider for a moment an airline route map in which dots represent cities and two dots are connected by a segment whenever there are flights between the corresponding cities. A portion of such an airline map is shown in Figure 3.1.   ■

FIGURE 3.1

EXAMPLE 3.2   Suppose we have four computers labeled A, B, C, and D, where there is a flow of information between computers A and B, C and D, and B and C. This situation can be represented by the diagram in Figure 3.2. This is usually referred to as a communication network.   ■

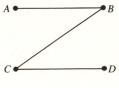

FIGURE 3.2

EXAMPLE 3.3   Suppose that there is a group of people and a set of jobs where some people can do only some of the jobs. For example, for individuals A, B, and C and jobs D, E, and F, suppose A can do only job D, B can do jobs D and E, and C can do jobs E and F. This type of situation can be represented by the diagram in Figure 3.3, where line segments are drawn between an individual and the jobs that person can do.   ■

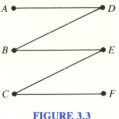

FIGURE 3.3

The general idea in the three examples is to represent by a picture a set of objects in which some pairs are related. We will now describe this type of representation more carefully.

A **graph** is a nonempty finite set V along with a set E of 2–element subsets of V. The elements of V are called **vertices** and the elements of E are called **edges.**

Figure 3.2 depicts a graph with vertices A, B, C, and D and edges $\{A, B\}$, $\{B, C\}$, and $\{C, D\}$. Thus, a graph can be described either by the use of sets or by the use of a diagram, where segments between the vertices in V describe which 2–element subsets are being included. Figure 3.3 shows a graph with vertices A, B, C, D, E, and F and with edges $\{A, D\}$, $\{B, D\}$, $\{B, E\}$, $\{C, E\}$, and $\{C, F\}$.

Whenever we have an edge $e = \{U, V\}$, we say that the edge e **connects** or **joins** the vertices U and V and that U and V are **adjacent.** It is also said that edge e is **incident** on the vertex U and that the vertex U is **incident** on the edge e. With the graph in Figure 3.2 we see that vertices A and B are adjacent, whereas vertices A and C are not because there is no segment between them (that is, the set $\{A, C\}$ is not an edge). In Figure 3.3 the edge $\{B, E\}$ is incident with the vertex B.

Note that the diagram in Figure 3.2 can be drawn in a different way and still represent the same graph. Another representation of this graph is given in Figure 3.4.

FIGURE 3.4

FIGURE 3.5

The way our picture is drawn is not important although one picture may be much easier to understand than another. What is important in the picture is which vertices are connected by edges, for this describes what relationships exist between the vertices. In Figure 3.5 we have redrawn the graph from Figure 3.2 in such a way that the edges meet at a place other than a vertex. It is important not to be misled into believing that there is now a new vertex. Sometimes it is not possible to draw a picture of a graph without edges meeting in this way, and it is important to understand that such a crossing does not generate a new vertex of the graph. It is often very difficult to determine if a graph can be drawn without any of the edges crossing except at vertices. Also it is generally a very hard problem to tell if two different pictures represent the same graph.

We caution the reader that the use of terminology in graph theory is not consistent among users, and so in consulting other books definitions should always be checked to see how words are being used. In our definition of a graph the set of vertices is required to be a finite set. Some authors do not make this restriction, but we find it convenient to do so. Also our definition of a graph does not allow an edge to go from a vertex to itself, or different edges between the same two vertices. Some authors do allow such edges but we do not.

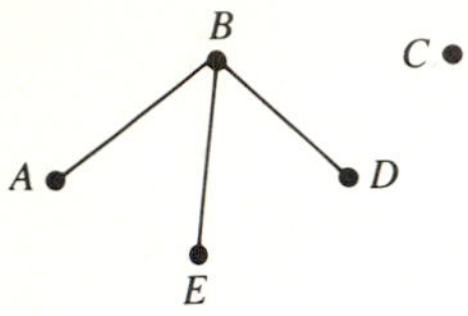

In a graph the number of edges incident with a vertex V is called the **degree** of V and is denoted as $\deg(V)$. In Figure 3.6 we see that $\deg(A) = 1$, $\deg(B) = 3$, and $\deg(C) = 0$.

One special graph that is encountered frequently is the **complete graph** on n vertices, where every vertex is connected to every other vertex. This graph is denoted by K_n. Figure 3.7 shows K_3 and K_4.

FIGURE 3.6

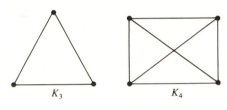

K_3 K_4

FIGURE 3.7

In Figures 3.6 and 3.7 notice that adding the degrees of the vertices in each graph yields a number that is twice the number of edges. This result is true in general.

THEOREM 3.1   In a graph the sum of the degrees of the vertices equals twice the number of edges.

Proof.   The key to understanding why Theorem 3.1 is true is that each edge is incident on two vertices. When we take the sum of the degrees of the vertices, each edge is counted twice in this sum. Thus, the sum of the degrees is twice the number of the edges. Look again at Figures 3.5, 3.6, and 3.7 to see how this double counting of edges takes place.   ▪

Matrix Representations

It is often necessary to analyze graphs and perform a variety of procedures and algorithms upon them. When they have large numbers of vertices and edges, it is often essential to use a computer to perform these algorithms. Thus, it is necessary to communicate to the computer the vertices and edges of a graph. One way to do so is to represent a graph by means of numbers, which are easy to store and manipulate.

An $m \times n$ **matrix** is a rectangular array of numbers in which there are m horizontal rows and n vertical columns.

EXAMPLE 3.4

Array A below is a 3×2 matrix because there are 3 horizontal rows and 2 vertical columns of numbers. In the same way, B is a 4×3 matrix.

$$A = \begin{bmatrix} 1 & 2 \\ 5 & 0 \\ 6 & 7 \end{bmatrix} \qquad B = \begin{bmatrix} 1 & 2 & 1 \\ 3 & 4 & 0 \\ 5 & 9 & 1 \\ 7 & 8 & 3 \end{bmatrix}$$

The number in the ith row and jth column is called the i,j **entry** of the matrix. Thus, in matrix A the 2,1 entry is 5 because the entry lying in row 2 and column 1 is 5. For matrix B the 3,2 entry is 9.

Two matrices A and B are said to be **equal** whenever A and B have the same number of rows and the same number of columns, and the i,j entry of A is equal to the i,j entry of B for all possible i and j. In other words, two matrices are equal when they have the same size and all pairs of corresponding entries are equal.

EXAMPLE 3.5

The matrix C is not equal to the matrix D because C has 2 columns and D has 3 columns. Also, the matrices C and E are not equal because their corresponding entries are not the same. In particular, the 1,2 entries are different.

$$C = \begin{bmatrix} 1 & 2 \\ 3 & 4 \end{bmatrix} \qquad D = \begin{bmatrix} 1 & 2 & 0 \\ 3 & 4 & 0 \end{bmatrix} \qquad E = \begin{bmatrix} 1 & 3 \\ 2 & 4 \end{bmatrix}$$

Suppose we have a graph G with n vertices labeled $V_1, V_2, \ldots, V_n$. Such a graph is called a **labeled graph.** To represent the labeled graph G by a matrix, we form an $n \times n$ matrix in which the i,j element is 1 if there is an edge between the vertices V_i and V_j and 0 if there is not. This matrix is called the **adjacency matrix** of G (with respect to the labeling) and is denoted by $A(G)$.

EXAMPLE 3.6

Figure 3.8 contains two graphs and their adjacency matrices. For (a) the 1,2 entry is 1 because there is an edge between vertices V_1 and V_2, and the 3,4 element is 0 because there is no edge between V_3 and V_4. For (b) we see that the 1,2 and 1,3 entries are 1 because of the edges between V_1 and V_2 and between V_1 and V_3.

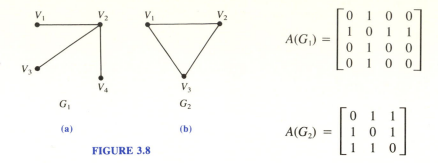

$$A(G_1) = \begin{bmatrix} 0 & 1 & 0 & 0 \\ 1 & 0 & 1 & 1 \\ 0 & 1 & 0 & 0 \\ 0 & 1 & 0 & 0 \end{bmatrix}$$

$$A(G_2) = \begin{bmatrix} 0 & 1 & 1 \\ 1 & 0 & 1 \\ 1 & 1 & 0 \end{bmatrix}$$

FIGURE 3.8

Note that in Figure 3.8(a) the sum of the entries in row 1 is 1, which is the degree of V_1, and likewise the sum of the entries in row 2 is the degree of V_2. This illustrates a more general result.

THEOREM 3.2   The sum of the entries in row i of the adjacency matrix of a graph is the degree of the vertex V_i in the graph.

Proof.   We recall that each 1 in row i corresponds to an edge on the vertex V_i. Thus, the number of 1's in row i is the number of edges on V_i, which is the degree of V_i.   ▨

Matrices are not the only way to represent graphs in a computer. Although an adjacency matrix is easy to construct, this form of representation requires $n \cdot n = n^2$ units of storage for a graph with n vertices and can be quite inefficient if the matrix contains lots of zeros. This means that if an algorithm to be performed on the graph requires a lot of searching of vertices and adjacent vertices, then the matrix representation can require a lot of unnecessary time. A better representation for such a matrix is an **adjacency list.**

The basic idea of an adjacency list is to list each vertex followed by the vertices adjacent to it. This provides the basic information about a graph: the vertices and the edges. To form the adjacency list, we begin by labeling the vertices of the graph G. Then we list the vertices in a vertical column, and after each one we write down the adjacent vertices. Thus, we see that the 1's in a row of an adjacency matrix tell what vertices are listed in the corresponding row of an adjacency list.

EXAMPLE 3.7   For the graph in Figure 3.9 there are 6 labeled vertices, and we list them in a vertical column as in (b). Beside vertex V_1 we list the adjacent vertices, which are V_2 and V_3. Then proceeding to the next vertex V_2, we list the vertices adjacent to it, V_1 and V_4. This process is continued until we get the adjacency list in (b).   ▨

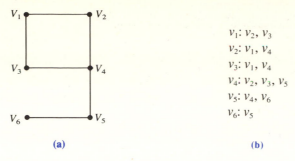

v_1: v_2, v_3
v_2: v_1, v_4
v_3: v_1, v_4
v_4: v_2, v_3, v_5
v_5: v_4, v_6
v_6: v_5

(a)   **(b)**

FIGURE 3.9

Matrix Operations

We will define only the two most common matrix operations, namely addition and multiplication.

Suppose A and B are two $m \times n$ matrices. The **sum** of A and B, denoted by $A + B$, is the $m \times n$ matrix in which the i,j entry of $A + B$ is the sum of the i,j entry of A and the i,j entry of B. In other words, we add two matrices of the same size by adding together the corresponding entries. Thus, only matrices of the same size can be added.

EXAMPLE 3.8   With the matrices given below, note how the sum of the two matrices is formed by adding together the 1,1 entries of the two matrices, then the 1,2 entries, then the 2,1 entries and continuing until the complete sum is formed.

$$\begin{bmatrix} 3 & 4 \\ 2 & 1 \\ 0 & 5 \end{bmatrix} + \begin{bmatrix} 2 & 6 \\ 3 & 0 \\ 2 & 1 \end{bmatrix} = \begin{bmatrix} 3+2 & 4+6 \\ 2+3 & 1+0 \\ 0+2 & 5+1 \end{bmatrix} = \begin{bmatrix} 5 & 10 \\ 5 & 1 \\ 2 & 6 \end{bmatrix}$$

The multiplication of two matrices is, unfortunately, not quite so direct. Suppose A is an $m \times k$ matrix with the i,j entry represented by a_{ij} and B is a $k \times n$ matrix with the i,j entry represented by b_{ij}. Then the **product** of A and B, denoted by AB, is the $m \times n$ matrix where the i,j entry is given by

$$a_{i1}b_{1j} + a_{i2}b_{2j} + \ldots + a_{ik}b_{kj}.$$

Note that the number of columns of A is required to be equal to the number of rows of B in order to multiply A and B. We will see later in an exercise that AB need not be equal to BA, unlike the ordinary multiplication of real numbers. Next note that the i,j entry of the product AB is formed by using entries from row i of A and column j of B, each of which has k entries. We multiply corresponding entries of row i and column j to get $a_{i1}b_{1j}$, $a_{i2}b_{2j}$, $\ldots$, $a_{ik}b_{kj}$, and then we add these products together. This process must be repeated for all possible i and j in order to compute AB.

EXAMPLE 3.9   For the matrices A and B listed below,

$$A = \begin{bmatrix} 1 & 2 & 3 \\ 4 & 5 & 6 \end{bmatrix} \qquad B = \begin{bmatrix} 1 & 3 & 5 & 8 \\ 0 & 1 & 6 & 9 \\ 2 & 4 & 7 & 1 \end{bmatrix}$$

the 1,3 element in the product AB is

$$a_{11}b_{13} + a_{12}b_{23} + a_{13}b_{33} = 1 \cdot 5 + 2 \cdot 6 + 3 \cdot 7 = 38.$$

The 2,4 element in AB is

$$a_{21}b_{14} + a_{22}b_{24} + a_{23}b_{34} = 4 \cdot 8 + 5 \cdot 9 + 6 \cdot 1 = 83.$$

Continuing in this fashion the product AB is found to be

$$\begin{bmatrix} 1\cdot1+2\cdot0+3\cdot2 & 1\cdot3+2\cdot1+3\cdot4 & 1\cdot5+2\cdot6+3\cdot7 & 1\cdot8+2\cdot9+3\cdot1 \\ 4\cdot1+5\cdot0+6\cdot2 & 4\cdot3+5\cdot1+6\cdot4 & 4\cdot5+5\cdot6+6\cdot7 & 4\cdot8+5\cdot9+6\cdot1 \end{bmatrix}$$

$$= \begin{bmatrix} 7 & 17 & 38 & 29 \\ 16 & 41 & 92 & 83 \end{bmatrix}.$$

EXERCISES 3.1

In Exercises 1–6 list the edges and vertices for each graph.

1.

2.

3.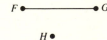

4.

5.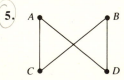

6.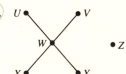

In Exercises 7–12 draw a diagram representing the graph with vertices V and edges E.

7. $V = \{A, B, C, D\}, E = \{\{B, C\}, \{C, A\}, \{B, D\}\}$

8. $V = \{X, Y, Z, W\}, E = \{\{X, Y\}, \{X, Z\}, \{Y, Z\}, \{Y, W\}\}$

9. $V = \{G, H, J\}, E = \varnothing$

10. $V = \{A, X, B, Y\}, E = \{\{A, X\}, \{X, B\}, \{B, Y\}, \{Y, A\}\}$

11. $V = \{A, B, C, D\}, E = \{\{B, C\}, \{C, D\}, \{D, B\}\}$

12. $V = \{X, Y, Z, W, U\}, E = \{\{X, Y\}, \{W, U\}\}$

In Exercises 13–20 determine if a graph is indicated.

13. A •

14.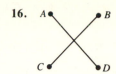

15. A •

(graph with vertices A, C, B)

16. A•   •B

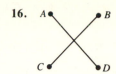

C •   •D

17. $V = \{A, B, C, D\}, E = \{\{A, B\}, \{A, A\}\}$

18. $V = \{A, B\}, E = \{\{A, B\}, \{B, C\}\}$

19. $V = \{A, X, B, Y, Z\}, E = \varnothing$

20. $V = \{A, B, C, D\}, E = \{\{A, B\}, \{C, B\}\}$

21. Construct the graph where the vertices are you, your parents, and your grandparents with a relationship of ''born in the same state.''

22. Construct the graph using ''are the same sex'' in place of ''born in the same state'' in Exercise 21.

23. There is a group of 6 students, Alice, Bob, Carol, Dean, Santos, and Tom, where Alice and Carol are always feuding, likewise for Dean and Carol, and for Santos, Tom, and Alice. Draw the graph to represent this situation.

24. Draw the graph with vertices $V = \{1, 2, \ldots, 10\}$ and edges $E = \{\{x, y\} : x, y$ in V, $x \neq y$, and x divides y or y divides $x\}$.

In Exercises 25–30 list the vertices adjacent to A and give the degree of A. Repeat for the vertex B.

25. *(graph with vertices A, B, F, E, C, D)*

26. *(graph with vertices A, B, C, D)*

27. A•————————•B

28. C•————————•B

A •

29. *(graph with vertices B, C, A, D, E, F)*

30. 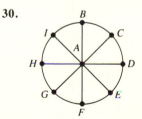 *(circular graph with vertices B, C, A, D, E, F, G, H, I)*

31. Draw the graph with vertices $X, Y, Z, W, R,$ and S, where X and R are adjacent, $W, R,$ and S are adjacent to each other, and Y and Z are adjacent.

32. Draw a graph where
  (a) there are 4 vertices, each with degree 1.
  (b) there are 4 vertices, each with degree 2.

33. Show that there are an even number of vertices with odd degree in any graph.

34. Can there be a graph with exactly 5 vertices, each with degree 3?

35. How many edges does K_3 have? K_4? K_5? and K_n in general?

36. Can there be a graph with 8 vertices and 29 edges?

37. How many vertices are there in a graph with 10 edges if each vertex has degree 2?

In Exercises 38–43 find the adjacency matrix and the adjacency list for each graph.

38. $V_1 \bullet\!\!-\!\!\!-\!\!\!-\!\!\!-\!\!\bullet V_2$

39. K_4

40.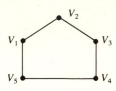

41.

42. K_5

43. $V_1 \bullet\!\!-\!\!\!-\!\!\!-\!\!\!-\!\!\bullet V_2$

$V_3 \bullet$

In Exercises 44–47 construct the graph for each adjacency matrix. Label the vertices V_1, V_2, V_3, ...

44.

$$\begin{bmatrix} 0 & 1 & 1 & 0 & 1 \\ 1 & 0 & 1 & 1 & 1 \\ 1 & 1 & 0 & 1 & 0 \\ 0 & 1 & 1 & 0 & 0 \\ 1 & 1 & 0 & 0 & 0 \end{bmatrix}$$

45.

$$\begin{bmatrix} 0 & 1 & 1 & 1 \\ 1 & 0 & 1 & 1 \\ 1 & 1 & 0 & 1 \\ 1 & 1 & 1 & 0 \end{bmatrix}$$

46.

$$\begin{bmatrix} 0 & 1 & 1 & 0 & 1 & 1 & 0 \\ 1 & 0 & 1 & 0 & 1 & 0 & 1 \\ 1 & 1 & 0 & 1 & 1 & 1 & 1 \\ 0 & 0 & 1 & 0 & 0 & 0 & 0 \\ 1 & 1 & 1 & 0 & 0 & 1 & 0 \\ 1 & 0 & 1 & 0 & 1 & 0 & 1 \\ 0 & 1 & 1 & 0 & 0 & 1 & 0 \end{bmatrix}$$

47.

$$\begin{bmatrix} 0 & 1 & 1 \\ 1 & 0 & 1 \\ 1 & 1 & 0 \end{bmatrix}$$

In Exercises 48–51 construct the graph for each adjacency list.

48. $V_1: V_2, V_4, V_5$
$V_2: V_1, V_3$
$V_3: V_2, V_5$
$V_4: V_1$
$V_5: V_1, V_3$

49. $V_1: V_2, V_3$
$V_2: V_1, V_3$
$V_3: V_1, V_2$

50. $V_1: V_2, V_3$
$V_2: V_1, V_4$
$V_3: V_1, V_4$
$V_4: V_1, V_3$

51. $V_1: V_2, V_3, V_5$
$V_2: V_1, V_4, V_5$
$V_3: V_1, V_4, V_5$
$V_4: V_2, V_3, V_5$
$V_5: V_1, V_2, V_3, V_4$

52. What does it mean when the adjacency matrix of a graph is all zeros?

In Exercises 53–55 can each matrix be an adjacency matrix?

53.

$$\begin{bmatrix} 0 & 1 & 1 & 0 \\ 1 & 0 & 1 & 0 \\ 1 & 1 & 0 & 1 \\ 0 & 0 & 1 & 0 \end{bmatrix}$$

54.

$$\begin{bmatrix} 0 & 1 & 0 & 1 & 0 \\ 1 & 0 & 0 & 1 & 1 \\ 0 & 0 & 0 & 0 & 1 \\ 1 & 1 & 0 & 0 & 0 \\ 0 & 0 & 0 & 0 & 0 \end{bmatrix}$$

55.

$$\begin{bmatrix} 1 & 1 & 1 \\ 1 & 1 & 1 \\ 1 & 1 & 1 \end{bmatrix}$$

56. For the matrices A, B, C, and Z, compute $A + B$, $B + A$, $B + C$, $A + (B + C)$, $(A + B) + C$, and $A + Z$.

$$A = \begin{bmatrix} 4 & 3 & 6 \\ -1 & 6 & 8 \end{bmatrix}, B = \begin{bmatrix} 0 & 3 & 4 \\ -1 & 4 & -2 \end{bmatrix}, C = \begin{bmatrix} -3 & 1 & 2 \\ 4 & 7 & -5 \end{bmatrix},$$

$$Z = \begin{bmatrix} 0 & 0 & 0 \\ 0 & 0 & 0 \end{bmatrix}$$

57. For the matrices A, B, C, and Z, compute $A + B$, $B + A$, $B + C$, $A + (B + C)$, $(A + B) + C$, and $A + Z$.

$$A = \begin{bmatrix} 5 & 8 & 2 \\ -7 & 1 & 0 \\ 8 & 0 & 0 \end{bmatrix}, B = \begin{bmatrix} 0 & 4 & 1 \\ 0 & 6 & -2 \\ 9 & 8 & 3 \end{bmatrix}, C = \begin{bmatrix} -2 & -3 & 1 \\ 5 & -2 & -2 \\ 7 & -1 & 4 \end{bmatrix},$$

$$Z = \begin{bmatrix} 0 & 0 & 0 \\ 0 & 0 & 0 \\ 0 & 0 & 0 \end{bmatrix}$$

58. For the matrices A, B, C, and S, compute AB, BC, $A(BC)$, $(AB)C$, AS, and SA.

$$A = \begin{bmatrix} 2 & 1 \\ -1 & 0 \end{bmatrix}, B = \begin{bmatrix} 2 & 3 & 4 \\ -1 & 0 & 5 \end{bmatrix}, C = \begin{bmatrix} 1 & 0 & 1 \\ 2 & 1 & 1 \\ 0 & 1 & 0 \end{bmatrix}, S = \begin{bmatrix} 3 & 0 \\ 0 & 3 \end{bmatrix}$$

59. For the matrices A, B, C, and S, compute AB, BC, $A(BC)$, $(AB)C$, AS, and SA.

$$A = \begin{bmatrix} 0 & 2 & 5 \\ -1 & 0 & 2 \\ 3 & 4 & 1 \end{bmatrix}, B = \begin{bmatrix} 3 & -1 \\ -2 & 2 \\ 4 & 0 \end{bmatrix}, C = \begin{bmatrix} 1 & 0 \\ 1 & 1 \end{bmatrix}, S = \begin{bmatrix} 4 & 0 & 0 \\ 0 & 4 & 0 \\ 0 & 0 & 4 \end{bmatrix}$$

60. For the matrices A, B, and C, compute AB, AC, $AB + AC$, $B + C$, $A(B + C)$, BC, CB, and C^2.

$$A = \begin{bmatrix} 2 & 1 & 4 \\ 4 & 1 & 7 \\ 2 & 0 & 5 \\ 3 & 0 & 2 \end{bmatrix}, B = \begin{bmatrix} 2 & 2 & 4 \\ -2 & 0 & 1 \\ 1 & 1 & -1 \end{bmatrix}, C = \begin{bmatrix} 2 & 2 & 1 \\ -1 & 2 & 2 \\ -3 & 1 & -3 \end{bmatrix}$$

61. For the matrices A, B, and C, compute AB, AC, $AB + AC$, $B + C$, $A(B + C)$, BC, CB, and B^2.

$$A = \begin{bmatrix} 3 & 2 & -1 \\ 0 & 1 & 2 \end{bmatrix}, B = \begin{bmatrix} 2 & 3 & 4 \\ -1 & 0 & 5 \\ 1 & 1 & -1 \end{bmatrix}, C = \begin{bmatrix} -1 & 0 & 1 \\ 2 & 1 & 2 \\ -3 & 1 & -2 \end{bmatrix}$$

62. Consider the closed intervals $[m, n]$, where m and n are distinct integers between 1 and 4, inclusive, as the vertices of a graph. Let two distinct vertices be adjacent if the corresponding intervals have at least one point in common. Draw the graph.

63. A few years ago the National Football League had two conferences each with 13 teams. It was decided by the league office that each team would play a total of 14 games, 11 of which were to be with teams in their own conference and the other 3 games with teams outside their own conference. Show that this is not possible.

64. Suppose a graph has n vertices, each with degree at least 1. What is the smallest number of edges the graph can have? $\left\lceil \dfrac{n}{2} \right\rceil$

65. Suppose a graph has n vertices, each with degree at least 2. What is the smallest number of edges it can have?

66. A graph has m edges with $m \geq 2$. What is the smallest number of vertices it can have?

67. Let $V = \{1, 2, \ldots, n\}$ and $E = \{\{x, y\} : x, y \text{ in } V, x \neq y, x \text{ divides } y \text{ or } y \text{ divides } x\}$ with $n > 3$. What vertices of this graph have degree 1?

68. Suppose Mr. and Mrs. Lewis attended a bridge party one evening. There were three other married couples in attendance and several handshakes took place. No one shook hands with himself or herself, no spouses shook hands, and no two people shook hands more than once. When each other person told Mr. Lewis how many hands he or she shook, the answers were all different. How many handshakes did Mr. and Mrs. Lewis each make?

69. Prove that if a graph has at least two vertices, then some pair of distinct vertices have the same degree.

Suppose a graph G is drawn in the plane so that the edges of G intersect only at the vertices of G. Then G partitions the plane into a finite number of parts, called regions. An illustration is given below where the regions are labelled $A, B, C, D, E, F, H,$ and I.

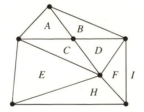

70. If G is such a graph in which e is the number of edges, v is the number of vertices, and f is the number of regions, prove that $f - e + v = 2$. (This result is called Euler's formula.)

3.2   Paths and Circuits

As we have seen, graphs can be used to describe a variety of situations. In many cases we want to know whether it is possible to go from one vertex to another by following a route using the edges. In other cases it may be necessary to perform a test that involves finding a route through all the vertices or over all the edges. While many situations can be described by graphs as we have defined them, there are others where it may be necessary to allow an edge from a vertex to itself or to allow more than one edge between vertices. For example, when a road system is being described, there can be two roads, an interstate highway and an older 2–lane road, between the same two towns. There could even be a scenic route starting and ending at the same town.

A **multigraph** is a generalization of a graph that consists of a finite set of vertices and a finite set of edges. Unlike a graph, a multigraph can contain edges from a vertex to itself and several edges between the same two vertices. An edge from a vertex to itself is called a **loop.** When there is more than one edge between two vertices, these edges are called **parallel edges.** It is important to note that a graph is a special kind of multigraph. Thus, all the definitions given for multigraphs apply to graphs as well.

EXAMPLE 3.10
The diagram in Figure 3.10 represents a multigraph but not a graph because there are two parallel edges k and m between the vertices Y and Z and a loop h at vertex X. ▆

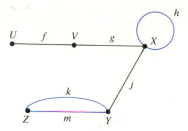

FIGURE 3.10

In a multigraph the number of edges incident with a vertex V is called the **degree** of V and is denoted as $\deg(V)$. A loop on a vertex V is counted twice in $\deg(V)$. Thus, in Figure 3.10 $\deg(Y) = 3$ and $\deg(X) = 4$.

Suppose G is a multigraph and U and V are vertices, not necessarily distinct. A **U–V path** or a **path from U to V** is an alternating sequence $V_1, e_1, V_2, e_2, V_3, \ldots, V_n, e_n, V_{n+1}$ of vertices and edges, where the first vertex V_1 is U and the last vertex V_{n+1} is V and the edge e_i connects V_i and V_{i+1} for $i = 1, 2, \ldots, n$. The **length** of this path is n, the number of edges listed. We note that U is a path to itself of length 0.

In a path the vertices need not be distinct from each other and likewise some of the edges can be the same. When there can be no chance of confusion a path can be represented by the vertices $V_1, V_2, \ldots, V_{n+1}$ only or by the edges $e_1, e_2, \ldots, e_n$ only.

EXAMPLE 3.11
In Figure 3.10 U, f, V, g, X is a path of length 2 from U to X. This path can also be written as f, g. Likewise f, g, h is a path of length 3 from U to X, and U, f, V, f, U is a path of length 2 from U to U. The path Z, m, Y cannot be described by just listing Z, Y since it would not be clear which edge between Z and Y, k or m, is part of the path. ▆

A path provides a way of describing how to go from one vertex to another by following edges. A $U-V$ path need not be an efficient route; that is, it may repeat vertices or edges. However, a $U-V$ **simple path** is a path from U to V in which no vertex is repeated.

There are no simple paths of length 1 or more from a vertex to itself. Furthermore, a simple path does not have loops or pairs of parallel edges in it. In some sense, a simple path is an efficient route between vertices, whereas a path allows wandering back and forth, repeating vertices and edges.

EXAMPLE 3.12

For the graph in Figure 3.11 the edges a, c, d, j form a simple path from U to Z, whereas a, c, m, d, j is a path from U to Z that is not a simple path because the vertex W is repeated. Similarly, e, i is a simple path from X to Z, but f, i, j is a path from X to Z that is not a simple path. Note also that c, p, f, i, e, n is a path from V to U that is not simple, but deleting f, i, e produces a simple path c, p, n from V to U. This illustrates the following result.

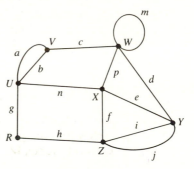

FIGURE 3.11

THEOREM 3.3

Every $U-V$ path contains a $U-V$ simple path.

Proof.   Let us suppose that $U = V_1, e_1, V_2, \ldots, e_n, V_{n+1} = V$ is a $U-V$ path. In the special case that $U = V$ we can choose our $U-V$ simple path to be just the vertex U. Now suppose U and V are different. If all of the vertices $V_1, \ldots, V_{n+1}$ are different initially, then our path is already a $U-V$ simple path. Thus, let us suppose that at least two of the vertices are the same, say that $V_i = V_j$ where $i < j$. See Figure 3.12 for an illustration of how a path from V_i to V_j is found.

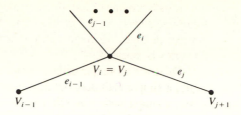

FIGURE 3.12

We delete e_i, V_{i+1}, . . . , e_{j-1}, V_j from the original path. What has been deleted is the part that is between vertex V_i and edge e_j. This still leaves a path from U to V. If there are only distinct vertices left after this deletion, then we are done. If there are still repetitions among the remaining vertices, the above process is repeated. Because the number of vertices is finite, this process will eventually end and give a U–V simple path from U to V. ■

Vertices U and V in a multigraph are called **connected** if there is a path between them. A multigraph G is **connected** if every two vertices of G are connected. Thus, in a connected multigraph we can go from any one vertex to another by following some route along the edges.

EXAMPLE 3.13   The multigraph in Figure 3.11 is connected since a path can be found between any two vertices. However, the graph in Figure 3.13 is not connected since there is no path from vertex U to vertex W. ■

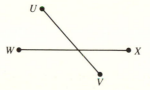

FIGURE 3.13

A **cycle** is a path V_1, e_1, V_2, e_2, . . . , V_n, e_n, V_{n+1}, where $n > 0$, $V_1 = V_{n+1}$, and all the vertices V_1, . . . , V_n and all the edges e_1, . . . , e_n are distinct. Thus, a cycle of length 3 or more cannot have loops or parallel edges as part of it.

EXAMPLE 3.14   For the multigraph in Figure 3.11 the edges *a, c, p, n* form a cycle. Likewise, the edges *g, b, c, p, f, h* form a cycle. Furthermore, the edges *f, p, d, e, n, g, h* do not form a cycle because the vertex *X* is used twice. ■

Euler Circuits and Paths

In testing a communication network it is often necessary to test each link (edge) in the system. In order to minimize the cost of such a test it is desirable to devise a route that goes through each edge exactly once. Similarly, when a multigraph is used to describe the street system in a town (corners are the vertices and the individual lanes are the edges), it may be necessary to devise a street sweeping route that will go over each edge (lane). Again to minimize cost and time, it is desirable not to send the street sweeper over an edge twice. Thus, a path is needed that includes every edge exactly once. Because the mathematician Leonhard Euler was the first one known to consider this question, a path in a multigraph *G* that includes exactly once all the edges of *G* and has different first and last vertices is called an **Euler path.** A path that includes exactly once all the edges of *G* and has the same initial and terminal vertices is called an **Euler circuit.**

EXAMPLE 3.15   For the graph in Figure 3.14(a) the path *a, b, c, d* is an Euler circuit since all the edges are included and each edge is included exactly once. However, the graph in Figure 3.14(b) has neither an Euler path nor circuit because to include all the three edges in a path we would have to backtrack and use an edge twice. For the graph in Figure 3.14(c) there is an Euler path *a, b, c, d, e, f* but not an Euler circuit. ■

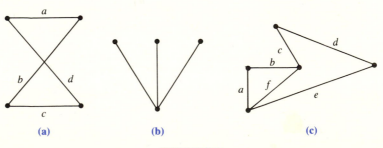

FIGURE 3.14

As we proceed along an Euler path or circuit, each time an intermediate vertex is reached along some edge there must be another edge for us to exit that vertex. In fact, whenever each vertex has even degree the following algorithm produces an Euler circuit.

Euler Circuit Algorithm

This algorithm finds an Euler circuit for a connected multigraph G such that every vertex of G has even degree.

Step 1 (start path). Select a vertex V and an edge on V.

Step 2 (get V–V path). If the other vertex on the last chosen edge is not V, then choose an unused edge on this other vertex. Repeat step 2.

Step 3 (start new path at A). If all of the edges have been used, then stop (an Euler circuit has been constructed). Otherwise choose an unused edge on a vertex already visited, and give this previously visited vertex a temporary name A.

Step 4 (get A–A path). If the other vertex on the last chosen edge is not A, then choose an unused edge on this other vertex. Repeat step 4.

Step 5 (join paths together). Insert these newly chosen edges at the vertex A. Go to step 3.

EXAMPLE 3.16   For a graph such as the one in Figure 3.15, it may be possible to look at the graph and with minimal trial and error construct an Euler circuit. However, for the purpose of illustration let us use the Euler circuit algorithm.

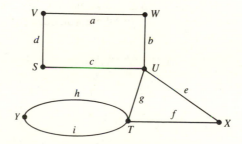

FIGURE 3.15

First choose some vertex at which to begin, say V. Next choose an edge on V, say a. Then the other vertex on the edge a is W and $W \neq V$. Since there is only one unused edge on W, namely the edge b, we choose b. The other vertex on the edge b is the vertex U and $U \neq V$. Since there are three unused edges on U, arbitrarily choose an edge, say edge c. Then edge d must be chosen next, returning our path to vertex V. These edges a, b, c, d do not form an Euler circuit because some of the edges of the graph are missing.

Since edge e has not been used and is on a vertex U which has been visited, we start with U and the edge e. Thus, we assign U the temporary name A. Since the vertex X is on the edge e and $X \neq A$, we choose the other edge f on X. Next we arbitrarily choose the edge g, returning us to A. Now these two groups of edges are put together by joining them at the vertex U to form a, b, e, f, g, c, d. Note how the edges are put together so that a path with no repetitions of edges is formed. Again this is not an Euler circuit. The process is now started again with vertex T and edge h. Then the edges h, i are obtained and now these are joined into the previously chosen edges at vertex T to get a, b, e, f, h, i, g, c, d, which is an Euler circuit. ■

The following theorem gives necessary and sufficient conditions for a connected multigraph to have an Euler circuit or path and justifies the Euler circuit algorithm.

THEOREM 3.4   Suppose a multigraph G is connected. Then G has an Euler circuit if and only if every vertex of G has even degree. Furthermore, G has an Euler path if and only if every vertex of G has even degree except for two distinct vertices, which have odd degree. When this is the case, the Euler path starts at one and ends at the other of these two vertices of odd degree.

Proof.   We will give a proof only in the case that G contains no loops. An easy modification establishes the result when there are loops in G.

Suppose the multigraph G has an Euler circuit. Every time this Euler circuit passes through a vertex, it enters along an edge and leaves along a different edge. Thus, the Euler circuit contributes twice to the degree of the vertex each time the Euler circuit goes through the vertex. Since every edge is used in an Euler circuit, every edge through a vertex can be paired as one of two, either coming in or going out. Thus, each vertex has even degree.

Conversely, suppose each vertex has even degree. The Euler circuit algorithm constructs a path starting at V. This path must return to V since when we enter a different vertex along one edge, another edge must leave the vertex because its degree is even. Thus a V–V path is constructed. The algorithm proceeds by starting along unused edges on vertices on this path. Since the multigraph is connected, there always exists a path from any unused edge to the path already constructed; so each edge is eventually included.

If an Euler path exists between distinct vertices U and V, then clearly the degrees of U and V must be odd, while all the other vertices have even degree. Conversely, if only U and V have odd degrees in a connected multigraph, we could add an edge e between U and V. The new graph will have all degrees even, and so an Euler circuit will exist for it by what we have already proved. Removing e produces an Euler path between U and V. ■

From the last paragraph of the proof of Theorem 3.4 we see that the Euler circuit algorithm may be used to find an Euler path in a connected multigraph with exactly two vertices of odd degree by applying it to the multigraph formed by adding an edge between these two vertices.

EXAMPLE 3.17   For the multigraph in Figure 3.16(a) an edge e is added between the two vertices U and V of odd degree. This results in the multigraph in Figure 3.16(b) for which an Euler circuit, say e, a, d, c, b, can be found by using the Euler circuit algorithm. Deleting the edge e from this circuit gives the Euler path a, d, c, b between U and V for the multigraph in Figure 3.16(a).   ■

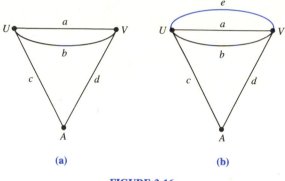

(a)   (b)

FIGURE 3.16

In analyzing the complexity of the Euler circuit algorithm we will use picking an edge as an elementary operation. Since each of the e edges is used once, this algorithm is of order at most e. For a graph with n vertices, $e \leq \frac{1}{2}n(n-1) = \frac{1}{2}(n^2 - n)$ since $\frac{1}{2}n(n-1)$ is the number of pairs of distinct vertices. (See Theorem 2.16.) Thus, for a graph this algorithm is of order at most n^2.

Hamiltonian Cycles

In Example 3.1 a graph was used to describe a system of airline routes. Suppose that a salesperson needs to visit each of the cities in this graph. In this situation time and money would be saved by visiting each city exactly once. What is needed for an efficient scheduling is a path that begins and ends at the same vertex and uses each vertex once and only once.

In the first part of this section paths that used each edge once and only once were considered. Now we want to find a cycle that uses each *vertex* of a multigraph exactly once. But since we want to avoid repetition of vertices, loops and parallel edges will not be of any assistance. Consequently, we may assume that we are working with a graph. In a graph a **Hamiltonian cycle** is a cycle that includes each vertex. These cycles are named after Sir William Rowan Hamilton, who developed a puzzle where the answer required the construction of this kind of cycle.

EXAMPLE 3.18   Suppose the graph in Figure 3.17 describes a system of airline routes where the vertices are towns and the edges represent airline routes. The vertex U is the home base for a salesperson who must periodically visit all of the other cities. To be economical the salesperson wants a path that starts at U, ends at U, and visits each of the other vertices exactly once. A brief examination of the graph shows that the edges a, b, d, g, f, e form a Hamiltonian cycle.

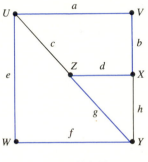

FIGURE 3.17

For the graph in Figure 3.1 there is no Hamiltonian cycle. For the only way to reach New York or St. Louis is from Chicago, and once in New York or St. Louis the only way to leave is to return to Chicago. ▪

Relatively easy criteria exist to determine if there is an Euler circuit or an Euler path. All that has to be done is to check the degree of each vertex. Furthermore, there is a straightforward algorithm to use in constructing an Euler circuit or path. Unfortunately, the same situation does not hold for Hamiltonian cycles. It is a major unsolved problem to determine necessary and sufficient conditions for a graph to have a Hamiltonian cycle. In general, it is very difficult to find a Hamiltonian cycle for a graph. There are, however, some conditions that guarantee the existence of a Hamiltonian cycle in a graph. We will provide an example of one of these theorems. The proof is found in Exercise 72.

THEOREM 3.5   Suppose G is a graph with n vertices, where $n > 2$. If each vertex has degree at least $\frac{n}{2}$, then G has a Hamiltonian cycle.

EXAMPLE 3.19 Theorem 3.5 can be used to say that the graph in Figure 3.18(a) has a Hamiltonian cycle because there are 6 vertices, each with degree 3. However, even though the theorem says there is a Hamiltonian cycle, it does not tell us how to find one. Fortunately in this case one can be found by a little bit of trial and error.

On the other hand, the graph in Figure 3.18(b) does not have a Hamiltonian cycle because no matter where we start we end up on the left side needing to go to a vertex on the left side, and there are no edges connecting the vertices on that side. Note that this graph does not satisfy the conditions of Theorem 3.5 because it has vertices of degree $2 < \frac{5}{2}$. However, the graph in Figure 3.18(c) also has 5 vertices, each with degree 2; yet it contains a Hamiltonian cycle. Thus, when some vertices have degree less than $\frac{n}{2}$, it is not possible to conclude anything in general about the existence or nonexistence of a Hamiltonian cycle.

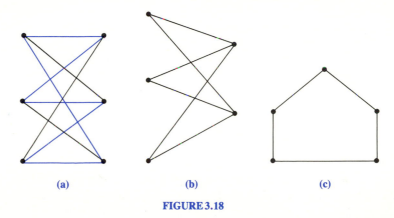

(a) (b) (c)

FIGURE 3.18

EXERCISES 3.2

In Exercises 1–8 determine if the multigraph is a graph.

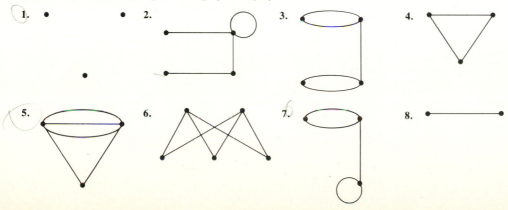

1. 2. 3. 4.

5. 6. 7. 8.

In Exercises 9–16 list the loops and parallel edges in the multigraph.

9. 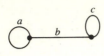   **10.** 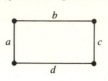   **11.** 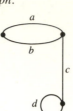   **12.**

13. 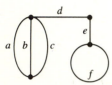   **14.**   **15.** 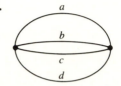   **16.**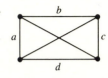

In Exercises 17–20

(i)   List at least 3 different paths from A to D. Give the length of each.
(ii) List the simple paths from A to D. Give the length of each.
(iii) For each path you listed in (i) find a simple path from A to D contained in it.
(iv) List the distinct cycles. (Two cycles are distinct if they do not contain the same edges.) Give the length of each.

17. 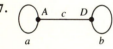   **18.** 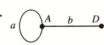   **19.** 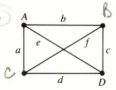   **20.**

21. Give an example of a multigraph satisfying each of the following.

   (a) There are exactly 2 cycles.
   (b) There is a cycle of length 1.
   (c) There is a cycle of length 2.

In Exercises 22–29 determine if the multigraph is connected.

22. 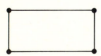   **23.**   **24.**   **25.**

26. 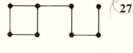   **27.**   **28.**   **29.**

In Exercises 30–37 determine if the multigraph has an Euler path. If it does, construct one using the Euler circuit algorithm.

30. *Euler circuit*

31. *neither*

32. *Euler path*

33. *Euler circuit*

34. *Euler path*

35. *neither*

36. *Euler path*

37. *Euler circuit*

In Exercises 38–45 determine if the multigraph in the indicated exercise has an Euler circuit. If it does, construct an Euler circuit using the Euler circuit algorithm.

38. Exercise 30

39. Exercise 31

40. Exercise 32

41. Exercise 33

42. Exercise 34

43. Exercise 35

44. Exercise 36

45. Exercise 37

46. The city of Königsberg, located on the banks of the Pregel River, had seven bridges which connected islands in the river to the shores as illustrated below. It was the custom of the townspeople to stroll on Sunday afternoons and, in particular, to cross over the bridges. The people of Königsberg wanted to know if it was possible to stroll in such a way that it was possible to go over each bridge exactly once and return to the starting point. Is it? (This problem was presented to the famous mathematician, Leonhard Euler, and his solution is often credited with being the beginning of graph theory.)

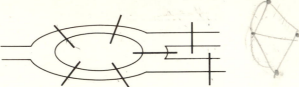

47. Could the citizens of Königsberg find an acceptable route by building a new bridge? no

48. Could the citizens of Königsberg find an acceptable route by building two new bridges? yes

49. Could the citizens of Königsberg find an acceptable route by tearing down a bridge? no

50. Could the citizens of Königsberg find an acceptable route by tearing down two bridges? yes

An old childhood game asks children to trace a figure with a pencil without either lifting the pencil from the figure or tracing a line more than once. Determine if this can be done for the figures in Exercises 51–54, assuming that you must begin and end at the same point.

51. yes    **52.** no 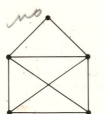   **53.** yes    **54.** no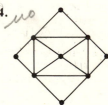

55. A street inspector wants to examine the streets in her region for potholes. If the map of her region is given below, is it possible for her to devise a route to examine each street once and return to her office? yes

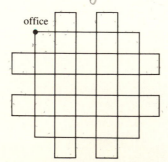

56. The following graph has 4 vertices of odd degree, and thus it has no Euler circuit or path. However, it is possible to find two distinct paths, one from *A* to *B*, the other

from C to D, such that the two paths use all the edges and the two paths have no edge in common. Find them.

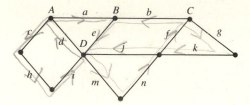

57. The following is the floor plan of the bottom level of a new home. Is it possible to enter the house at the front and exit at the rear and travel through the house going through each doorway exactly once?

58. In 1859 Sir William R. Hamilton, a famous Irish mathematician, marketed a puzzle which consisted of a regular dodecahedron made of wood in which each corner represented a famous city. The puzzle was to find a route that traveled along the edges of the dodecahedron visiting each city exactly once and returning to the original starting city. (To make the task somewhat easier, each corner had a nail in it and one was to use string while tracing out a path.) A representation of this puzzle drawn in the plane is given below. Can you find an answer to the puzzle?

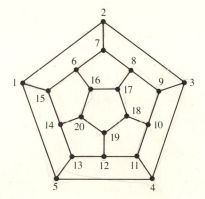

59. Give an example of a connected graph satisfying each of the following.
   (a) There is both an Euler circuit and a Hamiltonian cycle.
   (b) There is neither an Euler circuit nor a Hamiltonian cycle.
   (c) There is an Euler circuit but not a Hamiltonian cycle.
   (d) There is a Hamiltonian cycle but not an Euler circuit.

60. A **bipartite graph** is a graph in which the vertices can be divided into two disjoint nonempty sets A and B such that no two vertices in A are adjacent and no two vertices in B are adjacent. The **complete bipartite graph** $K_{m,n}$ is a bipartite graph in which the sets A and B contain m and n vertices, respectively, and every vertex in A is adjacent to every vertex in B. $K_{2,3}$ is given below. How many edges does $K_{m,n}$ have?

m·n

complete bipartite graph

61. For which m and n does $K_{m,n}$ have an Euler circuit? *m and n are even*

62. For which m and n does $K_{m,n}$ have a Hamiltonian cycle? *m or n > 1*

63. Prove that K_n has a Hamiltonian cycle when $n > 2$.

64. Find a Hamiltonian cycle for K_4 and for K_5.

65. In a multigraph with n vertices, what is the maximum length of a simple path? *n−1*

66. Prove that every $U-U$ path of positive length in a multigraph contains a cycle. *where none of the internal edges are repeated*

67. At a recent college party there were a number of young men and women present, some of whom had dated each other recently. This situation can be represented by a graph in which the vertices are the individuals in attendance with adjacency being defined by having dated recently. If this graph has a Hamiltonian cycle, show that the number of men is the same as the number of women.

68. Prove Theorem 3.3 by using induction.

69. Suppose that there are exactly 4 vertices of odd degree in a connected multigraph. Prove that there are two paths, one between two of these vertices and the other between the remaining two vertices such that every edge is in exactly one of these two paths.

70. Prove that if in a graph there is a $U-U$ path of odd length for some vertex U, then there is a cycle of odd length.

71. Prove that if a connected graph has n vertices, then it must have at least $n - 1$ edges.

72. Suppose G is a graph with n vertices, where $n > 2$. Prove that if each vertex has degree at least $\frac{n}{2}$, then G has a Hamiltonian cycle. (Hint: Consider a simple path with the largest number of vertices and show that there is a cycle containing the vertices of this simple path.)

3.3   Coloring a Graph

In Sections 3.1 and 3.2 we discussed several situations described by graphs or multigraphs. Sometimes the situation in which a graph can be used is somewhat unexpected. Two such examples follow.

EXAMPLE 3.20

Suppose that a chemical manufacturer needs to ship a variety of chemical products from a refinery to a processing plant. Shipping will be by rail, but according to EPA regulations not all of these chemical products can be shipped together in one railroad car because of the possibility of their mixing together and creating a violent reaction should an accident occur. How can these products be shipped? In order to minimize expenses the manufacturer wants to use the smallest possible number of railroad cars. What is this number?   ■■

EXAMPLE 3.21

The State Senate has a number of major standing committees with every senator on one or more of these. Each committee meets every week for an hour. Each senator must be able to attend each meeting of a committee he or she is on, and so no two committees can meet at the same time if they have a member in common. The Clerk of the Senate is responsible for scheduling these meetings. How should the Clerk schedule these committee meetings so that the senators can attend their major committee meetings and yet keep the number of meeting times as small as possible?   ■■

In these examples there are objects (chemical products or committee meetings) and relationships (cannot travel in the same railroad car or cannot meet at the same time) existing among them. Since this is the basic idea of a graph, it seems natural to describe each of these examples by a graph. In the first example the vertices are the chemical products and an edge is drawn between two vertices whenever they represent chemical products that cannot be in the same railroad car. In the second example the vertices are the committee meetings and an edge is drawn between two vertices whenever some senator is on both of these committees.

To illustrate this idea further let us assume in Example 3.20 that there are six chemical products P_1, P_2, P_3, P_4, P_5, and P_6 and that P_1 cannot ride in the same railroad car as P_2, P_3, or P_4; P_2 cannot also ride with P_3 or P_5; P_3 also cannot be with P_4; and P_5 cannot be with P_6. The graph that is described is found in Figure 3.19, where the vertices represent the six products and the edges join pairs of products that cannot ride together.

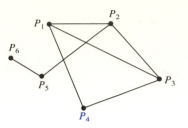

FIGURE 3.19

The question still remains: what is the smallest number of railroad cars needed? In the graph in Figure 3.19 products represented by adjacent vertices are to be in different rail cars. For example, product P_1 could be in car 1. Then because P_1 and P_2 are adjacent, a different car is needed for P_2, say car 2. Since P_3 is adjacent to both P_1 and P_2, there is still a need for another car for P_3, say car 3. But a new car is not needed for P_4; car 2 can be used again. Likewise, for P_5 a new car is not needed as either car 1 or car 3 can be used. Let car 1 be chosen. Then for P_6 car 2 or car 3 can be picked, say car 2. The graph in Figure 3.20 shows how the vertices are labeled so that incompatible chemical products travel in different cars. Furthermore, because P_1, P_2, and P_3 are adjacent to each other, at least three different railroad cars must be used. So three is the smallest number of railroad cars that can be used.

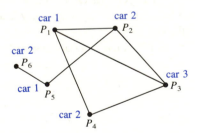

FIGURE 3.20

What has been done is to assign labels to the vertices of a graph so that adjacent vertices have different labels. This idea occurs frequently in graph theory, and for historical reasons the labels are called colors. To **color a graph** means to assign a color to each vertex such that adjacent vertices have different colors. Asking what is the smallest number of railroad cars that is needed in Example 3.20 is the same as asking what is the smallest number of colors needed to color the graph in Figure 3.19, with a color corresponding to a railroad car.

When a graph G can be colored with n colors but not with a smaller number of colors, G is said to have **chromatic number** n. Thus, the graph G in Figure 3.19 has chromatic number 3.

EXAMPLE 3.22   The graph in Figure 3.21(a) has chromatic number 2 since the vertices V_1, V_3, and V_5 can be colored with one color (say red) and the other three vertices with a second color (blue) as shown in Figure 3.21(b). In general, if a cycle has an even number of vertices, then it can be colored using 2 colors.   ■

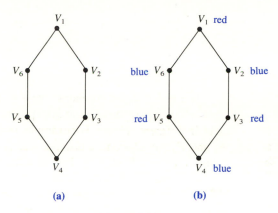

(a)   (b)

FIGURE 3.21

EXAMPLE 3.23   When a cycle has an odd number of vertices, such as that in Figure 3.22(a), then 3 colors must be used. If we try to alternate colors as was done in Figure 3.21 with the color red assigned to vertices V_1 and V_3 and the color blue assigned to vertices V_2 and V_4, then it is not possible to use either red or blue for V_5. Using 3 colors to color a cycle with an odd number of vertices is illustrated in Figure 3.22(b).   ■

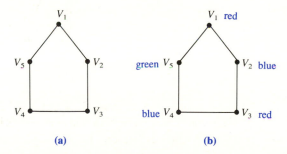

(a)   (b)

FIGURE 3.22

EXAMPLE 3.24   The complete graph K_n with n vertices can be colored using n colors. But since every vertex is adjacent to every other vertex, a smaller number of colors will not work. Thus K_n has chromatic number n.   ■

EXAMPLE 3.25   The graph in Figure 3.23(a) can be colored with two colors as indicated in Figure 3.23(b). ■

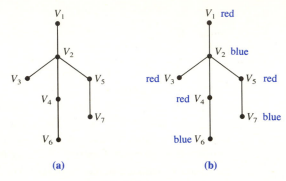

(a)   (b)

FIGURE 3.23

EXAMPLE 3.26   The graph in Figure 3.24 has chromatic number 2 since the vertices on the left can be colored with one color and the vertices on the right can be colored with a second. ■

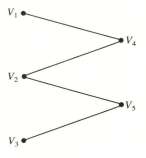

FIGURE 3.24

In general, it is very difficult to find the smallest number of colors needed to color a graph. One method is to list all the different ways to assign colors to the vertices of a graph, then go through these ways one at a time to see which of them is a coloring, and then finally determine which colorings have the smallest number of colors. Unfortunately, even if the graph has a relatively small number of vertices, this becomes an extraordinarily time-consuming process, measured in centuries rather than minutes even with the use of a supercomputer.

Nevertheless, there are a number of results that describe the chromatic number of a graph. For example, as seen in Example 3.23, a cycle with an odd length has chromatic number 3. Thus, any graph containing a cycle of this type needs at least 3 colors. The graph in Figure 3.20 is an example of this. When there are no cycles of odd length in a graph, then 2 colors are enough.

THEOREM 3.6 A graph G has no cycles of odd length if and only if G can be colored with two colors.

Proof. As noted above, when G has a cycle of odd length, then coloring G requires at least 3 colors.

Now suppose G has no cycles of odd length. A vertex V is selected and the color red is assigned to it. Then to each vertex adjacent to V the color blue is assigned. Now the vertices adjacent to the recently colored blue vertices are assigned the color red. Can one of these new red vertices, say W, be adjacent to the vertex V, which also has color red? The diagram in Figure 3.25 illustrates this situation.

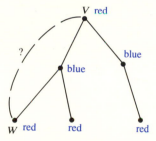

FIGURE 3.25

We see that if V and W were adjacent, then there would be a cycle of length 3. In a similar fashion, any other vertex just colored red is not adjacent to any other vertex with color red, for otherwise there would be a cycle of odd length. Next the vertices which are adjacent to those just colored red are assigned the color blue. This is illustrated in Figure 3.26.

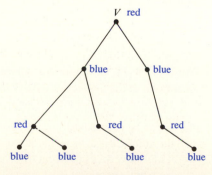

FIGURE 3.26

Again, if any two vertices colored blue were adjacent, then there would be a cycle of odd length. Continue by coloring red the vertices adjacent to these recently colored blue vertices. As before none of these newly colored red vertices can be adjacent to a previously colored red vertex. This process is repeated until there are no uncolored vertices adjacent to colored vertices.

If the graph is not connected, then there will be vertices that are not adjacent to any of the colored vertices and hence are uncolored. For these vertices the above process is repeated again using the colors red and blue. Eventually all the vertices can be colored with two colors. ■

EXAMPLE 3.27   For the graph in Figure 3.27(a) the coloring process is started by selecting a vertex V and coloring it red. Then since F, B, and A are the vertices adjacent to V, the color blue is assigned to them. The uncolored vertices adjacent to the blue colored vertices are C, D, and E, and so the color red is assigned to them. Finally, the vertex G is an uncolored vertex adjacent to red vertices and so is assigned the color blue. Now X is an uncolored vertex not adjacent to any colored vertex, and so X is given the color red. Then Y is given the color blue, and finally Z is assigned the color red. The coloring is given in Figure 3.27(b). ■

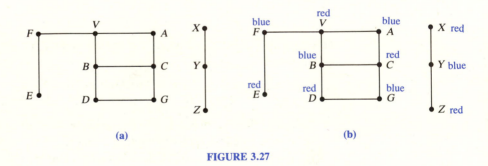

(a)   **(b)**

FIGURE 3.27

The following result gives an upper bound on the number of colors needed to color a graph.

THEOREM 3.7   For a graph G, the chromatic number of G cannot exceed one more than the maximum of the degrees of the vertices of G.

Proof.   Let k be the maximum of the degrees of the vertices of G. We will show that G can be colored using $k + 1$ colors $C_0, \ldots, C_k$. First a vertex V is selected and the color C_0 is assigned to it. Next some other vertex W is picked. Since there are at most k vertices adjacent to W and there are at least $k + 1$ colors available to choose from, there is at least one color (possibly many) that has not been used on a vertex adjacent to W. Choose such a color. This process can be continued until all the vertices of G are colored. ▄▄

EXAMPLE 3.28   The procedure described in Theorem 3.7 may use more colors than are really necessary. The graph in Figure 3.28 has a vertex of degree 4, which is the maximum degree, and so by Theorem 3.7 can be colored using $1 + 4 = 5$ colors. However, by using the procedure described in Theorem 3.6, it can be colored using 2 colors. ▄▄

FIGURE 3.28

We will present one additional algorithm, due to Welsh and Powell, for coloring the vertices of a graph.

Welsh and Powell Algorithm

This algorithm gives a coloring of a graph by labeling the vertices according to their degrees.

Step 1   (label vertices by degree). Label the vertices $V_1, V_2, \ldots, V_n$ such that $\deg(V_1) \geq \deg(V_2) \geq \ldots \geq \deg(V_n)$. (Ties can be broken arbitrarily.)

Step 2   (color first uncolored vertex and uncolored vertices not adjacent to it). Assign an unused color to the first uncolored vertex in the list of vertices. Go through the list of vertices in order, assigning this new color to any vertex not adjacent to any other vertex with this color.

Step 3   (graph colored?). If some of the vertices are uncolored, then return to step 2.

Step 4   (done). A coloring of the graph has been constructed.

EXAMPLE 3.29   For the graph in Figure 3.29(a) vertex F has the largest degree, namely 4, and so F is given the label V_1. Vertices A, D, and E have degree 3 and so are assigned labels V_2, V_3, and V_4 in some random manner. Likewise, vertices B and C with degree 2 are assigned labels V_5 and V_6. Vertex G is the only remaining vertex and is assigned V_7. This is illustrated in Figure 3.29(b).

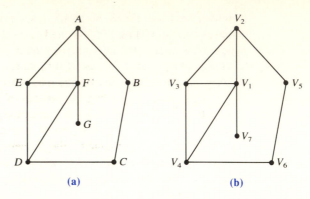

(a)   (b)

FIGURE 3.29

The adjacency list form of representation is very convenient to use with the Welsh and Powell algorithm. For the graph in Figure 3.29(b) the adjacency list representation is given below.

$$V_1: V_2, V_3, V_4, V_7$$
$$V_2: V_1, V_3, V_5$$
$$V_3: V_1, V_2, V_4$$
$$V_4: V_1, V_3, V_6$$
$$V_5: V_2, V_6$$
$$V_6: V_4, V_5$$
$$V_7: V_1$$

In the Welsh and Powell algorithm the first uncolored vertex in the list, V_1, is given the color red. Then we go down the list looking for the next vertex which is not adjacent to V_1, that is, for the lowest-numbered vertex that is not in the list following V_1. This is the vertex V_5, which is assigned the color red. We continue down the list looking for the next vertex which is adjacent to neither V_1 nor V_5. Since there are none, we go back to the top of the list and find the first uncolored vertex, which is V_2, and assign it the color blue. Then the next uncolored vertex is found which is not adjacent to V_2. This is V_4 and it is given the color blue. Continuing in this same manner, we find that V_7 is the next uncolored vertex adjacent to neither V_2 nor V_4, and so V_7 is assigned the color blue. Since all the uncolored vertices are adjacent to blue vertices, we go back to the top to find the next uncolored vertex, which is V_3, and assign it the color green. Since V_6 is uncolored and not adjacent to V_3, it is assigned the color green. So the graph in Figure 3.29(a) can be colored with 3 colors.

The adjacency list representation makes the Welsh and Powell algorithm easy to use because the vertices can be marked off as they are colored and thus easily avoided in the rest of the coloring process. ▮

One of the most famous problems of the 19th century concerned how many colors were necessary to color a map. It is understood that, when coloring a map, countries with a common boundary other than a point are to be colored with different colors. The map is assumed to be drawn on a flat surface or globe as opposed to a more complicated surface such as a doughnut. The usual approach to this problem is to let each country be a vertex of a graph and to connect vertices representing countries with a common boundary other than a point. Then coloring a map is the same as coloring the vertices of this graph so that no two adjacent vertices have the same color. It was conjectured in 1852 that four colors would be enough to color any such map, but it was not until 1976 that K. Appel and W. Haken, two mathematicians at the University of Illinois, verified this conjecture. Their verification required an exhaustive analysis of over 1900 cases that took over 1200 hours on a high-speed computer.

EXAMPLE 3.30 In Figure 3.30(a) is a portion of a map of the United States. The associated graph obtained as described above is shown in Figure 3.30(b). This graph can be colored using the Welsh and Powell algorithm as illustrated in Figure 3.30(c). ■

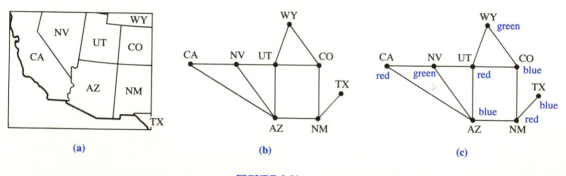

(a) (b) (c)

FIGURE 3.30

EXERCISES 3.3

In Exercises 1–8 find the chromatic number for the graph.

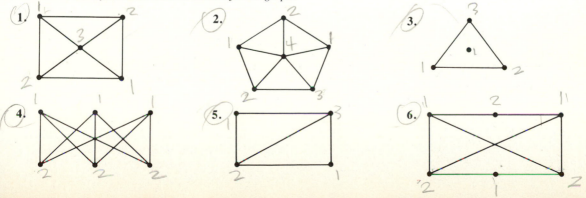

1. 2. 3.

4. 5. 6.

7.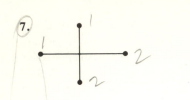

8.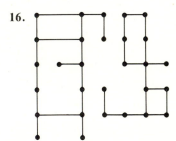

9. What does it mean for a graph to have chromatic number 1? *no edges*

10. What is the chromatic number for $K_{2,3}$? for $K_{7,4}$? for $K_{m,n}$? (See Exercise 60 of Section 3.2 for a definition of $K_{m,n}$.) *2   2   2*

11. Give an example of a graph where:
  (a) The chromatic number is one more than the maximum of the degrees of the vertices. *complete graph*
  (b) The chromatic number is not one more than the maximum of the degrees of the vertices.

12. It might be supposed that if a graph has a large number of vertices and each vertex has a large degree, then the chromatic number would have to be large. Show that this conclusion is incorrect by constructing a graph with at least 12 vertices, each of degree at least 3, that has chromatic number 2.

13. Using the process presented in the proof of Theorem 3.6, write an algorithm for coloring a graph with no cycles of odd length (no odd cycle coloring algorithm).

14. Show that when applied to a graph with n vertices and e edges the no odd cycle coloring algorithm (see Exercise 13) is of order at most $n + e$. Assume that coloring a vertex and using an edge are elementary operations in analyzing the algorithm.

In Exercises 15–18 color the graph using the no odd cycle coloring algorithm (see Exercise 13).

15.

16.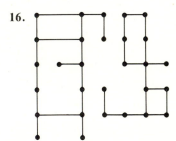

17.

18.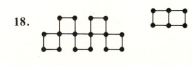

19. What is the chromatic number for the graph in Exercise 62 of Section 3.1?

In Exercises 20–23 color the graphs using the Welsh and Powell algorithm.

20.

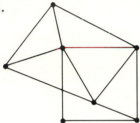

V_1	V_2	V_5
V_2	V_1	V_3
V_3	V_2	V_4
V_4	V_3	V_5
V_5	V_1	V_4

21.

22.

23.

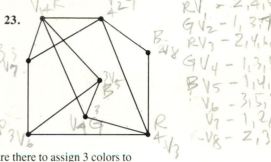

RV₁ = 2,4,5,7
GV₂ – 1,3,7,8
RV₃ – 2,4,6,8
GV₄ – 1,3,5
BV₅ – 1,4,6
V₆ – 3,5,7
V₇ – 1,2,6
–V₈ – 2,3

24. Suppose G is a graph with 3 vertices. How many ways are there to assign 3 colors to the vertices (this need not be a coloring of the graph)? Repeat using a graph with 4 vertices and 4 colors.

25. Generalize Exercise 24 to the case of a graph with n vertices and n colors.

26. Suppose G is a graph with n vertices, and there are n available colors to assign to the vertices. If one operation consists of assigning colors to the vertices and checking if a coloring has been made, how long would it take a supercomputer which can perform one billion operations per second to check all possible color assignments for a graph with 20 vertices? Would this be a good way to find a coloring using the least number of colors?

27. Color the following map using only 3 colors.

28. Color the following map using only 3 colors.

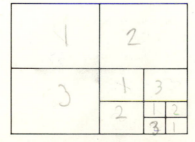

29. Color the following map using only 4 colors.

30. Color the following map using only 4 colors.

31. Solve Example 3.21 if there are 5 major committees: finance, budget, education, labor, and agriculture. The Clerk of the Senate needs only to consider State Senators Brown, Chen, Donskvy, Geraldo, Smith, and Wang. The finance committee has members Chen, Smith, and Wang; the budget committee has members Chen, Donskvy, and Wang; the education committee has members Brown, Chen, Geraldo, and Smith; the labor committee has only Geraldo; and the agriculture committee has Donskvy and Geraldo.

32. There are locations in a computer memory where stacks are stored during the execution of a computer program. Furthermore, a location can store only one stack at a time. Suppose stacks $S_1, \ldots, S_{10}$ are to be constructed during the execution of a computer program and stacks S_i and S_j will be in use at the same time if $i \equiv j \pmod 3$ or $i \equiv j \pmod 4$. What is the minimum number of locations for the stacks that will be needed during the execution of this computer program?

33. By representing the figure below by a graph, determine the minimum number of colors needed to color each circle so that touching circles have different colors.

34. The zookeeper of a major zoo wants to redo the zoo in such a way that the animals live together in their natural habitat. Unfortunately, it is not possible to put all the animals together in one location because some are predators of others. The dots in the chart below show which are predators or preys of others. What is the minimum number of locations the zookeeper needs?

	a	b	c	d	e	f	g	h	i	j
a		●			●					●
b	●			●			●			
c							●		●	
d		●				●				
e	●							●		
f				●						●
g		●								
h			●						●	
i					●			●		●
j	●		●			●			●	

35. There are 7 tour bus companies in the Los Angeles area, each visiting at most three different locations from among Hollywood, Beverly Hills, Disneyland, and Knott's Berry Farm during a day. It is understood that the same location cannot be visited by more than one tour company on the same day. The first tour company visits only Hollywood, the second only Hollywood and Disneyland, the third only Knott's Berry Farm, the fourth only Disneyland and Knott's Berry Farm, the fifth Hollywood and Beverly Hills, the sixth Beverly Hills and Knott's Berry Farm, and the seventh Disneyland and Beverly Hills. Can these tours be scheduled only on Monday, Wednesday, and Friday?

36. Prove that if a graph with n vertices has chromatic number n, then the graph has $\frac{1}{2}n(n-1)$ edges.

37. Show that it is possible to assign one of the colors red and blue to each edge of K_5 in such a way that no cycle of length 3 has all its edges the same color.

38. Show that the statement of Exercise 37 is incorrect if K_5 is replaced by K_6.

39. Prove Theorem 3.7 by induction on the number of vertices.

40. Suppose the no odd cycle coloring algorithm (see Exercise 13) is applied to a connected graph with no cycles of odd length and the vertex V is picked first and colored red. Prove that the paths from V to any red-colored vertex have even length and the paths from V to any blue-colored vertex have odd length.

41. Suppose for the graph G that whenever a vertex V and the edges incident on V are removed from G, the resulting graph has a smaller chromatic number. Prove that if the chromatic number of G is k, then the degree of each vertex of G is at least $k-1$.

3.4   Directed Graphs

In previous applications of graphs, an edge was used to represent a two-way or symmetric relationship between two vertices. However, there are situations where relationships hold in only one direction. In these cases the use of a line segment is not descriptive enough, and a directed line segment is used.

EXAMPLE 3.31   In many urban downtown areas the city streets are one-way. It is necessary to use a directed line segment to indicate the legal flow of traffic. In Figure 3.31 major downtown locations are represented by dots, and two dots are connected by an

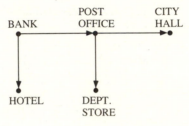

FIGURE 3.31

arrow when it is possible to go from the first location to the second by means of a one-way street. The arrow from BANK to HOTEL says that there is a one-way street from BANK to HOTEL. ■

EXAMPLE 3.32   Although in a communication network there are routes where the flow of information can travel either way, there are also some where the flow is in just one direction. Within a microcomputer system data usually can travel in either direction between the CPU and the Memory, but only from the Input to the Memory and from the Memory to the Output. This type of situation can be represented by the diagram in Figure 3.32, where the arrows indicate how the data can flow. ■

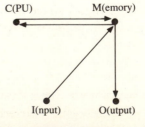

FIGURE 3.32

A **directed graph** is a finite nonempty set V and a set E of ordered pairs of distinct elements of V. The elements of V are called **vertices** and the elements of E are called **directed edges.**

Figure 3.32 depicts a directed graph with vertices $C, I, M,$ and O and directed edges $(C, M), (M, C), (I, M),$ and (M, O). As was true for graphs, a directed graph can be described either by the use of sets or by the use of a diagram, where arrows between the vertices in V describe which ordered pairs of vertices are being included.

Because the definition of a directed graph is similar to that of a graph, many of the ideas introduced for graphs can be easily reinterpreted for a directed graph. Whenever there is a directed edge $e = (A, B)$, it is said that e is **a directed edge from A to B.** In Figure 3.32 there is a directed edge from M to O but no directed edge from O to M. Similarly, there is a directed edge from M to C and one from C to M.

Just as for graphs, two directed edges crossing in a diagram do not create a new vertex. Likewise, in this book the set of vertices is to be a finite set (although not all authors require this). Finally, a directed edge cannot go from a vertex to itself, nor can there be two or more directed edges from one vertex to another.

In a directed graph the number of directed edges *from* vertex A is called the **outdegree** of A and is denoted as outdeg(A). Similarly, the number of directed edges *to* vertex A is called the **indegree** of A and is denoted by indeg(A). In Figure 3.32 we see that outdeg(M) = 2, indeg(C) = 1, and outdeg(O) = 0. Theorem 3.1 states that in a graph the sum of the degrees is equal to twice the number of edges. Because each directed edge leaves one vertex and enters a second vertex, there is the following similar theorem for directed graphs.

THEOREM 3.8   In a directed graph the following three numbers are equal: the sum of the indegrees of the vertices, the sum of the outdegrees of the vertices, and the number of directed edges.

Representations of Directed Graphs

As for graphs, a directed graph can be represented by a matrix. Suppose we have a directed graph D with n vertices labeled $V_1, V_2, \ldots, V_n$. Such a directed graph is called **labeled.** An $n \times n$ matrix is formed where the i,j entry is 1 if there is a directed edge from the vertex V_i to the vertex V_j and 0 if there is not. This matrix is called the **adjacency matrix** of D (with respect to the labeling) and is denoted by $A(D)$.

EXAMPLE 3.33   Figure 3.33 contains a directed graph and its adjacency matrix. The 1,4 entry is 0 because there is no directed edge from V_1 to V_4, but the 4,1 entry is 1 because there is a directed edge from V_4 to V_1. Row 3 contains all zeros because there are no directed edges from the vertex V_3. Since there are no directed edges to the vertex V_4, column 4 also contains all zeros. ▪

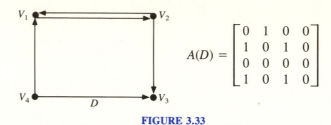

$$A(D) = \begin{bmatrix} 0 & 1 & 0 & 0 \\ 1 & 0 & 1 & 0 \\ 0 & 0 & 0 & 0 \\ 1 & 0 & 1 & 0 \end{bmatrix}$$

FIGURE 3.33

The last two observations in the previous example suggest the following version of Theorem 3.2 for directed graphs. The proof is omitted.

THEOREM 3.9   The sum of the entries in row i of the adjacency matrix of a directed graph equals the outdegree of the vertex V_i, and the sum of the entries in column j equals the indegree of the vertex V_j.

Directed graphs can also be represented by adjacency lists. To form an adjacency list, we begin by labeling the vertices of the directed graph. Then we list the vertices in a column, and after each vertex we list the vertices to which there is a directed edge from the given vertex.

EXAMPLE 3.34   For the directed graph in Figure 3.33 the adjacency list is given below. Since V_2 is the only vertex to which there is a directed edge from V_1, V_2 is the only vertex listed after V_1. Similarly, since there are directed edges from V_4 to V_1 and V_3 and only these, these are the two vertices listed after V_4.

$$V_1: V_2$$
$$V_2: V_1, V_3$$
$$V_3: \text{(none)}$$
$$V_4: V_1, V_3 \quad ▪$$

Directed Multigraphs

In Section 3.2 we introduced the concepts of multigraph, path, simple path, and cycle. These concepts have analogs using directed edges. To illustrate these definitions we will consider the diagram in Figure 3.34.

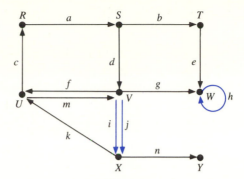

FIGURE 3.34

The diagram in Figure 3.34 represents a **directed multigraph.** Note there is a **directed loop** h at the vertex W and **parallel directed edges** i and j from V to X. Because a directed graph is also a special kind of directed multigraph, definitions developed for a directed multigraph also apply to a directed graph.

The alternating sequence of vertices and directed edges R, a, S, b, T, e, W is a **directed path** from R to W, which can also be written R, S, T, W or a, b, e. This directed path is said to have length 3, which is the number of directed edges listed. The directed path $V, i, X, k, U, m, V, j, X$ cannot be described by just using the vertices because there are two directed edges from V to X. However, this directed path can be described by just listing the directed edges i, k, m, j. Also T, b, S, d, V is not a directed path since the directed edge b is not from T to S but from S to T. Likewise, there is no directed path of positive length starting from Y. Again a vertex is a directed path of length 0.

The directed path a, d, g is a **simple directed path** from R to W, that is, a directed path with no vertex repeated. The directed path a, d, i, k, m, g is not a simple directed path because the vertex V is repeated. It is easily seen that a directed path contains a simple directed path. The proof follows very closely the one for graphs and is omitted.

THEOREM 3.10   Every directed path contains a simple directed path.

The directed path a, d, f, c is a **directed cycle** because it is a directed path from R to R in which no other vertex is visited twice. But b, e, g, d is not a directed cycle because the directed edges g and d go in the wrong direction. Both h and f, m are considered to be directed cycles. The directed path k, m, f, c, a, d, j is not a directed cycle because the vertices U and V appear twice.

Directed paths provide a way of describing how to move from one vertex to another in a directed multigraph. A directed multigraph D is **strongly connected** if for every pair A and B of vertices in D there is a directed path from A to B. Thus, in a strongly connected directed multigraph we can go from any vertex to any other by following some route along the directed edges.

EXAMPLE 3.35   The directed multigraph in Figure 3.34 is not strongly connected since there is no directed path from Y to any other vertex. The directed graph in Figure 3.35(a) is strongly connected since a directed path can be found from any vertex to any other. On the other hand, the directed graph in Figure 3.35(b) is not strongly connected since there is no directed path from A to C.

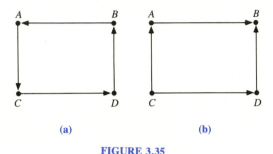

(a)   (b)

FIGURE 3.35

EXAMPLE 3.36   Suppose the city council of a middle-sized city is concerned about traffic congestion in the downtown area. It instructs the city traffic engineer to turn each two-way street into a one-way street in the downtown area in such a way that there still will be a route from each downtown location to any other.

   If a graph is used to represent the current downtown street system (where corners are vertices and streets are edges), then the city traffic engineer is to assign a direction to each edge, transforming the graph into a directed graph. Since there is to be a route from any place to any other, we want this new directed graph to be strongly connected. For example, if the graph in Figure 3.36 represents the downtown streets, then assigning directions as in Figure 3.35(a) produces a directed graph which is strongly connected. Thus, this assignment of directions satisfies the city council's requirement. On the other hand, the assignment of directions as in Figure 3.35(b) yields a directed graph that is not strongly connected, and so it does not satisfy the city council's requirement.

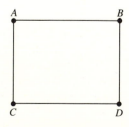

FIGURE 3.36

An important question: When can directions be assigned to the edges of a graph to yield a directed graph that is strongly connected? In Figure 3.37 there is an example of a graph that cannot be transformed into a strongly connected directed graph. The source of difficulty is the edge joining A and B. For if we direct this edge from A to B, then we cannot find a route from a place on the right side to any place on the left side. A similar problem occurs if we direct this edge from B to A. This edge joining A and B possesses an interesting property: if it is removed from the graph, the graph is no longer connected.

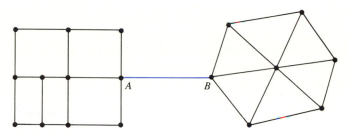

FIGURE 3.37

It can be proved that the absence of an edge whose removal disconnects a connected graph is equivalent to the existence of an assignment of directions to the edges to create a strongly connected directed graph. A proof can be found in suggested reading [9]. Thus, for the city traffic engineer in Example 3.36 to decide if there is an acceptable pattern of one-way streets, it suffices for him to find if there is an edge whose removal will disconnect the graph. We will return to the problem of assigning directions to edges in Section 4.3.

Directed Euler Circuits and Paths

The ideas of a directed Euler path and a directed Euler circuit in a directed multigraph are similar to the corresponding ones in a multigraph. A directed path in a directed multigraph D that includes exactly once all the directed edges of D and has different initial and terminal vertices is called a **directed Euler path.** A directed path that includes exactly once all the directed edges of D and has the same first and last vertices is called a **directed Euler circuit.**

Recall that in the proof of Theorem 3.4 constructing an Euler circuit required that each time we entered a vertex along an edge, there was another edge for us to leave on. This translated into the requirement that each vertex be of even degree. In constructing a directed Euler circuit we require similarly that for each directed edge going into a vertex, there must be another directed edge leaving from that vertex. This implies that the indegree of each vertex is the same as the outdegree. These observations are summarized in the following theorem.

THEOREM 3.11   Suppose the directed multigraph D has the property that whenever the directions are ignored on the directed edges, the resulting graph is connected. Then D has a directed Euler circuit if and only if for each vertex of D, the indegree is the same as the outdegree. Furthermore, D has a directed Euler path if and only if every vertex of D has its indegree equal to its outdegree except for two distinct vertices B and C, where the outdegree of B exceeds its indegree by 1 and where the indegree of C exceeds its outdegree by 1. When this is the case, the directed Euler path begins at B and ends at C.

The algorithm for constructing an Euler circuit in a graph may be modified in the obvious way (by choosing an unused *directed* edge leaving the vertex) to construct directed Euler circuits and paths in directed graphs that satisfy the hypotheses of Theorem 3.11.

EXAMPLE 3.37   In telecommunications there is an interesting application of directed Euler circuits which is due to Liu. (See [7] in the suggested readings.) Suppose there is a rotating drum with 8 different sectors, where each sector contains either a 0 or a 1. There are three detectors which are placed so that they can read the contents of 3 adjacent sectors. (See Figure 3.38.)

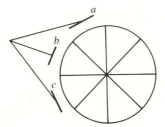

FIGURE 3.38

The task is to assign 1's and 0's to the sectors so that a reading of the detectors describes the exact position of the rotating drum. Suppose the sectors are assigned 1's and 0's as in Figure 3.39. Then a reading of the detectors gives 010. If the drum is moved 1 sector clockwise, the reading becomes 101. However, if the drum is moved still another sector clockwise, the reading becomes 010 again. Thus, two different positions of the rotating drum give the same reading. We want an assignment of 1's and 0's where this will not happen; that is, we want to arrange eight 1's and 0's in a circle so that every sequence of three consecutive entries is different.

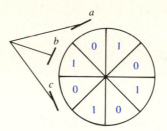

FIGURE 3.39

We construct a directed multigraph as follows. Use 00, 01, 11, and 10 as vertices. Next construct two directed edges from each vertex in the following manner. For the vertex *ab* consider the two vertices *b*0 and *b*1 (obtained from *ab* by dropping *a* and appending 0 and 1 at the end). Construct a directed edge from vertex *ab* to the vertex *bc* (where *c* is either 0 or 1) and assign this directed edge the label *abc*. For example, there is a directed edge from 01 to 10 with label 010 and from 01 to 11 with label 011. This directed multigraph is shown in Figure 3.40. Note that the labels assigned to the directed edges are all different and would be an acceptable set of readings for the detectors.

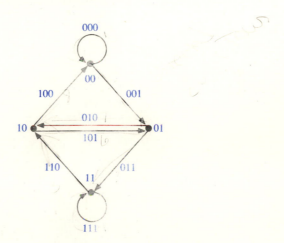

FIGURE 3.40

This is a directed multigraph such that when the directions on the directed edges are ignored, the resulting graph is connected. Furthermore, the indegree equals the outdegree for each vertex, and so a directed Euler circuit exists. Using the modifications indicated in the paragraph preceding this example, we start with vertex 01 and construct the directed Euler circuit 011, 111, 110, 101, 010, 100, 000, 001. For the directed edges in this directed Euler circuit, the last 2 digits in each label are the first 2 digits in the label of the next directed edge. Thus, if we select the first digit of the label (remember the labels are all different) of each directed edge in the directed Euler circuit, we get a sequence of 8 numbers where every sequence of 3 consecutive entries is different. For this example this selection process gives the sequence 01110100. When this sequence is placed in the sectors of the rotating drum, the 8 positions of the drum will give 8 different readings. ■

Directed Hamiltonian Cycles

A **directed Hamiltonian cycle (path)** is a directed cycle (path) that includes each vertex exactly once. Because directed loops and parallel directed edges are not needed for a directed Hamiltonian cycle or path, we will assume that we are working with directed graphs rather than directed multigraphs. As with graphs it is very difficult to decide if there is a directed Hamiltonian cycle and if so, to find one.

In a round robin contest each team plays every other team exactly once, and a tie between two teams is not permitted. Such a competition can be described by a directed graph in which the teams are represented by vertices, and there is a directed edge from one vertex to another if the first team beats the second team. A directed graph of this kind is called a **tournament directed graph** or more simply a **tournament.** An alternate way of thinking of a tournament is that it is the result of taking the complete graph K_n and assigning a direction to each edge.

EXAMPLE 3.38   Suppose there are three teams A, B, and C, where team A beats teams B and C, and team B beats team C. This is described in Figure 3.41(a). If instead, team C beats team A, the tournament is as in Figure 3.41(b). ■

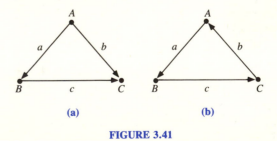

(a)   (b)

FIGURE 3.41

It may be desirable to find a ranking of the teams where the first team beats the second team, the second team beats the third team, and so forth. Finding a ranking of the teams is the same as finding a directed Hamiltonian path for the tournament. It can be shown that every tournament has a directed Hamiltonian path. In Figure 3.41(a) the directed path a, c is Hamiltonian and thus provides a ranking of the teams. However, an examination of Figure 3.41(b) shows that there can be more than one directed Hamiltonian path. In fact, there are three: a, c;

b, a; and *c, b*. This means that three separate rankings can be found. But also note in Figure 3.41(b) there is a directed cycle whereas in Figure 3.41(a) there is not. In general, if a tournament has no directed cycles, then there is only one directed Hamiltonian path, which provides a unique ranking of the teams.

Further illustration of these points is found in Figure 3.42. For the tournament in Figure 3.42(a) there are no directed cycles and only one directed Hamiltonian path, namely *a, f, d,* which gives the ranking *ABCD* of the teams. In Figure 3.42(b) the tournament has a directed cycle, for example, *a, f, d, e,* and so there are several directed Hamiltonian paths, such as *a, f, d* and *e, a, f*.

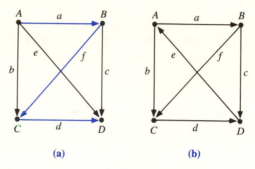

(a)   (b)

FIGURE 3.42

EXERCISES 3.4

In Exercises 1–4 list the vertices and directed edges for the graph.

1.

2.
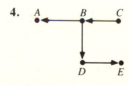

3.
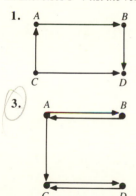

4.
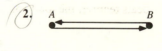

In Exercises 5–10 draw a diagram representing the directed graph with vertices V and directed edges E.

5. $V = \{A, B, C, D\}, E = \{(A, B), (C, D), (D, A), (D, C)\}$

6. $V = \{X, Y, Z, W, U\}, E = \{(X, Y), (Z, U), (Y, X), (U, Z), (W, X), (Z, X)\}$

7. $V = \{A, B, C, D\}, E = \varnothing$

8. $V = \{A, B, C, D\}, E = \{(A, C), (B, D)\}$

9. $V = \{A, B, C\}, E = \{(A, B), (B, C), (C, A), (B, A), (C, B)\}$

10. $V = \{A, B, C, D\}, E = \{(A, D), (D, B), (D, A)\}$

In Exercises 11–16 construct the directed graph for the adjacency matrix.

11.

$$\begin{bmatrix} 0 & 1 & 0 & 1 \\ 1 & 0 & 1 & 0 \\ 0 & 0 & 0 & 0 \\ 0 & 1 & 1 & 0 \end{bmatrix}$$

12.

$$\begin{bmatrix} 0 & 0 & 0 \\ 0 & 0 & 0 \\ 0 & 0 & 0 \end{bmatrix}$$

13.

$$\begin{bmatrix} 0 & 1 & 1 & 1 \\ 1 & 0 & 1 & 1 \\ 1 & 1 & 0 & 1 \\ 0 & 1 & 0 & 0 \end{bmatrix}$$

14.

$$\begin{bmatrix} 0 & 1 & 0 & 1 & 1 \\ 1 & 0 & 0 & 0 & 0 \\ 1 & 0 & 0 & 1 & 1 \\ 1 & 1 & 0 & 0 & 0 \\ 1 & 0 & 0 & 1 & 0 \end{bmatrix}$$

15.

$$\begin{bmatrix} 0 & 0 & 1 & 1 \\ 0 & 0 & 0 & 0 \\ 1 & 0 & 0 & 1 \\ 1 & 0 & 1 & 0 \end{bmatrix}$$

16.

$$\begin{bmatrix} 0 & 1 & 1 \\ 1 & 0 & 0 \\ 0 & 1 & 0 \end{bmatrix}$$

In Exercises 17–20 list for the directed graph the other vertices on the directed edges to A, the other vertices on the directed edges from A, the indegree of A, and the outdegree of A. Repeat for the vertex B.

17.

18.

19.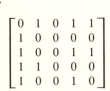

20.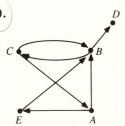

In Exercises 21–24 find the adjacency matrix and adjacency list for the directed graph in the indicated exercises.

21. Exercise 17   **22.** Exercise 18   **23.** Exercise 19   **24.** Exercise 20

25. Draw the directed graph with vertices $V = \{1, 2, \ldots, 10\}$ and directed edges $E = \{(x, y) : x, y \text{ are in } V, x \neq y, \text{ and } x \text{ divides } y\}$.

26. What does it mean if a row in an adjacency matrix for a directed graph is all zeros? What if a column is all zeros?

27. Can there be a directed graph with 6 vertices where the outdegrees of the vertices are 2, 3, 4, 1, 0, and 5 and the indegrees of the vertices are 2, 4, 1, 1, 5, and 2?

outdegree = indegree

28. Construct all the possible directed graphs with vertices $V = \{A, B\}$.

29. Construct the directed graph where the vertices are you, your parents, and your grandparents using the relationship "is a child of."

30. Construct the directed graph using "is a parent of" in place of "is a child of" in Exercise 29. How do the two directed graphs in Exercises 29 and 30 compare?

31. Susan has a fondness for chocolate desserts, in particular, pudding, pie, ice cream, eclairs, and cookies. Her preference is for pie over ice cream and cookies, eclairs over pie and cookies, cookies over pudding and ice cream, and pudding over eclairs with no other preferences. Draw a directed graph to represent this situation.

32. In a large corporation the chief executive officer communicates with her vice presidents but only they can communicate with her. Furthermore, the vice presidents can communicate with the directors, field managers, and division heads but only the directors can communicate back. Also the field managers and division heads can communicate with salespersons, but they can communicate back only with the field managers. Draw a directed graph to represent this situation.

For the graphs in Exercises 33–36
(i)   List at least 3 different directed paths from A to B. Give the length of each.
(ii) List the simple directed paths from A to B. Give the length of each.
(iii) For each directed path you listed in (i), find a simple directed path from A to B contained in it.
(iv) List the distinct directed cycles. (Two directed cycles are distinct if they do not contain the same directed edges.) Give the length of each.

33.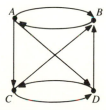

34.

35.

36.

In Exercises 37–40 determine if the directed graphs in the indicated exercises are strongly connected.

37. Exercise 33   **38.** Exercise 34   **39.** Exercise 35   **40.** Exercise 36

41. Give an example of a directed graph with 4 vertices where every directed path of positive length has length 1.

42. Give an example of a directed graph with 6 vertices where every directed path is a simple directed path.

43. In a directed multigraph with n vertices what is the maximum length of a simple directed path?

In Exercises 44–49 determine if a direction can be assigned to each edge of the graph resulting in a directed graph that is strongly connected.

44. *yes*

45. *no*

46.

47.

48. *no*

49.

In Exercises 50–53 assign a direction to each edge of the graph so that the resulting directed graph is strongly connected.

50.

51.

52.

53.

54. If a directed graph has a directed Hamiltonian cycle, then why is it strongly connected?

In Exercises 55–60 determine if the directed multigraph has a directed Euler path or circuit. If there is one, construct it using the appropriate algorithm as discussed in this section.

55.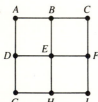

56.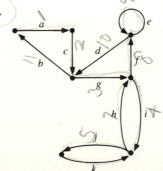

57.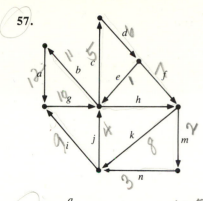

58.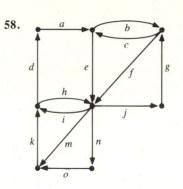

59. 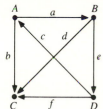 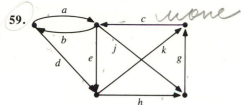 more

60. 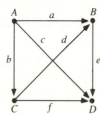 more

61. Suppose in Example 3.37 the rotating drum has only 4 different sectors on the drum and 2 detectors. Using the procedure described in that example, find a sequence of four 0's and 1's to be used on the rotating drum so that every sequence of two consecutive entries is different.

62. Do Example 3.37 if there are 16 different sectors and 4 detectors.

63. Show that in a tournament with n vertices the sum of the outdegrees is $\frac{1}{2}n(n-1)$.

64. Show that in a round robin contest with 7 players there cannot be 23 winners.

In Exercises 65–68 find all the directed Hamiltonian paths in the tournament.

65.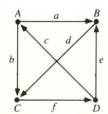

66.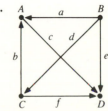

67.

68.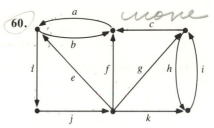

69. Suppose that Susan in Exercise 31 has established a preference between any two chocolate desserts (pie, pudding, ice cream, cookies, and eclairs). She prefers cookies over all of the others, ice cream over all but cookies, pie over pudding, and eclairs over pie and pudding. Is there a ranking to her preferences? How many?

70. In a tournament the outdegree of a vertex is called the **score** (the number of wins for that team). In the following tournament find a vertex with a maximum score and show that there is a directed path of length 1 or 2 from that vertex to any other.

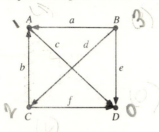

71. Repeat Exercise 70 for the following tournament.

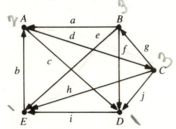

72. Can a tournament have 2 teams that lose every time?

73. Suppose that the teams in the NFL Central Division play a round robin contest in which the Bears beat every other team, the Lions lose to every other team, the Packers beat only the Lions and Bucs, and the Vikings beat everyone but the Bears. Is there a ranking of the teams? Is this ranking unique?

74. Prove Theorem 3.8.

75. Prove Theorem 3.9.

76. Prove Theorem 3.10.

77. Write an algorithm for finding a directed Euler circuit.

78. Prove Theorem 3.11.

    Prove that every tournament has a directed Hamiltonian path.

    Prove that if A is a vertex of maximum score (see Exercise 70) in a tournament, then there is a directed path of length 1 or 2 from A to any other vertex.

*A tournament is **transitive** if whenever (A, B) and (B, C) are directed edges in the tournament, then so is (A, C).*

    Prove that a tournament is transitive if and only if there are no directed cycles.

    Prove that the scores (see Exercise 70) in a transitive tournament with n vertices are $0, 1, 2, 3, \ldots, n - 1$.

3.5   Shortest Paths and Distance

In this section we will be looking at ways to find a shortest path between vertices in a graph or directed graph. The need to find shortest paths in graphs arises in many different situations.

We want to find a path of minimal length between two vertices S and T, that is, a path from S to T that has the smallest possible number of edges. This smallest possible number of edges in a path from S to T is called the **distance** from S to T. To find the distance from S to T, the general approach is to first look at S, then at the vertices adjacent to S, then at the vertices adjacent to these vertices, and so forth. By keeping a record of the way in which vertices are examined, it is possible to backtrack and find a shortest path from S to T. A formal description of this process is presented below.

Breadth-First Search Algorithm

This algorithm will determine the distance and a shortest path from vertex S to vertex T in a graph G. In the algorithm, L denotes the set of labeled vertices and the *predecessor* of vertex A is a vertex in L that is used in labeling A.

Step 1 (start with S). Assign S the label 0, let $L = \{S\}$, and let S have no predecessor.

Step 2 (check for completion). If T is not in L, go to step 3. If T is in L, then stop. The label for T is the distance from S to T. A shortest path from S to T is formed by using in reverse order the vertices T, the predecessor T_1 of T, the predecessor T_2 of T_1, and so forth until S is reached.

Step 3 (find next vertex). If T is not in L, find the unlabeled vertices in G that are adjacent to vertices in L with the largest label number k. If there are no such vertices, then there is no path from S to T. Otherwise, assign these newly found vertices the label $k + 1$ and put them in L. If B is one of these newly found vertices and B is adjacent to a vertex C in L with label k, let C be the predecessor of B. (If there is more than one choice for a predecessor, choose one at random.) Return to step 2.

It can be shown that the label assigned to each vertex by the breadth-first search algorithm is the distance from S (see Exercise 35).

We will regard labeling a vertex and using an edge to find an adjacent vertex as the elementary operations in analyzing this algorithm. For a graph with n vertices and e edges each vertex is labeled exactly once and each edge will be used at most once to find an adjacent vertex. Hence, there will be at most $n + e$ operations. But since $n + e \le n + \frac{1}{2}n(n - 1)$, we see that this algorithm is of order at most n^2.

EXAMPLE 3.39   For the graph in Figure 3.43(a) there is more than one path from S to T. One path is S, A, C, B, D, G, J, T and another is S, A, E, F, H, T. We will use the breadth-first search algorithm to find the distance and a path of minimal length between S and T. First label S with 0, set $L = \{S\}$, and let S have no predecessor. Since T is not in L, we look at the vertices adjacent to S, which are A and B. To them the label 1 is assigned and L then becomes $\{S, A, B\}$. A and B are given the predecessor S. Again since T is not in L, we find the unlabeled vertices adjacent to vertices with label 1. These are the vertices $E, C,$ and D, which are then assigned the label 2. Now $L = \{S, A, B, E, C, D\}$. Since E and C are adjacent to A which has label 1, E and C each have predecessor A, and similarly, D has predecessor B. Again because T is not in L, the unlabeled vertices adjacent to those with label 2 are determined. These are $F, H,$ and G, and they are assigned the label 3 and included in L. So L is now $\{S, A, B, E, C, D, F, G, H\}$. Because F is adjacent to E with label 2, F has predecessor E. Similarly, H has predecessor C, and G has predecessor D. Again T is not in L, and so the unlabeled vertices adjacent to those with label 3 are located. They are J and T, which are then assigned the label 4. Finally, the predecessor of T is H and that of J is G. We indicate all this on the graph in Figure 3.43(b) by writing the label and predecessor (in parentheses) by each vertex.

Since the label 4 has been assigned to T, the distance from S to T is 4. By using the vertices T, the predecessor of T (which is H), the predecessor of H (which is C), the predecessor of C (which is A), and the predecessor of A (which is S) in reverse order, we find that S, A, C, H, T is a shortest path from S to T. Note that it is not the only one, as the path S, B, C, H, T is also a path of minimal length from S to T. ◼

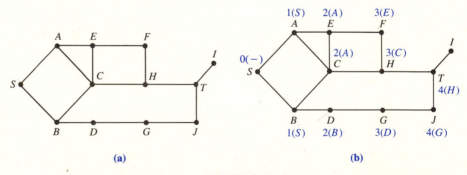

(a)   (b)

FIGURE 3.43

Although the algorithm and discussion above are for graphs, analogous ideas and procedures can be used for directed graphs. The only change in the algorithm is in Step 3 where we must find the unlabeled vertices on directed edges coming from those in L. In other words, we must follow the directions on the directed edges as we look for unlabeled vertices to include in the set L.

Weighted Graphs

Frequently when graphs or directed graphs are used to describe relationships between objects, a number is associated with each edge. For example, if a graph is being used to represent a highway system in the usual way, then a number can be assigned to each edge indicating the mileage between the two cities. If a directed graph is being used to represent possible energy states, a number may be assigned to each directed edge to describe the change in energy from one state to another. This idea of assigning numbers to the edges or directed edges is a very important one in applications. We shall discuss only graphs, but the reader should keep in mind that similar considerations apply to directed graphs.

A **weighted graph** is a graph in which a number called the **weight** is assigned to each edge. The **weight** of a path is the sum of weights of the edges in the path. In the case that a weighted graph describes a highway system with vertices representing cities and weights representing mileage between cities, the weight of a path is simply the total mileage between the cities representing the start and end of the path.

EXAMPLE 3.40   The graph in Figure 3.44 is a weighted graph since each edge has a number assigned to it. For example, the weight of the edge on A and B is 3 and the weight of the edge on D and F is 5. The weight of the path A, C, D, F is $4 + 2 + 5 = 11$, and the weight of the path F, D, B, E, D is $5 + 1 + 2 + 1 = 9$. ■

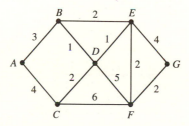

FIGURE 3.44

In many applications we need to find a path of smallest weight. However, there need not always be one. This kind of situation can occur if there is a cycle with negative weight.

EXAMPLE 3.41   For the weighted graph in Figure 3.45 the path A, B, D, E has weight 2, and the path A, B, D, C, B, D, E has weight -2, which is a smaller weight than that for the first path. Note that as the cycle B, D, C is repeated, the weight of the path gets smaller and smaller. Thus, there is no path of smallest weight between A and E. ■

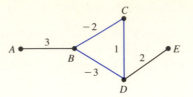

FIGURE 3.45

Consequently, we shall assume, unless explicitly stated otherwise, that weighted graphs do not have a cycle with negative weight. This assumption assures the existence of a path of smallest weight between two vertices if there is any path between them. Furthermore, a path of smallest weight between two vertices may be assumed to be simple since any cycle of weight 0 could be removed as in Theorem 3.3. A path of smallest weight is called a **shortest path** between those two vertices, and the weight of that path is called the **distance** between them.

When the weights assigned to edges or directed edges are positive, as is the case with highway or airline mileage, there is an algorithm which finds the distance and a shortest path between two vertices S and T. In fact, it can be used to find the distance and a shortest path between S and all other vertices at the same time. We will discuss this procedure in detail for graphs and indicate the minor changes needed for directed graphs.

The idea of this algorithm, which is due to Dijkstra, one of the pioneers in the study of computer programming, is to find first the vertex closest to S, then the second closest vertex to S, and so forth. In addition, by keeping a record of the vertices used in determining distances, it is possible to backtrack and find a shortest path from S to any other vertex.

Dijkstra's Algorithm

Let G be a weighted graph in which there is more than one vertex and there are nonnegative weights on the edges. This algorithm will determine the distance and a shortest path from vertex S to every other vertex in G. In the algorithm P denotes the set of vertices with permanent labels. The *predecessor* of a vertex A is a vertex in P used to label A. The weight of the edge on vertices U and V will be denoted by $W(U,V)$, and if there is no edge on U and V, we will write $W(U,V) = \infty$.

Step 1 (start with S). Assign S the label 0, let $P = \{S\}$, and let S have no predecessor.

Step 2 (assign labels to all the other vertices). To each vertex V not in P, assign the (perhaps temporary) label $W(S,V)$, and let V have (perhaps temporary) predecessor S.

Step 3 (find nearest vertex to P and revise labels). Include in P a vertex U having the smallest label of the vertices not in P. (If there is more than one such vertex, arbitrarily choose any one of them.) For each vertex X not in P and adjacent to U, replace the label on X by the smaller of the old label on X and (label on U) + $W(U,X)$. If the label on X was changed, let U be the new (perhaps temporary) predecessor of X.

Step 4 (check for completion). If P does not contain all the vertices of G, then return to step 3. Otherwise, the label on a vertex Y is its distance from S. If the label on Y is ∞, then there is no path, hence, no shortest path, from S to Y. Otherwise, a shortest path from S to Y is formed by using in reverse order the vertices Y, the predecessor Y_1 of Y, the predecessor Y_2 of Y_1, and so forth until S is reached.

The proof that this algorithm actually computes the distance between S and every other vertex can be found in Exercises 38–41.

In analyzing this algorithm for a graph with n vertices, we will consider assignments involving one vertex as being just one operation. So in step 1 there is just one operation, and in step 2 there are $n - 1$ more. Step 3 is done $n - 1$ times. Each time at most $n - 2$ comparisons are done on the labels to find the smallest one, and then at most one assignment occurs. Also in revising the labels we examine at most $n - 1$ vertices and for each vertex there are an addition, two comparisons, and possible assignments for a total of 4 operations. So for step 3 there are at most $(n - 1)[n - 2 + 1 + 4(n - 1)] = (n - 1)(5n - 5)$ operations. In step 4 looking up the distance and tracing back at most $n - 1$ predecessors to find a shortest path takes at most n operations. From this we see that there are at most

$$1 + (n - 1) + (n - 1)(5n - 5) + n = 5n^2 - 8n + 5$$

operations, and so the algorithm is of order at most n^2.

EXAMPLE 3.42   For the weighted graph in Figure 3.46 we want to find a shortest path and the distance from S to every other vertex.

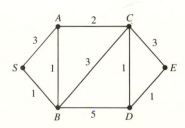

FIGURE 3.46

According to step 1, we start by putting S in P and assigning to S the label 0 and no predecessor. We indicate this on the graph by writing the label and predecessor (in parentheses) by S. We use an asterisk to show that S is in P. The graph now looks like Figure 3.47.

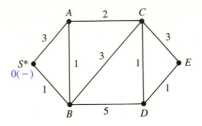

FIGURE 3.47

Next, according to step 2, we assign the label $W(S,V)$ and the predecessor S to the other vertices. Recall that $W(S,V) = \infty$ when there is no edge between S and V. Our graph now looks like Figure 3.48.

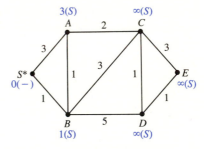

FIGURE 3.48

Now we apply step 3. The vertex not in P with the smallest label is B, and so we add it to P. The vertices not in P and adjacent to B are A, C, and D; and we replace the label on each such vertex X by the minimum of the old label and (label on B) + $W(B,X)$. These numbers are as follows.

Vertex X	Old label	(Label on B) + $W(B,X)$	Minimum
A	3	$1 + 1 = 2$	2
C	∞	$1 + 3 = 4$	4
D	∞	$1 + 5 = 6$	6

Since each label is changed, we also replace the predecessor of each of these vertices by B, producing Figure 3.49.

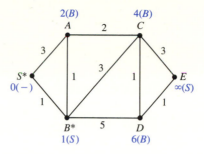

FIGURE 3.49

We continue in this way. Figure 3.50 shows the labels, predecessors, and vertices added to P at each stage. No entry in a column indicates no change from the previous stage.

Vertex		Labels and predecessors					Vertex added to P
	S	A	B	C	D	E	
	$0(-)$	$3(S)$	$1(S)$	$\infty(S)$	$\infty(S)$	$\infty(S)$	S
		$2(B)$		$4(B)$	$6(B)$		B
							A
				$5(C)$	$7(C)$		C
					$6(D)$		D
							E

FIGURE 3.50

The final graph is shown in Figure 3.51. In this figure the label on each vertex

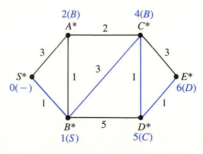

FIGURE 3.51

gives the distance between it and S, and a path of this length can be found by backtracking through the predecessors of the vertices. For example, the distance from S to E is 6, and the path S, B, C, D, E has this length. ▪

What has been done for weighted graphs can be repeated for directed weighted graphs with only two changes. First $W(U,V)$ now denotes the weight of the *directed* edge from U to V and $W(U,V) = \infty$ if there is no directed edge from U to V. Second, in step 3 we revise the labels on only those vertices X not in P for which there is a *directed* edge from U to X.

Number of Paths and Directed Paths

We conclude this section by considering the number of paths between two vertices, or alternatively how many paths of length m there are between a pair of vertices. One answer to these questions involves powers of the adjacency matrix of the graph.

THEOREM 3.12   For a graph G with vertices labeled $V_1, V_2, \ldots, V_n$ and adjacency matrix A, the number of paths of length m from V_i to V_j is the i,j entry of A^m.

Before we show how the theorem is proved for $m = 1, 2,$ and 3 (the general case is left as Exercise 18), an example is given to illustrate the theorem.

EXAMPLE 3.43   The graph in Figure 3.52(a) has the adjacency matrix A given in Figure 3.52(b).

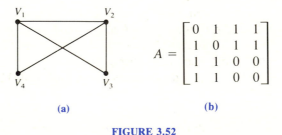

$$A = \begin{bmatrix} 0 & 1 & 1 & 1 \\ 1 & 0 & 1 & 1 \\ 1 & 1 & 0 & 0 \\ 1 & 1 & 0 & 0 \end{bmatrix}$$

(a)   (b)

FIGURE 3.52

To find the number of paths of length 2, we compute the product

$$A^2 = AA = \begin{bmatrix} 0 & 1 & 1 & 1 \\ 1 & 0 & 1 & 1 \\ 1 & 1 & 0 & 0 \\ 1 & 1 & 0 & 0 \end{bmatrix}\begin{bmatrix} 0 & 1 & 1 & 1 \\ 1 & 0 & 1 & 1 \\ 1 & 1 & 0 & 0 \\ 1 & 1 & 0 & 0 \end{bmatrix} = \begin{bmatrix} 3 & 2 & 1 & 1 \\ 2 & 3 & 1 & 1 \\ 1 & 1 & 2 & 2 \\ 1 & 1 & 2 & 2 \end{bmatrix}$$

That the 3,4 entry is 2 means there are 2 paths of length 2 between V_3 and V_4, namely, V_3, V_1, V_4 and V_3, V_2, V_4. Likewise the 1,3 entry being 1 means there is only one path of length 2 between V_1 and V_3, namely, V_1, V_2, V_3. The number of paths of length 3 is given by the product $A^2 \cdot A = A^3$ computed below.

$$A^3 = A^2 A = \begin{bmatrix} 3 & 2 & 1 & 1 \\ 2 & 3 & 1 & 1 \\ 1 & 1 & 2 & 2 \\ 1 & 1 & 2 & 2 \end{bmatrix} \begin{bmatrix} 0 & 1 & 1 & 1 \\ 1 & 0 & 1 & 1 \\ 1 & 1 & 0 & 0 \\ 1 & 1 & 0 & 0 \end{bmatrix} = \begin{bmatrix} 4 & 5 & 5 & 5 \\ 5 & 4 & 5 & 5 \\ 5 & 5 & 2 & 2 \\ 5 & 5 & 2 & 2 \end{bmatrix}$$

Since the 1, 2 entry of A^3 is 5, there are 5 paths of length 3 between V_1 and V_2, namely, V_1, V_2, V_1, V_2; V_1, V_2, V_4, V_2; V_1, V_2, V_3, V_2; V_1, V_3, V_1, V_2; and V_1, V_4, V_1, V_2. ■

We now return to the proof of Theorem 3.12 for $m = 1, 2$, and 3.

Let a_{ij} denote the i,j element of A. The number of paths of length 1 between V_i and V_j is either 0 or 1 depending on whether there is an edge on these vertices. But this is the same as a_{ij}, which is 1 when there is an edge on V_i and V_j and 0 otherwise. So the i,j entry of A gives the number of paths of length 1 from V_i to V_j.

For a path of length 2 between V_i and V_j there needs to be a vertex V_k such that there is an edge on V_i and V_k and an edge on V_k and V_j. In terms of the adjacency matrix A this is the same as saying that there is an index k such that both a_{ik} and a_{kj} are 1, or equivalently, $a_{ik}a_{kj} = 1$. Thus, the number of paths of length 2 between V_i and V_j is the number of k's where $a_{ik}a_{kj} = 1$. This number is the value of

$$a_{i1}a_{1j} + a_{i2}a_{2j} + a_{i3}a_{3j} + \ldots + a_{in}a_{nj}$$

since each term in the sum is 1 or 0. But this sum is also the i,j entry in A^2, the product of A with A, and thus the i,j entry in A^2 is the number of paths of length 2 between V_i and V_j.

For a path of length 3 between V_i and V_j there are vertices V_p and V_k with edges on V_i and V_p, on V_p and V_k, and on V_k and V_j. But this means there is a path of length 2 between V_i and V_k and an edge on V_k and V_j. If b_{ik} denotes the i,k entry of A^2, then the number of paths V_i, V_p, V_k, V_j of length 3 between V_i and V_j is $b_{ik}a_{kj}$. Thus the total number of paths of length 3 between V_i and V_j is the value of

$$b_{i1}a_{1j} + b_{i2}a_{2j} + \ldots + b_{in}a_{nj},$$

which is the same as the i,j entry in $A^2 \cdot A = A^3$. Hence the i,j entry in A^3 is the number of paths of length 3 on V_i and V_j.

There is a similar theorem for directed graphs. Because the proof is similar to that of Theorem 3.12, it is omitted.

THEOREM 3.13   For a directed graph D with vertices $V_1, V_2, \ldots, V_n$ and adjacency matrix A, the number of directed paths of length m from V_i to V_j is the i,j entry of A^m.

EXERCISES 3.5

In Exercises 1–6 determine the distance and a shortest path from S to T in the graph.

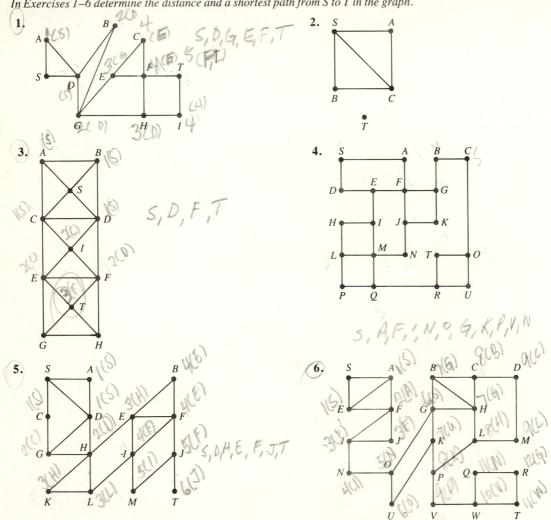

In Exercises 7–10 determine the distance and a shortest directed path from S to T in the directed graph.

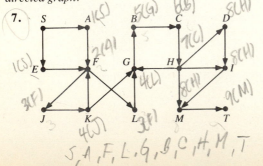

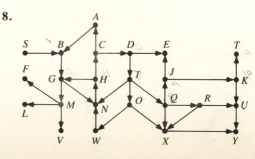

9.

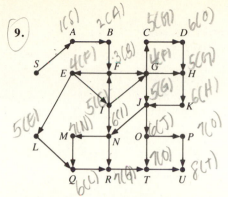

10.

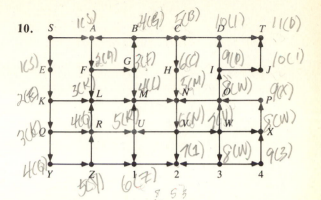

Solve Exercises 11 and 12 using graphs.

11. Suppose you have 3 jars, one holding 8 liters, another 5 liters, and the last one 3 liters. The 8–liter jar is full of water, and you can pour water from one jar to another. What is the smallest number of pourings that you need to make in order to get exactly 4 liters? (Hint: Let the vertices of a directed graph be triples (x, y, z), where there are x liters in the first jar, y in the second, and z in the third. For example, we can move from $(8, 0, 0)$ to $(3, 5, 0)$ by filling the second jar.)

12. There is a famous problem where there is a wolf, a goat, and a head of cabbage on one side of a river and a riverperson who is willing to take only one of them across at a time. However, the goat and the wolf cannot be left alone on the same side of the river nor can the goat and the head of cabbage be left on the same side. What is the smallest number of trips across the river needed to get all three on the other side?

In Exercises 13–16 determine the distance from S to all the other vertices in the graph.
Find a shortest path from S to A. Find a shortest path from S to B.

13.

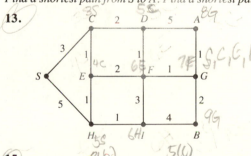

14.

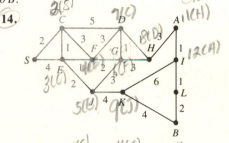

15.

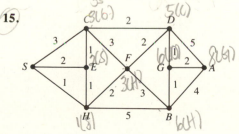

16.

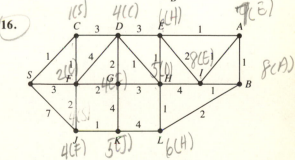

In Exercises 17–20 determine the distance from S to all the other vertices in the directed graph. Find a shortest directed path from S to A. Find a shortest directed path from S to B.

17.

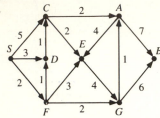

18.

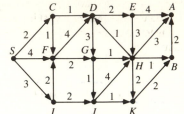

19.

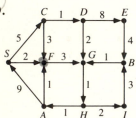

20.

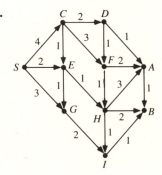

In Exercises 21–24 find a shortest path from S to T which goes through the vertex A in the graph.

21.

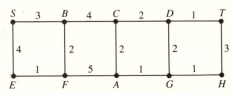

22.

23.

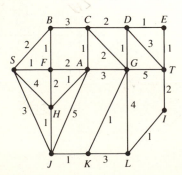

24.

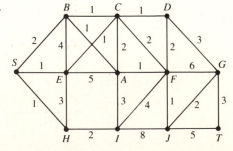

25. Suppose that the Illinois Electric Company has a power plant from which electrical power is sent along transmission lines to the surrounding communities. However, there is a continuing problem with power loss in these lines because of their deteriorated condition. The table below describes the loss along a transmission line from one community to another. What is the best route (one with least power loss) from the plant to the surrounding communities? (A dash in the table means there is no transmission line.)

To	Plant	Normal	Hudson	Ospur	Kenney	Lane	Maroa
Normal	—	—	4	4	—	3	—
Hudson	9	2	—	3	5	6	3
Ospur	3	—	—	—	6	9	4
Kenney	2	3	1	1	—	7	2
Lane	1	—	2	2	7	—	6
Maroa	6	2	3	4	2	2	—

Use Theorems 3.12 and 3.13 to solve Exercises 26–33.

26. For the graph determine the number of paths of length 1, 2, 3, and 4 from V_1 to V_2, and from V_2 to V_3.

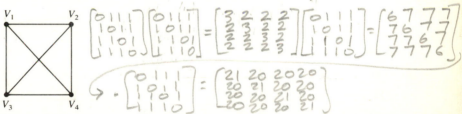

27. For the graph determine the number of paths of length 1, 2, 3, and 4 from V_1 to V_2, and from V_1 to V_3.

28. For the graph determine the number of paths of length 1, 2, 3, and 4 from V_1 to V_1, and from V_4 to V_3.

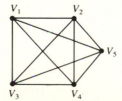

29. For the graph determine the number of paths of length 1, 2, 3, and 4 from V_1 to V_3, and from V_2 to V_4.

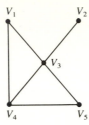

30. For the directed graph determine the number of directed paths of length 1, 2, 3, and 4 from V_1 to V_3, and from V_2 to V_4.

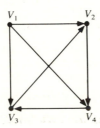

31. For the directed graph determine the number of directed paths of length 1, 2, 3, and 4 from V_1 to V_4, and from V_4 to V_1.

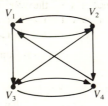

32. For the directed graph determine the number of directed paths of length 1, 2, 3, and 4 from V_1 to V_2, and from V_3 to V_5.

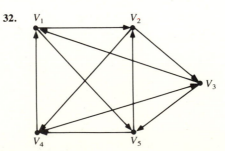

33. For the directed graph determine the number of directed paths of length 1, 2, 3, and 4 from V_1 to V_4, and from V_2 to V_5.

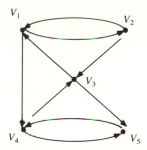

34. If A is the adjacency matrix of a graph G, what does the i,j element of $A + A^2 + A^3$ describe?

35. Prove that the label given to each vertex by the breadth-first search algorithm is the distance from S.

36. Prove Theorem 3.12 using induction.

37. Prove Theorem 3.13 using induction.

Exercises 38–41 provide a proof of the validity of Dijkstra's algorithm. Assume in them that G is a weighted graph with all weights W(U, V) positive, and that S is a vertex of G.

38. Suppose each vertex V of G is assigned a label $L(V)$ which is either a number or ∞. Assume that P is a set of vertices of G containing S such that (i) if V is in P, then $L(V)$ is the length of a shortest path from S to V and (ii) if V is not in P, then $L(V)$ is the length of a shortest path from S to V subject to the restriction that V is the only vertex of the path not in P. Let U be a vertex not in P with minimal label among such vertices. Show that a shortest path from S to U contains no element not in P except U.

39. Show that under the assumption of Exercise 38 the length of a shortest path from S to U is $L(U)$.

40. Assume the hypotheses of Exercise 38, and let P' be the set formed by U and the elements of P. Show that P' satisfies property (i) of Exercise 38; and show that if V is not in P', then the length of a shortest path from S to V, all of whose vertices except V are in P', is the minimum of $L(V)$ and $L(U) + W(U, V)$.

41. Prove that Dijkstra's algorithm gives the length of a shortest path from S to each vertex of G by induction on the number of elements in P. Let the induction hypothesis be that P is a set of vertices containing S satisfying properties (i) and (ii) of Exercise 38.

Suggested Readings

1. Bogart, Kenneth P. *Introductory Combinatorics*. Marsfield, MA: Pitman, 1983.
2. Bondy, J. A. and U. S. R. Murty. *Graph Theory with Applications*. New York: North Holland, 1976.
3. Chachra, Vinod, Prabhakar M. Ghare, and James M. Moore. *Applications of Graph Theory Algorithms*. New York: North Holland, 1979.
4. Chartrand, Gary. *Graphs as Mathematical Models*. Boston: Prindle, Weber & Schmidt, 1977.
5. Even, Shimon. *Graph Algorithms*. Rockville, MD: Computer Science Press, 1979.
6. Harary, Frank. *Graph Theory*. Reading: Addison-Wesley, 1969.
7. Liu, C. L. *Introduction to Combinatorial Mathematics*. New York: McGraw-Hill, 1968.
8. Ore, Oystein. *Graphs and Their Uses*. New York: L. W. Singer, 1963.
9. Polimeni, Albert D. and Joseph H. Straight. *Foundations of Discrete Mathematics*. Monterey, CA: Brooks/Cole, 1985.
10. Roberts, Fred S. *Graph Theory and Its Applications to Problems of Society*. Philadelphia: SIAM, 1978.

4

TREES

In Chapter 3 we studied several different types of graphs and their applications. A special class of graphs, trees, has been found to be very useful in computer science. Trees were first used in 1847 by Gustav Kirchhoff in his work on electrical networks and then later used by Arthur Cayley in the study of chemistry. Now trees are widely used in computer science as a way to organize and manipulate data.

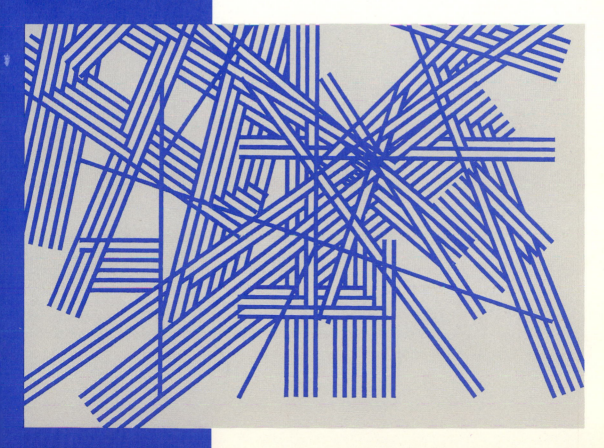

4.1   Properties of Trees

We begin this section by looking at some examples.

EXAMPLE 4.1   In 1857 Arthur Cayley studied hydrocarbons, chemical compounds formed from hydrogen and carbon atoms. In particular, he investigated saturated hydrocarbons, which have k carbon atoms and $2k + 2$ hydrogen atoms. He knew that a hydrogen atom was bonded (chemically kept together) with one other atom, and each carbon atom was bonded with four other atoms. These compounds are usually represented pictorially as in Figure 4.1, where a line segment between two atoms indicates a bonding.

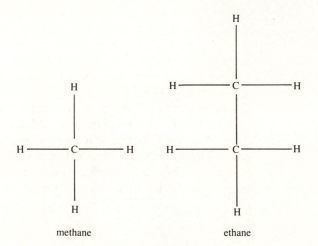

methane   ethane

FIGURE 4.1

These chemical diagrams can be redrawn as graphs, as illustrated in Figure 4.2. Note that in these graphs we have followed the customary practice of using the same chemical symbol on different vertices representing the same element. However, it is not really necessary to label the vertices with C and H since a vertex of degree 4 represents carbon and a vertex of degree 1 represents hydrogen. It was through the mathematical analysis of these graphs that Cayley predicted the existence of new saturated hydrocarbons. Later discoveries proved that his predictions were correct. ■

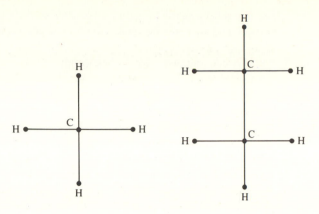

FIGURE 4.2

EXAMPLE 4.2   Suppose we are planning the telephone network for an underdeveloped area, where the goal is to link together five isolated towns. We can build a telephone line between any two towns, but time and cost limitations restrict us to building as few lines as possible. It is important that each town be able to communicate with each other town, but it is not necessary that there be a direct line between any pair of towns since it is possible to route calls through other towns. If we represent the towns by the vertices of a graph and the possible telephone lines by edges between the vertices, then the graph in Figure 4.3 represents all the possibilities we can have for the telephone lines. (This is merely the complete graph on five vertices.)

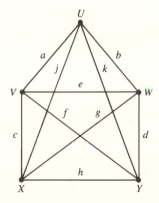

FIGURE 4.3

We need to select a set of edges that will give us a path between any two vertices and that has no more edges than necessary. One such set of edges is $\{a, b, c, d\}$, as illustrated in Figure 4.4. This choice of edges allows communication between any two towns. For example, to communicate between Y and X, we can use edges d, b, a, c in that order. Notice that if any edge from this set is deleted, then it is not possible to communicate between some pair of towns. For example, if we use only edges a, b, and c, towns U and Y cannot communicate. Another set of acceptable edges is $\{e, g, h, k\}$. Sets like $\{g, h, j, k\}$ or $\{a, b, e, h\}$ are not acceptable because not every pair of towns can communicate. Also the set $\{a, b, g, j, k\}$ is bigger than necessary. For example, the edge g can be left out without disrupting communication between any two towns. ▄▄

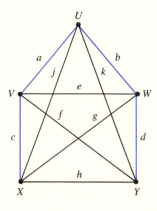

FIGURE 4.4

For the graphs in Figure 4.2 and the colored graph in Figure 4.4 (the one formed by the edges a, b, c, and d and the vertices U, V, W, X, and Y), we note two common characteristics; namely, these graphs are connected (there is a path between any two vertices) and have no cycles. Any graph which is connected and has no cycles is called a **tree.** Additional examples of trees follow.

EXAMPLE 4.3   Since each of the graphs in Figure 4.5 is connected and has no cycles, it is a tree. ▄▄

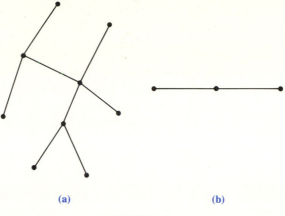

(a)   (b)

FIGURE 4.5

EXAMPLE 4.4   Each of the graphs in Figure 4.6 is not a tree. The graph in Figure 4.6(a) is not connected, and the graph in Figure 4.6(b) has a cycle. ■

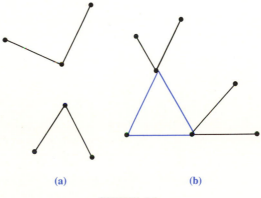

(a)   (b)

FIGURE 4.6

THEOREM 4.1   In a tree there is exactly one simple path between any two vertices.

Proof.   Suppose T is a tree. Since T is a connected graph, there is at least one path between any two vertices. Thus, by Theorem 3.3 there is a simple path between any two vertices. We will now show that there cannot be two distinct simple paths between any pair of vertices U and V. To do so, we will assume that there are two distinct simple paths P_1 and P_2 between U and V and show that this leads to a contradiction. Since P_1 and P_2 are different, there must be a vertex A (possibly $A = U$) lying on both P_1 and P_2 such that the vertex B following A on P_1 is not on P_2. In other words P_1 and P_2 separate at A. (See Figure 4.7.) Now follow path P_1 until we come to the first vertex C that is again on both paths. (The paths must rejoin because they meet again at V.) Consider the part of the simple path P_1 from A to C and the part of the simple path P_2 from C to A. These parts form a cycle. But trees contain no cycles, so we have a contradiction. It follows that there cannot be two distinct simple paths between any pair of vertices.   ▪

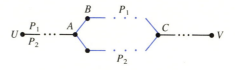

FIGURE 4.7

Looking at each of the previous examples of trees reinforces the idea that there is a *unique* simple path between any two vertices of a tree. Notice also that in all these examples every tree has at least two vertices of degree 1.

THEOREM 4.2   In a tree T with more than one vertex there are at least two vertices of degree 1.

Proof.   Since T is a connected graph with at least two vertices, there is a simple path with at least two distinct vertices. Thus, T contains a simple path with a maximal number of edges, say from U to V, where U and V are distinct. If U has degree more than 1, then since T has no cycles, a longer simple path would exist; likewise for V. Thus U and V have degree 1.   ▪

For the tree shown in color in Figure 4.4 there are 5 vertices and 4 edges, and for the tree in Figure 4.5(a) there are 9 vertices and 8 edges. In fact, in each of the previous examples of trees the number of vertices is one more than the number of edges. The next theorem establishes that this is always the case.

THEOREM 4.3

A tree with n vertices has exactly $n - 1$ edges.

Proof.   The proof will be done by induction on n, the number of vertices. We will first consider the case $n = 1$. Because a tree is a graph, there are no loops in a tree. Hence there can be no edges in a tree with only one vertex, and the theorem holds when $n = 1$.

Now assume the theorem holds for all trees which have k vertices. We will prove that the theorem holds for a tree T with $k + 1$ vertices. By Theorem 4.2 there is a vertex V which has degree 1. Remove the vertex V and the edge on V from the graph T to obtain a new graph T'. (See Figure 4.8.) This graph T' has k vertices and is still a tree. (Why?) Thus, by the induction assumption T' has $k - 1$ edges. But then T has k edges.

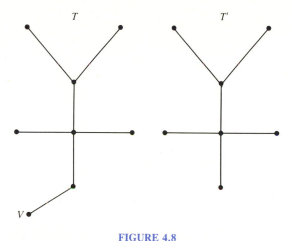

FIGURE 4.8

Thus, by mathematical induction the theorem holds for all positive integers n.   ▩

EXAMPLE 4.5

The KGB has established a network of 10 spies working in high technology industries. It is important that each spy be able to communicate with any other either directly or indirectly through a chain of others. Establishing secret locations to exchange messages is difficult, and the KGB wants to keep the number of these meeting places as small as possible.

Yet for reasons of secrecy no more than two spies should know about any particular meeting place. This communication network can be represented by a graph in which the vertices correspond to spies and an edge joins two vertices when the corresponding spies know about the same meeting place. In fact, this graph is a tree with 10 vertices, and so there will need to be 9 meeting places in all.   ▬

THEOREM 4.4   (a) When an edge is removed from a tree (leaving all the vertices), the resulting graph is not connected and hence is not a tree.

(b) When an edge is added to a tree (without adding additional vertices), the resulting graph has a cycle and hence is not a tree.

Proof.   If an edge is added or removed from a tree, the resulting new graph can no longer be a tree by Theorem 4.3. Since removing an edge cannot create a cycle nor adding an edge disconnect the graph, both parts of the theorem follow. ■

Theorem 4.4 shows that a tree has just the right number of edges to be connected and not have any cycles. By looking at the tree in Figure 4.5(a), we can see how the deletion of any edge produces a disconnected graph by breaking the tree into two parts. In addition, we can see how the addition of an edge between two existing vertices creates a cycle in the new graph.

The following theorem gives some other ways of characterizing a tree. Its proof will be left to the exercises.

THEOREM 4.5   The following are equivalent for a graph T:

(a) T is a tree.

(b) T is connected and the number of vertices is one more than the number of edges.

(c) T has no cycles and the number of vertices is one more than the number of edges.

(d) There is exactly one simple path between each two vertices in T.

(e) T is connected and the removal of any edge of T results in a graph that is not connected.

(f) T has no cycles and the addition of any edge results in a graph with a cycle.

It is the equivalence of parts (a) and (b) in Theorem 4.5 that helps in the mathematical analysis of saturated hydrocarbons of the type C_kH_{2k+2}. (See Example 4.1.) We know that there will be k carbon atoms and $2k + 2$ hydrogen atoms represented in the graph. Furthermore, since the atoms form a compound, the graph will be connected. Since a vertex represents an atom, there will be $k + 2k + 2 = 3k + 2$ vertices. Also, since each carbon atom has degree 4 and each hydrogen atom has degree 1, the sum of the degrees is $4k + 2k + 2 = 6k + 2$.

By Theorem 3.1 the number of edges is $\frac{1}{2}(6k + 2) = 3k + 1$, which is one less than the number of vertices. Hence, by part (b) of Theorem 4.5 the graph representing the chemical compound is always a tree. Knowing this, Cayley used information about trees to find new saturated hydrocarbons.

EXERCISES 4.1

In Exercises 1–10 determine if each graph is a tree.

1. • yes

2.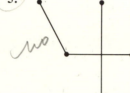

3. • • no

4.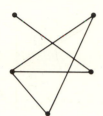

5. no

6.

7. yes

8.

9. yes

10.

11. How many vertices are there in a tree with 15 edges? *16*

12. How many edges are there in a tree with 21 vertices?

13. Seven farming communities in Iowa want to develop a computer telecommunication network to facilitate communication during a farm crisis. For reasons of economy they want to build as few lines as possible but still allow communication between any two towns. Indicate how this might be done for the map below.

Traer

Tama

Conrad

Lincoln

Garwin

Beaman

Gladbrook

14. As few trails as possible are to be built between houses in a primitive community so that it is possible for a resident to go from any house to any other. If there are 34 houses, how many trails need to be built? Since it is considered to be bad luck to live at the end of a trail, can the trails be constructed so that no house is so situated?

15. A farmer needs to irrigate the fields in which his crops are growing. (A map of the fields is given below in which the fields are the enclosed areas and edges represent earthen walls between the fields.) Because he lacks modern equipment, his method of irrigation is to break holes in the walls and let water from the outside cover the entire field. He wants to irrigate each field and to break as few walls as possible. In how many walls should he break holes?

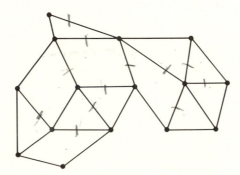

16. At the Illinois FBI office, Special Agent Jones is working with 7 informants who have infiltrated a gambling ring. She needs to arrange for the informants to communicate with each other in groups of two in such a way that messages can be passed on to others. For secrecy the number of meeting places must be kept as small as possible. How many meeting places must Agent Jones find?

17. Draw a graph which is not a tree for which the number of vertices is one more than the number of edges.

18. Draw a tree with at least 6 vertices that has exactly 2 vertices of degree 1.

19. What is the smallest number of edges in a connected graph with n vertices?

20. What is the largest number of vertices in a connected graph with n edges?

21. How many simple paths of nonzero length are there in a tree with n vertices, where $n \geq 2$?

22. What is the smallest number of colors needed to color a tree with n vertices, where $n \geq 2$?

23. Prove that the graph T' in the proof of Theorem 4.3 is a tree.

24. Prove that if an edge is deleted from a cycle in a connected graph, the graph remains connected.

25. For which n is K_n a tree? (K_n is defined in Section 3.1.)

26. If $m \geq 2$, for which n is $K_{m,n}$ a tree? ($K_{m,n}$ is defined in Exercise 60 of Exercises 3.2.)

27. Given a tree, can you always draw it on a sheet of paper so that the edges do not intersect except at vertices?

28. Ask a chemist about the chemical structure of benzene and draw a graph describing it. Is it a tree?

29. There are two saturated hydrocarbons of the type C_4H_{10}, butane and isobutane. Draw a tree representing the chemical structure of each.

30. Draw a graph representing a saturated hydrocarbon with 5 carbon atoms.

31. Can a tree with 13 vertices have 4 vertices of degree 3, 3 vertices of degree 4, and 6 vertices of degree 1?

32. How many vertices of degree 1 are there in a tree with 3 vertices of degree 4, 1 vertex of degree 3, 2 vertices of degree 2, and no vertices of degree more than 4?

33. Prove that if the maximum degree of a vertex in a tree is k, then there are at least k vertices of degree 1.

As trees on the vertices labeled A, B, and C, the two trees in figures (a) and (b) below are the same since they both have the same set of edges, namely, {A, B} and {B, C}. The trees in figures (a) and (c) are distinct since they do not have the same set of edges. For example, {A, C} is an edge of the tree in figure (c) but not of the tree in figure (a).

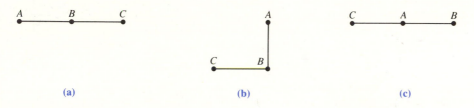

| (a) | (b) | (c) |

34. Draw the 3 distinct trees with 3 labeled vertices. (Use 1, 2, 3 as the labels.)

35. Draw the 16 distinct trees with 4 labeled vertices. (Use 1, 2, 3, 4 as the labels.)

In order to count the number of distinct trees with vertices $1, 2, \ldots, n$ we establish a one-to-one correspondence between each such tree and a list $a_1, a_2, \ldots, a_{n-2}$, where $1 \le a_i \le n$ for $i = 1, 2, \ldots, n - 2$. The following algorithm shows how to get such a list from a labeled tree T.

Prufer's Algorithm

This algorithm will construct a list $a_1, a_2, \ldots, a_{n-2}$ of numbers for a tree with n labeled vertices, where $n \ge 3$ and the labels are $1, 2, \ldots, n$.

Step 1 (start). Let $k = 1$, and let T be the given tree.

Step 2 (find degree 1 vertex). Select the vertex X of degree 1 in T that has the smallest label.

Step 3 (find the edge). Find the edge e on X. Let W be the other vertex on e, and let a_k be the label on W.

Step 4 (make new tree). Delete the edge e and vertex X from T to form a new tree T'.

Step 5 (repeat?). If T' has more than two vertices, then let $T = T'$, increase k by 1, and return to step 2. Otherwise we are done.

For example, the list for the following tree is 6, 5, 1, 5, 6.

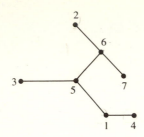

In Exercises 36–41 use Prufer's algorithm to find the list for each tree in the indicated exercise or graph.

36. Exercise 34 $(2, 2, 5, 5, 5)$

37. Exercise 35

38.

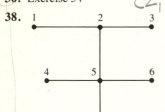

39. $(2, 2, 1, 3, 3)$

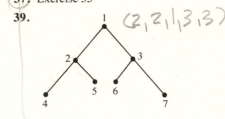

40. $(1, 2, 4, 4)$

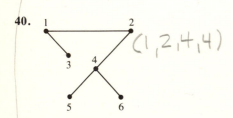

41. $(1, 5, 6, 5, 4, 5, 9$

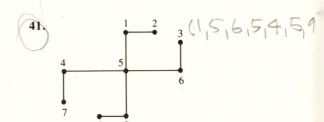

42. We can construct a tree from a list L of $n - 2$ numbers taken from $N = \{1, 2, \ldots, n\}$ as follows. (Here we assume that the vertices of the tree are labeled $1, 2, \ldots, n$.) Pick the smallest number k from N that is not in the list L and construct an edge on that number and the first number in the list L. Then delete the first number in L and delete k from N and repeat the process. When L is exhausted, join the two numbers remaining in N. For example, the tree generated by the list 6, 5, 1, 5, 6 is the one pictured before Exercise 36. Construct the tree for the list 2, 2, 2, 2.

In Exercises 43–46 repeat Exercise 42 for each list.

43. 1, 2, 3, 4   **44.** 1, 2, 3, 2, 1   **45.** 4, 3, 2, 1   **46.** 3, 5, 7, 3, 5, 7

47. Use Prufer's algorithm to prove that the number of distinct trees with vertices $1, 2, \ldots, n$ is n^{n-2} for $n > 1$.

48. Prove that part (a) implies part (b) in Theorem 4.5.

49. Prove that part (b) implies part (c) in Theorem 4.5.

50. Prove that part (c) implies part (d) in Theorem 4.5.

51. Prove that part (d) implies part (e) in Theorem 4.5.

52. Prove that part (e) implies part (f) in Theorem 4.5.

53. Prove that part (f) implies part (a) in Theorem 4.5.

54. Give an inductive proof of Theorem 4.3 which does not use Theorem 4.2. (*Hint:* Use induction on the number of edges.)

55. Use Theorem 4.3 to give an alternate proof of Theorem 4.2.

56. Suppose $d_1, d_2, \ldots, d_n$ for $n \geq 2$ are positive integers with sum $2n - 2$. Prove that there is a tree with n vertices having degrees $d_1, d_2, \ldots, d_n$. (*Hint:* Use induction on n.)

4.2   Rooted Trees and Spanning Trees

People have always been interested in learning about the descendants of historically important individuals. To assist in these investigations a genealogical chart is often drawn. An example is given in Figure 4.9, where for simplicity only first names are used. It is understood that the downward lines represent the "is a parent of" relationship.

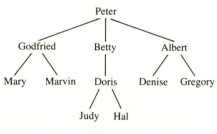

FIGURE 4.9

This chart can also be represented by a directed graph in which vertices represent individuals and directed edges begin at a parent and end at a child. Such a directed graph is shown in Figure 4.10.

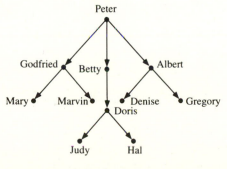

FIGURE 4.10

Since all the arrows in Figure 4.10 point downward, it is not really necessary to draw the arrowheads on the edges as long as the directions are understood to be downward. Figure 4.11 shows the corresponding directed tree without these arrowheads.

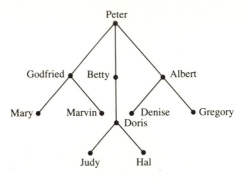

FIGURE 4.11

For the directed graph in Figure 4.10 there is one vertex with indegree 0, and all the other vertices have indegree 1. Furthermore, when the directions on the edges are ignored, we have a tree. A **rooted tree** is a directed graph T satisfying two conditions: (1) When the directions of the edges in T are ignored, the resulting undirected graph is a tree; and (2) there is a unique vertex R such that the indegree of R is 0 and the indegree of any other vertex is 1. This vertex R is called the **root** of the rooted tree. The directed graph in Figure 4.10 is a rooted tree with Peter as its root. We will follow the customary practice of drawing rooted trees with the roots at the top and omitting arrowheads on the directed edges with the understanding that edges are directed downward.

EXAMPLE 4.6   The graph in Figure 4.12(a) is a rooted tree with root A since (1) when the directions on the edges are ignored, the resulting graph is a tree; and (2) A has indegree 0, and all the other vertices have indegree 1. The usual way of drawing this tree is shown in Figure 4.12(b). ■

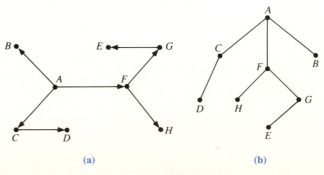

(a)   (b)

FIGURE 4.12

Rooted trees are often used to describe hierarchical structures. One such example occurs with the family tree of Peter. Another example is given below.

EXAMPLE 4.7

A rooted tree can be used to describe the organization of a book by using ''book'' as the root and other vertices as subdivisions. In some books there are subsections of a section, and so another level of vertices could be added in this case. See Figure 4.13 for an illustration. ▬

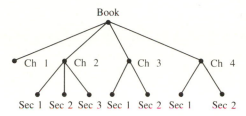

FIGURE 4.13

THEOREM 4.6

In a rooted tree with root R:

(a) The number of vertices is one more than the number of directed edges.

(b) There are no directed cycles.

(c) There is a unique directed simple path from R to every other vertex.

Proof. The proofs of (a) and (b) follow immediately since T becomes a tree when the directions on the directed edges are ignored. Next we will show that there is a directed path (and, hence, there will be a simple directed path) from R to any other vertex $V \neq R$. Since the indegree of V is 1, there is a vertex $V_1 \neq V$ and a directed edge from V_1 to V. If $V_1 = R$, we are finished. If not, since the indegree of V_1 is 1, there is a vertex $V_2 \neq V_1$ and a directed edge from V_2 to V_1. Since there are no directed cycles, $V_2 \neq V$. If $V_2 = R$, then we are done. Otherwise, this process can be repeated with each iteration generating a new vertex. Since the number of vertices is finite, we must eventually reach R. Thus, we create a directed path from R to V. The uniqueness of a simple directed path from R to V follows immediately as in parts (a) and (b). ▬

Family terms are used to describe the relationships among vertices in a rooted tree just as they describe relationships in a genealogical chart. If in a rooted tree there is a directed edge from a vertex U to a vertex V, we say U is a **parent** of V or V is a **child** of U. For a vertex V the vertices other than V on the directed simple path from the root to V are called the **ancestors** of V or, equivalently, we say that V is a **descendant** of these vertices. A **terminal vertex** is a vertex that has no children, and an **internal vertex** is one that has children. For the rooted tree in

Figure 4.12, E is a child of G, and A, F, and G are ancestors of E. Also, F has H, G, and E as its descendants. Vertices B, D, E, and H are terminal vertices, and the rest are internal vertices. Note that in any rooted tree the root has no ancestors, and every other vertex is a descendant of the root. A terminal vertex is a vertex with outdegree 0, and an internal vertex has nonzero outdegree. As the name suggests, a terminal vertex is at the end of a directed path from the root.

We will now consider two examples where a rooted tree is used to obtain a solution to a problem.

EXAMPLE 4.8

In Chapter 2 the concept of a partition of a set was introduced. To list all the partitions of $\{1, 2, \ldots, n\}$ requires a systematic approach so as not to miss any possibility. The rooted tree in Figure 4.14 shows one such approach for $n = 4$. Here the terminal vertices are the partitions. Do you recognize the pattern? The children of $\{1, 3\}, \{2\}$ are $\{1, 3, 4\}, \{2\}; \{1, 3\}, \{2, 4\};$ and $\{1, 3\}, \{2\}, \{4\}$. ■

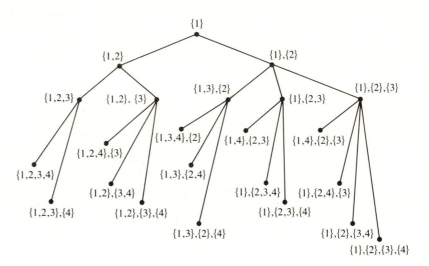

FIGURE 4.14

EXAMPLE 4.9

Suppose we have seven identical coins and an eighth that looks the same but is heavier. With the use of a balance scale we want to identify the counterfeit coin in the smallest number of weighings. Let us label the coins 1, 2, . . . , 8. Note that when coins are placed on the two sides of the balance scale either the left side will go down, the two sides will balance, or the right side will go down. We can construct a rooted tree as in Figure 4.15 giving a systematic approach to doing the weighings. The label beside each vertex indicates which coins are being weighed on each side of the balance scale. For example, $\{1,2\} - \{3,4\}$ means that coins 1 and 2 are weighed on the left side and coins 3 and 4 on the right. If the right side goes down, we proceed to the child on the right side for the next weighing, and similarly when the left side goes down. The terminal vertex indicates the heavy coin. For example, we begin by comparing the weight of coins 1, 2, 3, and 4 on the left with the weight of coins 5, 6, 7, and 8 on the right. If the balance tips to the left, we then compare coins 1 and 2 against coins 3 and 4. If in this weighing the right side goes down, we next compare coins 3 and 4. If this weighing shows the right side going down again, we reach the terminal vertex indicating that 4 is the counterfeit coin. Since each terminal vertex is at the end of a simple directed path of length 3 from the root, we see that this scheme requires three weighings to find the counterfeit coin.

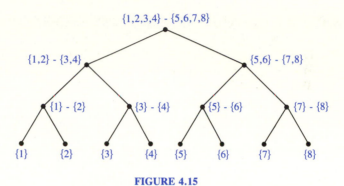

FIGURE 4.15

Could there be a different approach that will find the counterfeit coin with fewer weighings? Since a balance scale has three possible outcomes, we can build a rooted tree in which there are three children rather than just two as was done above. Figure 4.16 gives one such possibility, where we proceed to the middle child when the two sides balance. Here because each terminal vertex is at the end of a simple directed path of length two from the root, we can find the counterfeit coin with just two weighings.

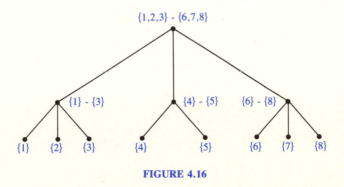

FIGURE 4.16

The trees in Figures 4.15 and 4.16 are called **decision trees** because of the way they structure a decision-making process.   ■

Spanning Trees

In Example 4.2 in Section 4.1 we found a tree which contained all the vertices of the original graph. This is an idea that appears in many applications, including power lines, pipeline networks, and road construction.

EXAMPLE 4.10   Suppose an oil company wants to build a series of pipelines between six storage facilities in order to be able to move oil from one storage facility to any of the other five. Because the construction of a pipeline is very expensive, the company wants to construct as few pipelines as possible. Thus, the company does not mind if oil has to be routed through a second or even a third facility to go from one to another. For environmental reasons it is not possible to build a pipeline between each pair of storage facilities. The graph in Figure 4.17(a) shows the pipelines that can be built.

The task is to find a set of edges which, together with the vertices incident on these edges, form a connected graph containing all the vertices and having no cycles. This will allow oil to go from any storage facility to any other without unnecessary duplication of routes and, hence, unnecessary building costs. Thus, again a tree containing all the vertices of a graph is being sought. One selection of edges is b, e, g, i, and j, as illustrated in Figure 4.17(b). ▪

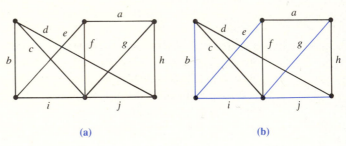

(a)   (b)

FIGURE 4.17

A **spanning tree** of a graph G is a tree (formed by using edges and vertices of G) containing all the vertices of G. Thus, in Figure 4.17 the edges b, e, g, i, and j along with the vertices on them form a spanning tree for the graph. We shall follow the customary practice of describing a tree by listing only its edges with the understanding that its vertices are those incident on the edges. Thus, in Figure 4.17 we would say that the edges b, e, g, i, and j form a spanning tree for the graph.

If the graph G is a tree, then its only spanning tree is G itself. A graph may have more than one spanning tree. For example, the edges a, b, c, d, and e also form a spanning tree for the graph in Figure 4.17(a).

There are several ways to find a spanning tree for a graph. One way is to remove an edge from each cycle. This method is used to find the fundamental system of circuits in an electrical network. This process is illustrated in the following example.

EXAMPLE 4.11   The graph in Figure 4.18(a) is not a tree because it contains cycles such as a, b, e, d. In order to obtain a tree our procedure will be to delete an edge in each cycle. Deleting b from the cycle a, b, e, d gives the graph in Figure 4.18(b), which is still not a tree because of the cycle c, e, d. So we delete an edge in this cycle, say e. The resulting graph in Figure 4.18(c) is now a tree. This, then, is a spanning tree for the original graph. ▰

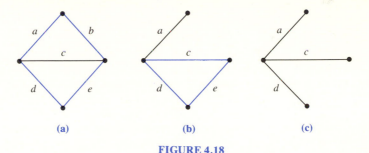

FIGURE 4.18

If a connected graph has n vertices and e edges, with $e \geq n$, we must perform this deletion process $e - n + 1$ times in order to obtain a tree. For by performing these deletions we change the number of edges from e to $n - 1$, which is the number of edges in a tree with n vertices.

The Breadth-First Search Algorithm

The method described above is not the only way to find a spanning tree. There are many others, and some of these are easier to program on a computer because they do not require that cycles be found. One of these methods is based on the idea of breadth-first search, which was discussed in Section 3.5.

Recall that in the breadth-first search algorithm we start with a vertex S. Then we find the vertices adjacent to S and assign them the label 1. Next we look at the unlabeled vertices that are adjacent to those with label 1. These newly found vertices are then given the label 2. We continue by finding unlabeled vertices adjacent to the vertices with label 2 and assigning these newly located vertices the label 3. This process is repeated until there are no more unlabeled vertices adjacent to labeled vertices. In this procedure there are edges that lead from a vertex with label k to a vertex with label $k + 1$ that form a spanning tree for the original graph.

Breadth-First Search Spanning Tree Algorithm

This algorithm will find a spanning tree, if it exists, for a graph G with n vertices. In the algorithm, L is the set of vertices with labels and T is a set of edges connecting the vertices in L.

Step 1 (start with a vertex). Pick a vertex U and assign U the label 0. Let $L = \{U\}$, $T = \varnothing$, and $k = 0$.

Step 2 (L has n vertices). If L contains all the vertices of G, then stop; the edges in T and the vertices in L form a spanning tree for G.

Step 3 (L has fewer than n vertices). If L does not contain all the vertices of G, find the vertices not in L that are adjacent to the vertices in L with largest label number k. If there are no such vertices, G has no spanning tree. Otherwise, assign these newly found vertices the label $k + 1$ and put them in L. For each new vertex with label $k + 1$, place in T one edge connecting this vertex to a vertex with label k. If there is more than one such edge, choose one arbitrarily. Return to step 2.

The construction process in step 3 of this algorithm guarantees that the edges in T and the vertices in L form a connected graph. Furthermore, each edge in T joins two vertices labeled with consecutive integers, and no vertex in L is connected by an edge in T to more than one vertex with a smaller label. Therefore, no collection of edges in T forms a cycle. Thus, after each iteration of step 3 the edges in T and the vertices in L form a tree. We will follow the usual practice of referring to this tree as simply T with the understanding that the vertices of the tree are those incident on the edges (or equivalently, those in L).

EXAMPLE 4.12   We shall apply the breadth-first search spanning tree algorithm to find a spanning tree for the graph in Figure 4.19. For convenience names have been given to the vertices and edges. In particular, one vertex has been selected to be the starting vertex U.

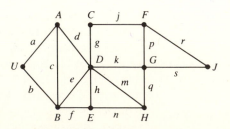

FIGURE 4.19

First label U with 0, and set $L = \{U\}$ and $T = \varnothing$. The unlabeled vertices adjacent to the vertex with label 0 are A and B. To them the label 1 is assigned, and L becomes $\{U, A, B\}$. Furthermore, the edges a and b are added to T, so that $T = \{a, b\}$. Since L does not contain all the vertices of the graph, we find the unlabeled vertices adjacent to those with label 1. These are the vertices D and E, which are then assigned the label 2. Now $L = \{U, A, B, D, E\}$. Although there are 3 edges between the vertices with label 1 and those with label 2, we are to choose only 2 of them, one incident with D and the other incident with E. Arbitrarily choose e and f, so that $T = \{a, b, e, f\}$. Again because L does not contain all the vertices of the graph, the unlabeled vertices adjacent to those with label 2 are determined. These are C, G, and H, and they are assigned the label 3 and included in L. So L is now $\{U, A, B, C, D, E, G, H\}$. Moreover, since 3 vertices were added to L, we need to select 3 edges to add to T. Suppose we choose g, k, and m, so that $T = \{a, b, e, f, g, k, m\}$. (Note that it is not possible to choose both m and n.) Again L does not contain all the vertices, and so the unlabeled vertices adjacent to those with label 3 are located. They are F and J, which are then assigned the label 4. Now L is $\{U, A, B, C, D, E, F, G, H, J\}$. We also add the edges p and s to T, and so $T = \{a, b, e, f, g, k, m, p, s\}$. Since L contains all the vertices of the graph, T is a spanning tree as illustrated in Figure 4.20. The labels on the vertices are shown in parentheses. ∎

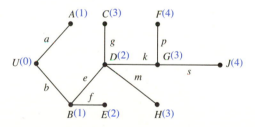

FIGURE 4.20

We should note that when using this algorithm, there are places where edges are chosen arbitrarily. Different choices lead to different spanning trees. In Example 4.12, for instance, instead of choosing the edges m and p, we could have chosen the edges n and j. This would give the spanning tree in Figure 4.21.

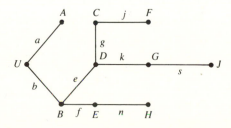

FIGURE 4.21

A spanning tree constructed by means of the breadth-first search algorithm is sometimes called a **shortest path tree** because the path from U to any other vertex using edges in the spanning tree is a shortest path in the original graph. (Recall from Section 3.5 that the label on each vertex is its distance from the vertex U.)

In the examples so far, the graphs have had spanning trees. However, this is not always the case, as the next example shows.

EXAMPLE 4.13   The graph G given in Figure 4.22 does not have a spanning tree because it is not possible to choose edges that connect all the vertices of G. In particular, we cannot find edges of G that can be used to make a path from A to E. ▪

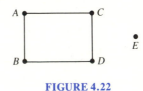

FIGURE 4.22

In the examples we have seen that the existence of a spanning tree is related to the connectedness of the graph. This relationship is made explicit in the following theorem.

THEOREM 4.7   A graph G is connected if and only if G has a spanning tree.

Proof.   Suppose that the graph G has a spanning tree T. Since T is a connected graph containing all the vertices in G, for any two vertices U and V in G there is path between U and V using edges from T. But since the edges of T are also edges of G, we have a path between U and V using edges in G. Hence, G is connected.

Conversely, suppose G is connected. Applying the breadth-first search spanning tree algorithm to G yields a set L of vertices with labels and a set T of edges connecting the vertices in L. Moreover, T is a tree. Since G is connected, each vertex of G is labeled. Thus, L contains all the vertices of G and T is a spanning tree for G. ▪

EXERCISES 4.2

In Exercises 1–8 determine if each directed graph is a rooted tree.

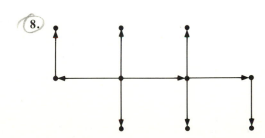

In Exercises 9–12 draw the rooted trees in the indicated exercises in the usual way with the root at the top and without the arrowheads.

9. Exercise 1

10. Exercise 4

11. Exercise 6

12. Exercise 8

13. LISP is the primary programming language used in artificial intelligence. There are seven objects manipulated by LISP: S–expressions, atoms, lists, numbers, symbols, fixed-point numbers, and floating-point numbers. An S–expression can be an atom or a list, an atom can be a number or a symbol, and a number can be either fixed-point or floating-point. Draw a rooted tree describing these relationships.

14. Draw a rooted tree for your mother and her descendants.

15. Tom and Sue are first cousins living in a state that allows first cousins to marry. If a child is born to this marriage, what effect would this have upon a genealogical chart in which the root is Tom and Sue's common grandfather?

16. It is known that a male bee has only a mother and that a female bee has both a mother and a father. Draw a rooted tree giving the ancestors of a male bee for four generations back, assuming no mating between ancestors.

17. Write an algorithm describing how a tree with a vertex labeled R can be transformed into a rooted tree with root R. Illustrate your algorithm on the tree below.

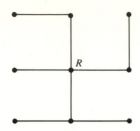

18. Repeat the second part of Exercise 17 for the following tree.

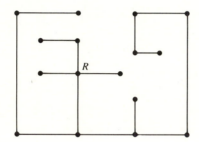

19. In how many ways can a tree with a vertex labeled R be transformed into a rooted tree with root R?

In Exercises 20–23 list for each rooted tree:
(i)   *the root.*
(ii)   *the internal vertices.*
(iii) *the terminal vertices.*
(iv)   *the parent of G.*
(v)   *the children of B.*
(vi)   *the descendants of D.*
(vii) *the ancestors of H.*

20.

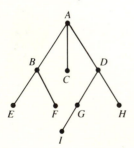

21.

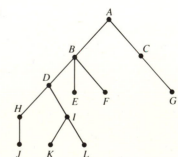

22.

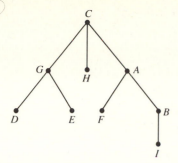

23.

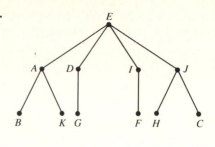

24. Draw a rooted tree with 7 vertices having as many terminal vertices as possible.

25. Draw a rooted tree with 7 vertices having as many internal vertices as possible.

26. Use Figure 4.14 to determine the number of partitions of $\{1, 2, 3, 4, 5\}$.

27. Draw a rooted tree describing all the possible outcomes for a two-game match between two chess players. (Remember that a chess game can end in a win, draw, or loss.)

28. Draw a rooted tree showing how to sort letters having a three-digit zip code in which the digits are 1 and 2.

29. Suppose we have three identical coins and a fourth that looks the same but is lighter. Construct a decision tree that will find the counterfeit coin by no more than two weighings on a balance scale.

30. Suppose we have eleven identical coins and a twelfth that looks the same but is lighter. Construct a decision tree that will find the counterfeit coin using no more than three weighings on a balance scale.

31. Suppose we have three identical coins and a fourth that looks the same but is heavier or lighter. Construct a decision tree that will find the counterfeit coin and determine if it is heavier or lighter using no more than three weighings on a balance scale.

32. Suppose we have seven identical coins and an eighth that looks the same but is counterfeit (can be heavier or lighter). Construct a decision tree that will find the counterfeit coin using no more than three weighings on a balance scale.

33. In a rooted tree the **level** of a vertex is defined to be the length of the simple directed path from the root to that vertex. What is the level of the root?

In Exercises 34–37 determine the level of each indicated vertex. (See Exercise 33 for the definition of "level.")

34. vertex F in the rooted tree of Exercise 20

35. vertex L in the rooted tree of Exercise 21

36. vertex H in the rooted tree of Exercise 22

37. vertex F in the rooted tree of Exercise 23

In Exercises 38–43 use the breadth-first search spanning tree algorithm to find a spanning tree for each connected graph. (Start with A and use alphabetical order when there is a choice for a vertex.)

38.

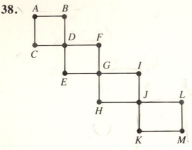

39.

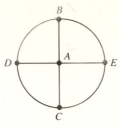

40.

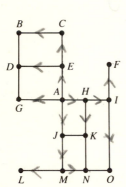

41.

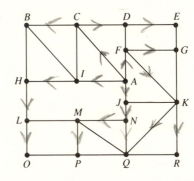

42.

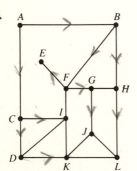

43.

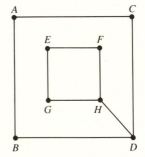

44. At its refinery an oil company has 7 major buildings that are connected by underground tunnels as illustrated below. Because of the possibility of a major explosion, there is a need to reinforce some of these tunnels to avoid a possible cave-in. The company wants to be able to go from any building to any other in the case of a major fire above ground, but it wants to avoid reinforcing more tunnels than necessary. How can this be done?

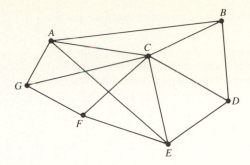

45. One of the primary responsibilities of the National Security Agency is to assist other governmental agencies in providing secure computer communications. The Department of Agriculture does not ordinarily need to be concerned about this, but when estimates of future crop productions arrive, it is important that these be kept secret until the time of public announcement. The map of computer links between reporting agencies for the Department of Agriculture is shown below. Realizing that there is a need for complete security only at certain times, the National Security Agency will make secure only the minimum number of lines. How can this be done?

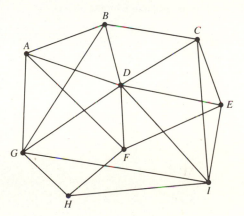

46. Will any two spanning trees for a connected graph always have an edge in common? If so, give a proof, and if not, give a counterexample.

47. How many different spanning trees are there for a cycle with n vertices, where $n \geq 3$?

48. Prove that any edge whose removal disconnects a connected graph is part of every spanning tree.

49. Draw the spanning tree formed by performing the breadth-first search spanning tree algorithm on K_n. (Name the vertices $1, 2, \ldots, n$ and start at 1.)

50. Draw the spanning tree formed by performing the breadth-first search spanning tree algorithm on $K_{m,n}$. (Name one vertex in the set of m vertices with 1, one vertex in the set of n vertices with $m + 1$, start at 1, and visit adjacent vertices in numerical order.)

51. Can the spanning tree formed by the breadth-first search spanning tree algorithm be a simple path?

52. The breadth-first search algorithm can be modified to work on a directed graph by labeling a vertex with label $k + 1$ when there is a directed edge to it from a vertex with label k. Write an algorithm for this modification. Apply to it the following directed graph starting at A.

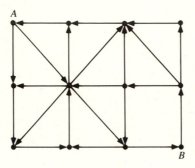

53. Repeat the second part of Exercise 52 using the directed graph in Exercise 52 and starting at B.

54. A graph with vertices labeled 1, 2, . . . , 9 is described below by giving its adjacency list. Determine if this graph is connected by using the breadth-first search spanning tree algorithm.
>   1: 2, 3, 5, 7, 9
>   2: 1, 3, 4, 5, 9
>   3: 1, 2, 4, 6, 8
>   4: 2, 3, 5, 6
>   5: 1, 2, 6, 7
>   6: 3, 4, 5, 7, 9
>   7: 1, 5, 6, 8, 9
>   8: 3, 7
>   9: 1, 2, 6, 7

55. Repeat Exercise 54 with the adjacency list below.
>   1: 2, 5
>   2: 1, 3
>   3: 2, 6
>   4: 5, 6
>   5: 1, 4
>   6: 3, 4
>   7: 8, 9
>   8: 7, 9
>   9: 7, 8

56. Suppose T is a rooted tree where every vertex has at most $k \geq 2$ children and the length of the longest path from the root to a terminal vertex is h. Prove that (i) T has at most $\dfrac{(h^{k+1} - 1)}{(k - 1)}$ vertices and (ii) if some vertex has k children, then there are at least $h + k$ vertices in T.

4.3   Depth-First Search

In Section 4.2 we saw how breadth-first search can be used to find a spanning tree in a graph. This algorithm starts from one vertex and spreads out to all the adjacent vertices. From each of these we spread out again to all the adjacent vertices that have not been reached and continue in this fashion until we can go no further. In this way we obtain both the distance from the initial vertex to each vertex and a spanning tree, if one exists.

Another algorithm for finding a spanning tree is the depth-first search algorithm. In this algorithm we label the vertices with consecutive integers. The underlying idea of the algorithm is that to find the vertex that should be labeled immediately after labeling vertex V, the first vertices to consider are the ones adjacent to V. If there is an unlabeled vertex W adjacent to V, W is assigned the next label number, and the process of searching for the next vertex to label is begun with W. If V has no unlabeled adjacent vertices, we back up to the vertex that was labeled immediately before V and continue backing up, if necessary, until we reach a vertex having an unlabeled adjacent vertex U. Vertex U is then assigned the next label number, and the process of searching for the next vertex to label is begun with U.

Depth-First Search Algorithm

This algorithm will assign labels to vertices and select edges of a graph. In the algorithm L is the set of vertices with labels and T is the set of edges selected.

Step 1 (start with a vertex). Pick a vertex U and assign U the label 1. Let $L = \{U\}$; $T = \varnothing$, and $k = 1$.

Step 2 (adjacent vertex?). If some vertex in L is adjacent to a vertex not in L, go to step 3. Otherwise, stop.

Step 3 (find next vertex). Let V be the vertex with largest label adjacent to a vertex W not in L. Add 1 to k, label W with k, include the edge on V and W in T, and go to step 2. *add W to the set L*

EXAMPLE 4.14   The first step in applying depth-first search to the graph in Figure 4.23 is to select some starting vertex arbitrarily, say A. Then A is assigned the label 1 and is put into L. (We will show the label in parentheses near the vertex as illustrated in Figure 4.24.) Since A is in L and the adjacent vertices B and E are not in L, step 3 is performed next. Now we choose an unlabeled vertex adjacent to A, say E, and assign E the label 2. Vertex E is included in L and the edge on A and E is placed in T. (The edge being placed in T is colored in Figure 4.25.) Since there are vertices

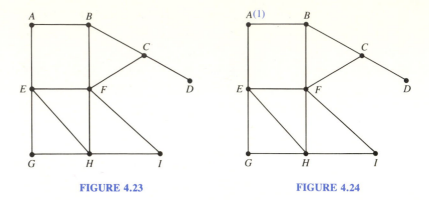

FIGURE 4.23   FIGURE 4.24

in L with adjacent vertices not in L, we proceed again to step 3. This time E is the vertex with largest label having adjacent vertices that are not in L. Choose one of these unlabeled adjacent vertices, say F. Assign F the label 3, place F in L, and include the edge on E and F in T. (See Figure 4.26.) Again step 3 needs to be repeated. Here F is the vertex with largest label having adjacent vertices that are not in L, and so we select one of these adjacent vertices, say B. Now B is assigned the label 4, and the edge on B and F is placed in T. Again step 3 is repeated by looking at the unlabeled vertices adjacent to B. Since the only such vertex is C, we assign the label 5 to C and place the edge on B and C in T. Likewise D is assigned the label 6 and the edge on C and D is placed in T. (See Figure 4.27.)

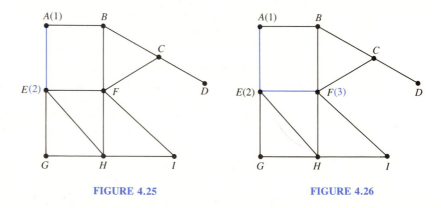

FIGURE 4.25   FIGURE 4.26

There are still vertices with labels that have adjacent vertices without labels. However, this time it is not the vertex D with the largest label 6 which has such adjacent vertices. Thus, we back up to the vertex with label 5 and find that it also has no adjacent vertices without labels. An examination of the vertex with label 4 shows the same thing. However, the vertex F with label 3 does have adjacent unlabeled vertices and we choose one, say H. Vertex H is then assigned the label 7 and the edge on F and H is placed in T. Step 3 is now repeated, with either G or I being chosen to receive label 8. Suppose we choose I; then the edge on H and I

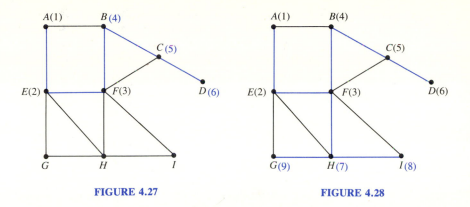

FIGURE 4.27   **FIGURE 4.28**

is included in T. Since I has no adjacent vertices without labels, we back up to H, which is adjacent to an unlabeled vertex G. Thus, G is given the label 9, and the edge on G and H is placed in T. Figure 4.28 illustrates the complete labeling and placement of edges in T. ▮

There is a fundamental difference between breadth-first search and depth-first search. With breadth-first search we fan out from each vertex to all the adjacent vertices, and this process is repeated at each vertex. Furthermore, at no time do we back up in order to continue the search. But with depth-first search we go out from a vertex as far as we can, and when unable to continue, we back up to the most recent vertex from which there was a choice; then we resume going out as far as we can.

An analogous situation can be found in two different ways to explore a cave with many tunnels. With the breadth-first search approach a posse searches the cave and whenever a tunnel branches off into several others, subgroups are formed to explore each of these simultaneously. With the depth-first search approach one person explores the cave by leaving a phosphorous trail to mark where she has been. When there is a choice of tunnels, an unexplored one is chosen at random to be explored next. Upon reaching a dead end, she backtracks using the marked trail to find the next unvisited tunnel.

THEOREM 4.8   Let the depth-first search algorithm be applied to a graph G.

    (a) The edges in T and the vertices in L form a tree.

    (b) Furthermore, if G is connected, this tree is a spanning tree.

Proof. (a) By the construction process of depth-first search, the edges of T and the vertices in L form a connected graph. In step 3 each time an edge is selected to be placed in T, one vertex is in L and the other is not in L. Thus, this selection does not create any cycles using the other edges in T. Consequently, at the end of the depth-first search algorithm the graph formed by the edges in T and the vertices in L contains no cycles and is, therefore, a tree.

The proof of part (b) is left as an exercise. ■

We will follow our convention and refer to the tree in Theorem 4.8 formed by the edges in T and the vertices in L as simply T. The tree T is called a **depth-first search tree.** The edges in T are called **tree edges** and the other edges are called **back edges.** The labeling of the vertices is called a **depth-first search numbering.** Thus, in Figure 4.28 vertex F has depth-first search number 3, the edge on vertices F and H is a tree edge, and the edge on vertices F and I is a back edge. Of course, the designation of edges as tree and back edges as well as the depth-first search numbering depends upon the choices made during implementation of the algorithm.

THEOREM 4.9 If the depth-first search algorithm is applied to a graph, then the edges in T, when oriented from the lower depth-first search number to the higher, form a rooted tree whose root is the vertex with depth-first search number 1.

Proof. Theorem 4.8 shows that T is a tree. Let R be the vertex with depth-first search number 1. Only during step 3 of depth-first search is a vertex assigned a depth-first search number and a tree edge going into it. This means that the root R has indegree 0, and each vertex in the tree other than R has indegree 1. ■

In Example 3.36 of Chapter 3 we investigated the problem of how to assign directions to the edges of a graph to create a strongly connected directed graph (a directed graph having a directed path between any two vertices). We stated that the absence of an edge whose removal disconnects the graph is necessary and sufficient to guarantee that there is a way to assign directions to edges so as to produce a strongly connected directed graph. However, no procedure was given for assigning directions to the edges when this can be done.

We can use depth-first search to obtain such a procedure. We begin by assigning directions to tree edges as in Theorem 4.9. See Figure 4.29 for such an orientation of the depth-first search tree in Figure 4.28. Next we assign directions to back edges by going from the larger depth-first search number to the smaller. With these assignments of directions to the tree and back edges, a directed graph D is formed.

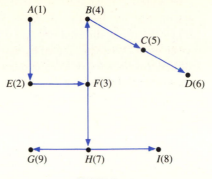

FIGURE 4.29

EXAMPLE 4.15 The depth-first search algorithm has been applied to a graph to yield the depth-first search numbering and tree edges in Figure 4.30(a). Now we assign directions to all the edges as described above, producing the directed graph shown in Figure 4.30(b). ■

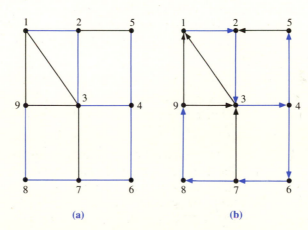

(a)   (b)

FIGURE 4.30

By Theorems 4.6 and 4.9 we see that there is a directed path in the depth-first search tree from the root to every other vertex. Thus, the fact that D is strongly connected is equivalent to there being a directed path from every other vertex to the root.

To determine whether there is a directed path in D from each vertex to the root R, we can use the following procedure: Reverse the directions of all the directed edges in D to obtain a new directed graph D', and apply directed breadth-first search to D' starting at R. (To apply breadth-first search to a directed graph, we modify the algorithm in the obvious way by only labeling a vertex with $k + 1$ when there is a directed edge to it from a vertex with label k.) Then if all the vertices of D' can be labeled by applying directed breadth-first search, this gives directed paths from R to each other vertex using the edges of D'. By reversing these directed paths, we have directed paths from each vertex to R using edges of D. When a graph can be made into a strongly connected directed graph, this assignment of directions to the edges is one way of doing it. The following theorem summarizes the method.

THEOREM 4.10   Let G be a connected graph. Apply depth-first search to G to make it into a directed graph D. (Directions are assigned to tree edges by going from the lower depth-first search number to the higher and to back edges by going from the higher to the lower.) Let D' be the directed graph formed by reversing the directions of the edges of D. Apply directed breadth-first search to D' starting at the vertex in D with depth-first search number 1. Then the edges of G can be oriented so as to produce a strongly connected directed graph if and only if this directed breadth-first search reaches every vertex of D'. Moreover, if such an orientation exists, the orientation of D is one.

Make a connected graph strongly connected if possible

EXAMPLE 4.16   Reversing the directed edges in Figure 4.30(b) yields the directed graph in Figure 4.31(a). Applying directed breadth-first search at R yields the labeling in Figure 4.31(b). Since all the vertices are labeled, the directed graph in Figure 4.30(b) is strongly connected. ■

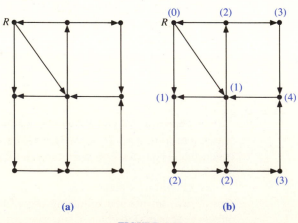

(a)   (b)

FIGURE 4.31

EXAMPLE 4.17

The depth-first search algorithm has been applied to a graph to yield the depth-first search numbering and tree edges shown in Figure 4.32(a). We assign directions to the edges yielding the directed graph in Figure 4.32(b). If the directions of the edges in the directed graph in Figure 4.32(b) are reversed, then the directed graph in Figure 4.32(c) results. When directed breadth-first search is applied at R to the directed graph in Figure 4.32(c), not all the vertices receive labels as shown in Figure 4.32(d). This shows that the original graph cannot be oriented so as to be strongly connected.

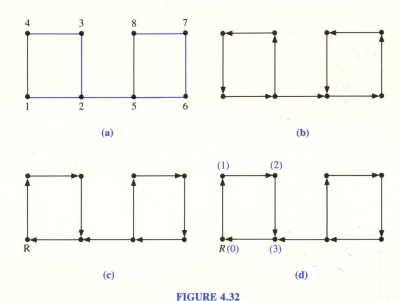

FIGURE 4.32

In order to analyze the complexity of the depth-first search algorithm, we will regard labeling a vertex and using an edge as the elementary operations. For a graph with n vertices and e edges, each vertex is labeled at most once and each edge is used at most twice, once in going from a labeled vertex to an unlabeled vertex and once in backing up to a previously labeled vertex. Hence, there will be at most $n + 2e \leq n + 2 \cdot \frac{1}{2}n(n-1)$ operations, and thus this algorithm is of order at most n^2.

Depth-first search can be used in many other ways to solve problems involving graphs and directed graphs. Interested readers should consult [8] in the suggested readings at the end of the chapter.

EXERCISES 4.3

Throughout these exercises, if there is a choice of vertices, choose the vertex that appears first in alphabetical order.

In Exercises 1–8 apply depth-first search to each graph to obtain a depth-first search numbering of the vertices.

1.

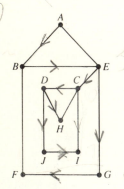

2.

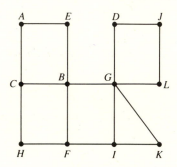

3.

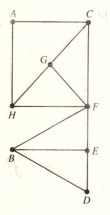

4.

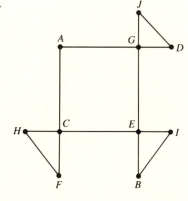

5.

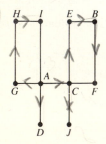

6.

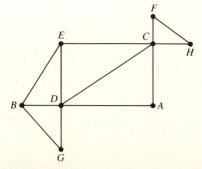

7.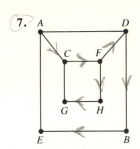

8.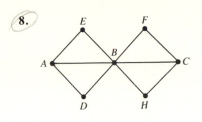

In Exercises 9–16 use the depth-first search numbering obtained for the graph in the indicated exercise to form the spanning tree described in Theorem 4.8.

9. Exercise 1   **10.** Exercise 2   **11.** Exercise 3   **12.** Exercise 4

13. Exercise 5   **14.** Exercise 6   **15.** Exercise 7   **16.** Exercise 8

In Exercises 17–24 use the depth-first search numbering obtained in the indicated exercises to list the back edges in the graphs.

17. Exercise 1   **18.** Exercise 2   **19.** Exercise 3   **20.** Exercise 4

21. Exercise 5   **22.** Exercise 6   **23.** Exercise 7   **24.** Exercise 8

In Exercises 25–32 use Theorem 4.10 to determine if there is an orientation of the edges of the graphs in the indicated exercises that will make each graph into a strongly connected directed graph. If so, assign directions to the edges as described in Theorem 4.10 using the depth-first search numbering obtained in the indicated exercise.

25. Exercise 1   **26.** Exercise 2   **27.** Exercise 3   **28.** Exercise 4

29. Exercise 5   **30.** Exercise 6   **31.** Exercise 7   **32.** Exercise 8

33. The city manager of a community with a large university believes that something needs to be done to handle the large influx of automobile traffic on those days that students are checking into the dormitories. She instructs the chief of police to transform the current two-way street system into a system of one-way streets to handle the extra traffic with the provision that students can still get from any place to any other. The campus area is given below. Is there a way for the chief of police to carry out these instructions? If so, how?

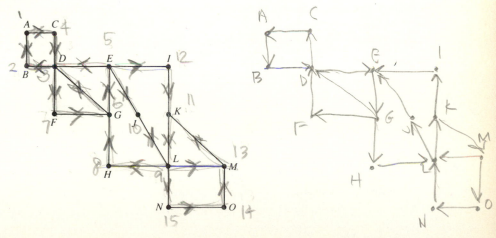

34. Can the spanning trees for a connected graph formed by breadth-first search and depth-first search be the same?

35. Prove part (b) of Theorem 4.8.

36. Show that every strongly connected directed graph with more than one vertex has at least one additional orientation of its edges under which it is strongly connected.

37. Label the vertices of K_3 as 1, 2, 3. Apply depth-first search to K_3 starting at 1. How many different depth-first search numberings are there?

38. Label the vertices of K_4 as 1, 2, 3, 4. Apply depth-first search to K_4 starting at 1. How many different depth-first search numberings are there?

39. Label the vertices of K_n as 1, 2, . . . , n. Apply depth-first search to K_n starting at 1. How many different depth-first search numberings are there?

40. Suppose breadth-first search and depth-first search are applied to a connected graph starting at the same vertex. If b is the label assigned to a vertex by breadth-first search and d is the label assigned to the same vertex by depth-first search, what is the relationship between b and d? Why?

41. For the tree obtained by applying depth-first search to a connected graph, prove that the descendants (relative to the depth-first search tree) of any vertex V have larger depth-first search numbers than V.

42. For the tree obtained by applying depth-first search to a connected graph, show that if a vertex with depth-first search number k has m descendants (relative to the depth-first search tree) with $m \geq 1$, then they have depth-first search numbers $k + 1$, $k + 2$, . . . , $k + m$.

43. Describe the algorithm that results from the replacement of "largest" by "smallest" in step 3 of the depth-first search algorithm.

44. Prove that when depth-first search is applied to a connected graph, one of the vertices on a back edge is an ancestor of the other (relative to the depth-first search tree).

45. Prove that when depth-first search is applied to a connected graph, one of the vertices on an edge is an ancestor of the other (relative to the depth-first search tree).

46. Let depth-first search be applied to a connected graph G. Prove that every cycle of G contains a back edge, and every back edge is contained in a cycle of G.

*A vertex A is called an **articulation point** of a connected graph G when the deletion of A and the edges incident on A creates a graph that is not connected. For example, the vertex A is an articulation point of the graph below.*

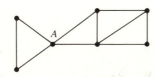

47. Prove that A is an articulation point of a connected graph G if and only if there exist vertices U and V such that U, V, and A are distinct and every path between U and V contains the vertex A.

For Exercises 48–49 let depth-first search be applied to a connected graph G and let A be a vertex in G.

48. If the depth-first search starts at A, prove that A is an articulation point of G if and only if A has more than one child.

49. If the depth-first search does not start at A, prove that A is an articulation point of G if and only if for some child C of A there is no back edge between C or any of its descendants and an ancestor of A.

4.4 Minimal Spanning Trees

When pipelines are to be constructed between oil storage facilities, it is likely that the cost of building each pipeline is not the same. Because of terrain, distance, and other factors it may cost more to build one pipeline than another. We can describe this problem by a weighted graph (discussed in Section 3.5), in which the weight of each edge is the cost of building the corresponding pipeline. Figure 4.33 depicts such a weighted graph. The problem is to build the cheapest set of pipelines. In other words, we want to find a spanning tree in which the sum of the costs of all the edges is as small as possible.

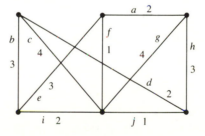

FIGURE 4.33

In a weighted graph the **weight of a tree** is the sum of the weights of the edges in the tree. A **minimal spanning tree** in a weighted graph is a spanning tree for which the weight of the tree is as small as possible. In other words, a minimal spanning tree is a spanning tree such that no other spanning tree has a smaller weight.

EXAMPLE 4.18   For the weighted graph in Figure 4.33 the edges b, c, e, g, and h form a spanning tree with weight $3 + 4 + 3 + 4 + 3 = 17$. The edges a, b, c, d, and e form another spanning tree with weight $2 + 3 + 4 + 2 + 3 = 14$. The edges a, d, f, i, and j form yet another spanning tree, which has weight 8. Since this spanning tree uses the five edges with the smallest weights, there can be no spanning tree with smaller weight. Thus, the edges a, d, f, i, and j form a minimal spanning tree for the weighted graph. ■

In the above example we were able to find a minimal spanning tree by trial and error. However, for a weighted graph with a large number of vertices and edges, this is not a very practical approach. One systematic approach would be to find all the spanning trees of a connected weighted graph, compute their weights, and then select a spanning tree with the smallest weight. Although this approach will always find a minimal spanning tree for a connected weighted graph, checking out all the possibilities can be a very time-consuming task, even for a supercomputer. A natural way to try to construct a minimal spanning tree is to build a spanning tree using edges of smallest weights. This approach is illustrated in Example 4.19.

EXAMPLE 4.19   For the weighted graph in Figure 4.34(a) we begin with the vertex A and select the edge of smallest weight on it, which is b. To continue building a tree, we look at the edges a, c, e, and f touching edge b. We select the one with the smallest weight, which is f. The next edges to look at are a, c, e, and g, the ones touching the edges already selected. There are two with the smallest weight, e and g, and we select one arbitrarily, say e. The next edges we consider are a, c, and d. (The edge g is not considered any longer, for its inclusion will form a cycle with e and f.) The edge with the smallest weight is a, and so it is added to the tree. These four edges a, b, e, f, form a spanning tree (see Figure 4.34(b)), which also turns out to be a minimal spanning tree. ▪

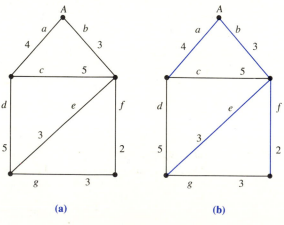

(a)   (b)

FIGURE 4.34

The method in Example 4.19 will always produce a minimal spanning tree, and is due to Prim. Prim's algorithm builds a tree by selecting any vertex and then an edge of smallest weight on that vertex. Next the tree is extended by choosing an edge of smallest weight that forms a tree with the previously chosen edge. The tree is extended still further by choosing an edge of smallest weight that forms a tree with the two previously chosen edges. This process is continued until a spanning tree is obtained, and this turns out to be a minimal spanning tree. This procedure can be formalized as follows.

Prim's Algorithm for Minimal Spanning Trees

This algorithm will find a minimal spanning tree, if one exists, for a weighted graph G with n vertices. In this algorithm S is a set of vertices and T is a set of edges.

Step 1 (start). Pick a vertex U, and let $S = \{U\}$ and $T = \varnothing$.

Step 2 (check for completion). If S contains all the vertices of G, then stop; the edges in T and the vertices in S form a minimal spanning tree for G.

Step 3 (pick next edge). If S does not contain all the vertices of G, find the edges which have one vertex in S and the other not in S. If there are no such edges, G is not connected and has no minimal spanning tree. Otherwise, choose one such edge of smallest weight (ties can be broken arbitrarily), and place it in T and its vertices in S (one of these is already in S). Return to step 2.

In step 3 of Prim's algorithm the selection of an edge with one vertex in S and the other not in S guarantees that there are no cycles formed by any collection of edges in T. Thus, at the end of each iteration of step 3 the edges in T and the vertices in S form a tree. Furthermore, when S contains all the vertices of G, a spanning tree is formed. As usual we will denote this tree by T. The proof that Prim's algorithm yields a minimal spanning tree is found in Exercises 22–24.

EXAMPLE 4.20 We shall apply Prim's algorithm to the weighted graph G in Figure 4.35. We start with vertex U, and so $S = \{U\}$ and $T = \varnothing$. Since S does not contain all the vertices

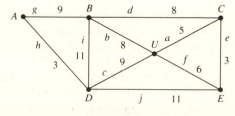

FIGURE 4.35

of G, we consider the edges on U, which are a, b, c, and f (see Figure 4.36), and pick the one of smallest weight. This is the edge a, and so a is added to T and its vertices are added to S. Thus, $S = \{U, C\}$ and $T = \{a\}$. Again we see S does not contain all the vertices of G, and so we look at the edges which have one vertex in S and one vertex not in S (see Figure 4.37). These are the edges b, c, d, e, and f.

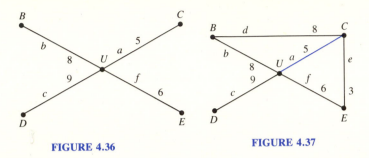

FIGURE 4.36   FIGURE 4.37

Of these the edge e has the smallest weight, and so it is included in the set T and the vertex E is added to the set S. Now $S = \{U, C, E\}$ and $T = \{a, e\}$. Since S does not contain all the vertices in G, we consider those edges with one vertex in S and one not in S (see Figure 4.38). These are the edges b, c, d, and j. Notice that the edge f is not considered, for it has both of its vertices in S. Of the edges b, c, d, and j there are two with the smallest weight, namely b and d. We arbitrarily pick

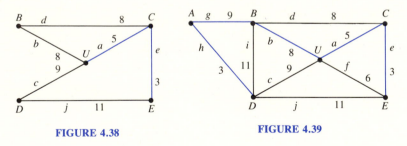

FIGURE 4.38   FIGURE 4.39

one, say b, and add it to T and B to S. Thus, $S = \{U, C, E, B\}$ and $T = \{a, e, b\}$. At this point the edges with one vertex in S and one out of S are c, g, i, and j, and of these both c and g have the smallest weight. So we arbitrarily choose g, making $S = \{U, C, E, B, A\}$ and $T = \{a, e, b, g\}$. Now we consider the edges c, h, i, and j. The one with smallest weight is h, and so it is chosen. Now after this choice, $S = \{U, C, E, B, A, D\}$ and $T = \{a, e, b, g, h\}$. Since S contains all the vertices of G, we are finished. We have found a minimal spanning tree of weight 28 as illustrated in Figure 4.39   ■

In two different places in the above example more than one edge of least weight could be chosen. The algorithm indicates that any edge of least weight can be chosen. Therefore, different minimal spanning trees could have been constructed depending upon these choices. For example, if in Example 4.20 we choose edge d instead of b, followed by the choices of c and h, the minimal spanning tree shown in Figure 4.40 results. Thus, we see that minimal spanning trees are not necessarily unique.

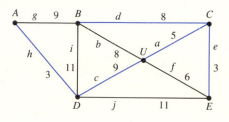

FIGURE 4.40

Prim's algorithm is an example of what is called a *greedy algorithm* since at each iteration we do the thing that seems best at that step (extending a tree by including an available edge of smallest weight). In Prim's algorithm this approach does lead to a minimal spanning tree, although in general a greedy algorithm need not produce the best possible result. (See Exercises 20 and 21.)

In analyzing the complexity of Prim's algorithm for a weighted graph with n vertices and e edges, we will consider comparing the weights of two edges as the basic operation. At each iteration of step 3 there will be at most $e - 1$ comparisons made in order to find an edge of smallest weight having one vertex in S and one vertex not in S. Step 3 is done at most n times, and so there are at most $n(e - 1)$ operations. Since $e \leq \frac{1}{2}n(n - 1)$, our implementation of Prim's algorithm is of order at most n^3.

Another algorithm that can be used to find a minimal spanning tree is due to Kruskal. Details are found in the exercises.

Let us now return to Figure 4.33. Suppose now that the weights of the edges measure the profit that results when oil is pumped through the corresponding pipelines. Our problem now is to find a spanning tree of pipelines that generates the most profit. Thus, we want a spanning tree for which the sum of the weights of the edges is not as small as possible but as large as possible.

A **maximal spanning tree** in a weighted graph is a spanning tree where the weight of the tree is as large as possible. In other words, there is no spanning tree where the weight is larger. Fortunately, finding a maximal spanning tree is very similar to finding a minimal spanning tree. All that is needed is to replace the word "smallest" by the word "largest" in Prim's algorithm.

EXAMPLE 4.21   For the weighted graph in Figure 4.35 we will construct a maximal spanning tree. We begin by picking the vertex U. Then $S = \{U\}$ and $T = \varnothing$. We now look at the edges on U (see Figure 4.41) and pick one with the largest weight. This is the edge c, and so S is now $\{U, D\}$ and $T = \{c\}$. Since S does not contain all the vertices of the graph, we find the edges with one vertex in S and one vertex not in S (see Figure 4.42). These are the edges $a, b, f, h, i,$ and j. Of these there are two with

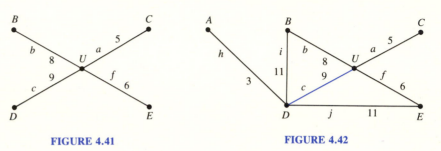

FIGURE 4.41   FIGURE 4.42

the largest weight, i and j. We choose one arbitrarily, say i. So now $T = \{c, i\}$ and $S = \{U, B, D\}$. Again the process is repeated (see Figure 4.43) by choosing edge j with largest weight having one vertex in S and one not in S. Now $T = \{c, i, j\}$ and $S = \{U, B, D, E\}$. Again we look at the edges with a vertex in S and one not in S. Of these g is the edge with largest weight, and so $T = \{c, g, i, j\}$ and $S =$

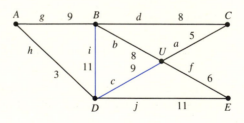

FIGURE 4.43

$\{U, A, B, D, E\}$. One more iteration yields the choice of the edge d; therefore, $T := \{c, d, g, i, j\}$ and $S = \{U, A, B, C, D, E\}$, which is the set of all vertices. Hence, T is a maximal spanning tree as illustrated in Figure 4.44.   ■

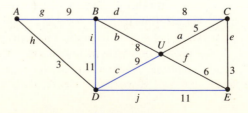

FIGURE 4.44

EXERCISES 4.4

Throughout these exercises, if there is a choice of edges to use in forming a minimal or maximal spanning tree, select edges according to alphabetical order.

In Exercises 1–4 use Prim's algorithm to find a minimal spanning tree for each weighted graph. (Start at A.)

1.

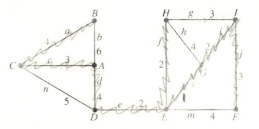

2.

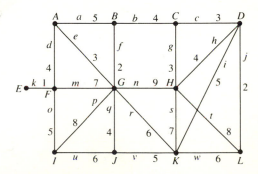

3.

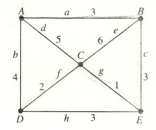

4.

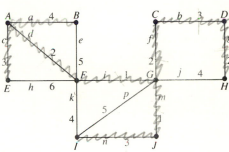

In Exercises 5–8 use Prim's algorithm to find a minimal spanning tree for the weighted graphs in the indicated exercises.

5. Exercise 1 (Start at E.)   **6.** Exercise 2 (Start at H.)

7. Exercise 3 (Start at G.)   **8.** Exercise 4 (Start at H.)

In Exercises 9–12 use Prim's algorithm to find a maximal spanning tree for the weighted graphs in the indicated exercises.

9. Exercise 1 (Start at A.)   **10.** Exercise 2 (Start at A.)

11. Exercise 3 (Start at F.)   **12.** Exercise 4 (Start at D.)

13. The Gladbrook Feed Company has 7 bins of corn which must be connected by grain pipes so that grain can be moved from one to another. To minimize the cost of construction they want to build as few grain pipes as possible. The cost (in thousands of dollars) of building a pipeline between two bins is given below, where a – indicates no pipeline can be built. How can the pipes be built at minimal cost?

Bin	1	2	3	4	5	6	7
1	–	4	–	6	2	–	3
2	4	–	5	2	–	3	1
3	–	5	–	7	–	2	2
4	6	2	7	–	4	1	–
5	2	–	–	4	–	1	–
6	–	3	2	1	1	–	2
7	3	1	2	–	–	2	–

14. FBI Special Agent Hwang is working with 5 informants who have infiltrated organized crime. She needs to make arrangements for the informants to communicate with each other, either directly or through others, but never in groups of more than two. For reasons of security the number of meeting places must be kept as small as possible. Furthermore, each pair of informants has been assigned a danger rating (given in table below) which indicates the risk involved in their being seen together. How can Special Agent Hwang arrange communication so as to minimize the danger? Assume the danger is proportional to the sum of the danger ratings of the individuals who meet directly.

	Jones	Brown	Hill	Ritt	Chen
Jones	–	3	4	5	2
Brown	3	–	3	1	4
Hill	4	3	–	2	3
Ritt	5	1	2	–	4
Chen	2	4	3	4	–

15. Give an example of a connected weighted graph (not a tree) where the same edge is part of every minimal spanning tree and every maximal spanning tree.

16. Modify Prim's algorithm to find a spanning tree that is minimal with respect to those containing a specified edge. Illustrate your modification with edge g in Exercise 1.

17. Repeat the second part of Exercise 16 with edge b in Exercise 3.

18. If the weights in a connected weighted graph correspond to distances, does a minimal spanning tree give the shortest distance between any two vertices? If so, give a proof. If not, give a counterexample.

19. Can a minimal spanning tree in a connected weighted graph (not a tree) contain an edge of largest weight? If so, give an example. If not, give a proof.

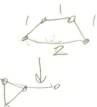

20. In the knapsack problem of Section 1.3 discuss why the greedy algorithm approach of using the highest rating does not give a good procedure.

21. Suppose that we want to mail a package and have stamps worth 1, 13, and 22 cents. If we want to make the necessary postage with the minimum number of stamps, the greedy algorithm approach is to use as many 22–cent stamps as possible, then as many 13–cent stamps as possible, and finally the necessary number of 1–cent stamps. Show that this approach need not result in the fewest stamps being used.

Exercises 22–24 provide a proof that Prim's algorithm yields a minimal spanning tree.

22. Let G be a connected weighted graph with vertex set V, let S be a set of vertices of G such that neither S nor $V - S$ is empty, and let T_0 be a minimal spanning tree for G. Among all edges $\{A, B\}$ with A in S and B not in S let e be one having minimal weight. Show that if e is not in T_0, then the set T_1 formed by e and the edges in T_0 contains a cycle which includes both e and another edge e' from a vertex in S to a vertex not in S.

23. Show with the hypotheses of Exercise 22 that $T_2 = T_1 - \{e'\}$ is a minimal spanning tree for G.

24. Prove Prim's algorithm gives a minimal spanning tree for a connected weighted graph G by induction on the number of edges in the set T. Let the induction hypothesis be that T is contained in some minimal spanning tree for G. (*Hint:* Use the previous two exercises.)

25. Prove that if the weights in a connected weighted graph are all different, then it has one minimal spanning tree.

Kruskal's Algorithm

This algorithm will find a minimal spanning tree, if one exists, for a weighted graph G with n vertices, where $n \geq 2$. In this algorithm S is a set of vertices and T is a set of edges.

Step 1 (start). If there are no edges, G is not connected and, hence, has no minimal spanning tree. Otherwise, pick an edge of smallest weight (ties can be broken arbitrarily). Place the edge in T and its vertices in S.

Step 2 (check for completion). If T contains $n - 1$ edges, then stop; the edges in T and the vertices in S form a minimal spanning tree. Otherwise, go to step 3.

Step 3 (pick next edge). Find the edges of smallest weight which do not form a cycle with any of the edges in T. If there are no such edges, G is not connected and has no minimal spanning tree. Otherwise, choose one such edge (ties can be broken arbitrarily), and place it in T and its vertices in S. Return to step 2.

In Exercises 26–29 use Kruskal's algorithm to find a minimal spanning tree for the weighted graphs in the indicated exercises.

26. Exercise 1

27. Exercise 2

28. Exercise 3

29. Exercise 4

30. Modify Kruskal's algorithm to find a spanning tree that is minimal with respect to all those containing a specified edge. Illustrate your modification with edge d in Exercise 1.

31. Repeat the second part of Exercise 30 with edge b in Exercise 3.

32. Prove that Kruskal's algorithm gives a minimal spanning tree for a connected weighted graph.

4.5   Binary Trees and Traversals

Expression Trees

In previous examples and applications of rooted trees it was not necessary to distinguish between the children of a parent. In other words, there was no need to designate a child as the first child or the second child. However, there are many situations where it is necessary to make such a distinction. For example, in an arithmetic expression such as $A - B$ the order of A and B is important. Thus, if we represent $A - B$ by a rooted tree in which the root represents the operation $(-)$ and the children represent the operands (A and B), then the order of the children is important.

A **binary tree** is a rooted tree in which each vertex has at most two children and each child is designated as being a **left child** or a **right child.** Thus, in a binary tree each vertex may have 0, 1, or 2 children. When drawing a binary tree, we will follow customary practice and draw a left child to the left and below its parent and a right child to the right and below its parent. The **left subtree** of a vertex V in a binary tree is the graph formed by the left child L of V, the descendants of L, and the edges connecting these vertices. The **right subtree** of V is defined in an analogous manner.

EXAMPLE 4.22   For the binary tree in Figure 4.45(a), A is the root. Vertex A has two children, a left child B and a right child C. Vertex B has one child, a left child D. Similarly, C has a right child E but no left child. The binary tree in Figure 4.45(b), in which A has a left child B, is different from the one in Figure 4.45(c), in which B is a right child. For the binary tree in Figure 4.45(d), the left subtree of V is shown in color in Figure 4.45(e) and the right subtree of W is shown in color in Figure 4.45(f). The right subtree of V consists of the vertex U alone.   ■

Binary trees are used extensively in computer science to represent ways to organize data and describe algorithms. For example, during the execution of a computer program it may be necessary to evaluate arithmetic expressions such as $(2 - 3 \cdot 4) + (4 + 8/2)$. Our knowledge of the conventions for the order of operations tells us how to proceed with this calculation: Scan from left to right, first doing multiplication and division and then addition and subtraction, with the understanding that parentheses have priority. However, when an expression needs to be evaluated frequently, this method cannot be used efficiently by a computer. An alternate approach is to represent an arithmetic expression by a binary tree and then process the data in some other way.

We will represent an arithmetic expression as a binary tree with the operations as internal vertices and the operands as terminal vertices. In this representation we let the root denote the final operation done in the expression and place the left

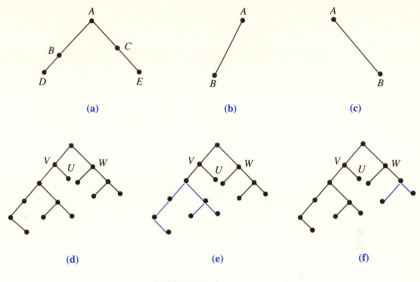

FIGURE 4.45

operand as its left child and the right operand as its right child. If necessary, this process is repeated on these operands. The binary tree created by this process is called an **expression tree.**

EXAMPLE 4.23 The expression $a * b$ (where $*$ denotes multiplication) is represented by the binary tree in Figure 4.46. Note that the operation $*$ is represented by an internal vertex and the operands a and b are represented by terminal vertices. ■

FIGURE 4.46

EXAMPLE 4.24 The expression $a + b * c$ means $a + (b * c)$. The last operation to be performed is addition. Thus, we first represent this expression by the binary tree in Figure 4.47(a). Repeating the process with the operand $b * c$ yields the expression tree in Figure 4.47(b). ■

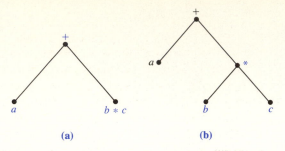

(a)   **(b)**

FIGURE 4.47

EXAMPLE 4.25   The expression tree for $a + d * (b - c)$ is created by the sequence of binary trees in Figure 4.48. ▮

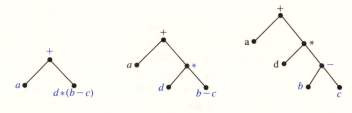

FIGURE 4.48

EXAMPLE 4.26   The expression $(a + b * c) - (f - d/e)$ is represented by the expression tree in Figure 4.49. ▮

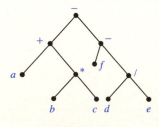

FIGURE 4.49

Preorder Traversal

We have seen that an arithmetic expression can be represented by an expression tree. Now we must process the expression tree in some way so as to obtain an evaluation of the original expression. We are looking for a systematic way to examine each vertex in the expression tree exactly once. Processing the data at a vertex is usually called **visiting a vertex,** and a search procedure that visits each vertex of a graph exactly once is called a **traversal** of the graph. For example, both breadth-first search and depth-first search are traversals of a connected graph because both are methods by which each vertex of the graph is visited (labeled) exactly once. Note that ''visit'' is used in a technical sense; merely considering a vertex in an algorithm does not necessarily constitute a visit.

Let us apply depth-first search to a binary tree by starting at the root and always choosing a left child of a vertex when there is a choice. We consider a vertex to be visited when it is labeled, and keep a list of the vertices in the order visited. This will give a traversal of the binary tree. This traversal is called a **preorder traversal** and the order in which the vertices are listed is called a **preorder listing.**

A characterization of the preorder traversal is that we visit a parent before its children and a left child before a right child. (This holds for all the vertices in the binary tree.) A systematic procedure for visiting the vertices in this order is provided by the depth-first search algorithm, which we state below in an alternate version that is consistent with the descriptions of the other traversals we will discuss. This is a **recursive** formulation of the preorder traversal, which means that in this description the algorithm refers to itself. This is analogous to the definition given for $n!$ in Section 2.5.

Preorder Traversal Algorithm

This algorithm gives a preorder listing of the vertices in a binary tree.

Step 1   (visit). Visit the root.

Step 2   (go left). Go to the left subtree, if one exists, and do a preorder traversal.

Step 3   (go right). Go to the right subtree, if one exists, and do a preorder traversal.

EXAMPLE 4.27   For the binary tree in Figure 4.50(a), we start by visiting the root A. (We use the word ''visit'' to indicate when a vertex should be listed, and in the figures we show the order of visiting in parentheses near the vertex.) Then we go to the left subtree of A (see Figure 4.50(b)) and start the preorder traversal again. Now we visit the root B and go to the left subtree of B (see Figure 4.50(c)), where we start another preorder traversal. Next we visit the root D. Since there is no left subtree of D, we go to the right subtree of D (which consists of just the vertex F), and again start a preorder traversal. Thus, we visit the root F. Since there are no subtrees of F, we have completed the preorder traversal of the left subtree of B. Consequently, we next begin a preorder traversal of the right subtree of B (see Figure 4.50(d)). To do this, we visit the root E and then go to the left subtree of E (which consists of only the vertex G) to begin another preorder traversal. Thus, we visit vertex G. Since G has no subtrees and E has no right subtree, both subtrees of B are traversed. This completes the traversal of the left subtree of A, and so we begin another preorder traversal on the right subtree of A. This consists only of visiting the root C and so completes the preorder traversal of the entire binary tree. The resulting preorder listing is A, B, D, F, E, G, C with a labeling of the vertices shown in Figure 4.50(e). ■

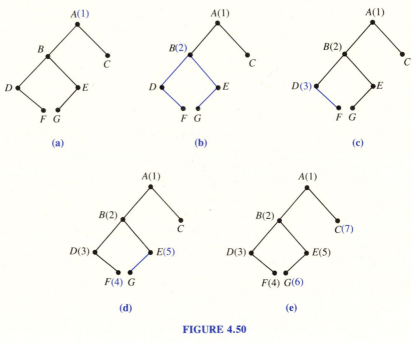

FIGURE 4.50

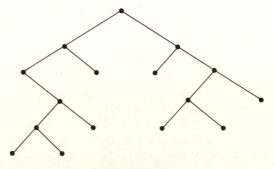

FIGURE 4.51

EXAMPLE 4.28   Applying preorder traversal to the binary tree in Figure 4.51 yields the order of visiting shown in Figure 4.52. ■

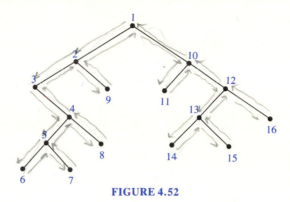

FIGURE 4.52

When a preorder traversal is performed on an expression tree, the resulting listing of operations and operands is called the **prefix form** or **Polish notation** for the expression. (The latter name is used in honor of the famous Polish logician Lukasiewicz.) For example, the four expressions in Examples 4.23, 4.24, 4.25, and 4.26 have as their Polish notations $* a\, b$, $+ a * b\, c$, $+ a * d - b\, c$, and $- + a * b\, c - f / d\, e$, respectively. An expression in Polish notation is evaluated according to the following rule: Scan from left to right until coming to an operation sign, say T, that is followed by two successive numbers, say a and b. Evaluate $T\, a\, b$ as $a\, T\, b$, and replace $T\, a\, b$ by this value in the expression. Repeat this process until the entire expression is evaluated.

EXAMPLE 4.29   The expression $(2 - 3 * 4) + (4 + 8/2)$ is represented by the expression tree in Figure 4.53. The Polish notation for this expression (found by doing a preorder traversal on the expression tree) is $+ - 2 * 3\, 4 + 4 / 8\, 2$. The evaluation is performed as follows.

First we evaluate $* 3\, 4$ and replace it by $3 * 4 = 12$. This substitution gives the new expression $+ - 2\, 12 + 4 / 8\, 2$.

Second we evaluate $- 2\, 12$ and replace $- 2\, 12$ by -10. This substitution yields the new expression $+ -10 + 4 / 8\, 2$, where we remember that the $-$ is part of -10 and is not a new operation.

Third we evaluate $/ 8\, 2$ and replace these symbols with 4. Thus, the current expression is $+ -10 + 4\, 4$.

Fourth we evaluate $+ 4\, 4$ as 8. The expression now has the form $+ -10\, 8$.

Fifth we evaluate $+ -10\, 8$ to obtain the final result, which is -2. ■

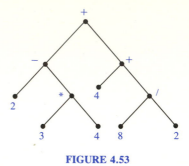

FIGURE 4.53

The Polish notation for an expression provides an unambiguous way to write it without the use of parentheses or conventions about the order of operations. Many computers are designed to rewrite expressions in this form.

Postorder Traversal

Readers who are familiar with hand calculators know that some require algebraic expressions to be entered in a form known as **reverse Polish notation** or **postfix form,** also introduced by Lukasiewicz. Unlike the Polish notation, in which the operation sign precedes the operands, in reverse Polish notation the operation sign follows the operands. The reverse Polish notation for the expression in Example 4.29 is $2\ 3\ 4\ * - 4\ 8\ 2\ / +\ +$. It is evaluated in a fashion similar to Polish notation except that as we scan from left to right we look for two numbers immediately followed by an operation sign. The steps in evaluating the expression above are

$2\ 3\ 4\ * - 4\ 8\ 2\ / +\ +$ (First $3\ 4\ *$ is evaluated and replaced.)

$2\ 12 - 4\ 8\ 2\ / +\ +$ (Next $2\ 12\ -$ is evaluated and replaced.)

$-10\ 4\ 8\ 2\ / +\ +$ (Then $8\ 2\ /$ is evaluated and replaced. Note that the first symbol is part of the number -10 and not an operation sign.)

$-10\ 4\ 4\ +\ +$ (Next $4\ 4\ +$ is evaluated and replaced.)

$-10\ 8\ +$ (Finally we evaluate $-10\ 8\ +$ to obtain the final result.)

-2

Again we see that we can evaluate an expression without the need for parentheses and without worrying about the order of operations. Thus, reverse Polish notation is an efficient method for use in hand calculators and computers. How can the reverse Polish notation for an expression be obtained from an expression tree?

By using a traversal called **postorder** we can obtain the reverse Polish notation for an expression. The postorder traversal is characterized by visiting children before the parent and a left child before a right child. (This holds for all the vertices in the binary tree.) A systematic way to do this is described in the following recursive algorithm.

Postorder Traversal Algorithm

This algorithm gives a postorder listing of the vertices of a binary tree.

Step 1 (go left). Go to the left subtree, if one exists, and do a postorder traversal.

Step 2 (go right). Go to the right subtree, if one exists, and do a postorder traversal.

Step 3 (visit). Visit the root.

EXAMPLE 4.30 For the binary tree in Figure 4.54(a), we begin by going to the left subtree of A (see Figure 4.54(b)) and begin postorder traversal again. Thus, we go to the left subtree of B (see Figure 4.54(c)) and start postorder traversal again. Since there is no left subtree of D, we go to the right subtree of D (which consists only of the vertex F) and begin postorder traversal again. Because F has no subtrees, we visit F. Since the right subtree of D has been traversed, we next visit D. (Again we use the word "visit" to indicate when a vertex should be listed, and in the figures show the order of visiting in parentheses near the vertex.) Now the left subtree of B has been traversed; so we go to the right subtree of B (see Figure 4.54(d)) and start postorder traversal again. Next we go to the left subtree of E (which is just G) and start postorder traversal again. Since G has no subtrees, G is visited. Because the left subtree of E has been traversed and there is no right subtree of E, we visit E. Now both the left and right subtrees of B have been traversed; so we visit B. This completes the traversal of the left subtree of A, and so we go to the right subtree of A and start postorder traversal again. Because C has no subtrees, C is visited. Since both subtrees of A have been traversed, we visit A. This completes the postorder listing F, D, G, E, B, C, A with a labeling of the vertices shown in Figure 4.54(e). ∎

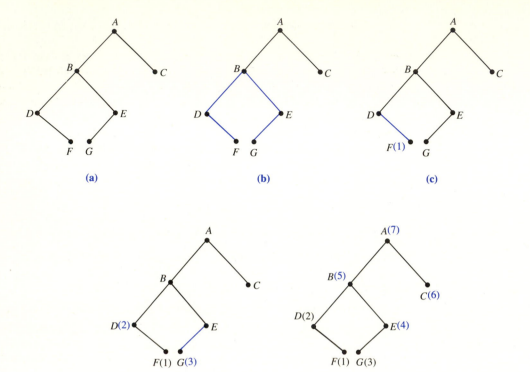

FIGURE 4.54

EXAMPLE 4.31   The postorder traversal applied to the binary tree in Figure 4.51 yields the order of visiting shown in Figure 4.55. ■

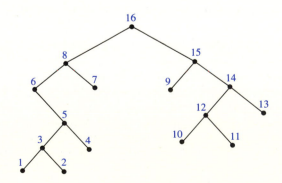

FIGURE 4.55

Inorder Traversal

We have seen how expression trees yield the Polish and reverse Polish notations for an expression. In these notations the operation sign precedes or follows the operands, respectively. With the use of the inorder traversal it is possible to obtain an expression with the operation sign between the operands. However, this traversal requires the careful insertion of parentheses in order to evaluate the expression properly.

The **inorder traversal** is characterized by visiting a left child before the parent and a right child after the parent. (This holds for all the vertices in the binary tree.) A systematic way to do this is described in the following recursive algorithm.

Inorder Traversal Algorithm

This algorithm gives an inorder listing of the vertices of a binary tree.

Step 1 (go left). Go to the left subtree, if one exists, and do an inorder traversal.

Step 2 (visit). Visit the root.

Step 3 (go right). Go to the right subtree, if one exists, and do an inorder traversal.

EXAMPLE 4.32   For the binary tree in Figure 4.56(a), we begin by going to the left subtree of *A* (see Figure 4.56(b)) and then start inorder traversal again. Next we go to the left subtree of *B* (see Figure 4.56(c)) and start inorder traversal again. Since there is no left subtree of *D,* we visit the root *D*. (Again we use the word "visit" to indicate when a vertex should be listed and in the figures show the order of visiting in parentheses near the vertex.) Then we go to the right subtree of *D* (which is just the vertex *F*) and start inorder traversal again. Since there is no left subtree of *F*, we visit the root *F*. Since *F* has no right subtree, we have traversed the left subtree of *B*. So we visit *B* and go to the right subtree of *B* (see Figure 4.56(d)) and do inorder traversal again. Thus, we go to the left subtree of *E* (which is just the vertex *G*) and start inorder traversal again. Since *G* has no left subtree, we visit *G*. Because *G* has no right subtree, we have traversed the left subtree of *E*. Thus, we visit *E*. Since *E* has no right subtree, we have traversed the right subtree of *B*. We have now traversed the left subtree of *A*. Thus, we visit the root *A* and go to the right subtree of *A* (which consists only of the vertex *C*) and begin inorder traversal again. Since *C* has no left subtree, we visit *C*. This step completes the inorder traversal giving the inorder listing *D, F, B, G, E, A, C* with the labeling of vertices shown in Figure 4.56(e). ■

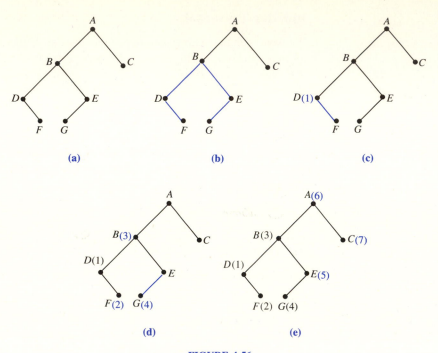

FIGURE 4.56

EXAMPLE 4.33   When inorder traversal is applied to the binary tree in Figure 4.51 the vertices are listed according to the numbering in Figure 4.57. ■

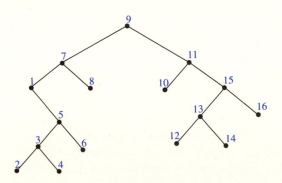

FIGURE 4.57

EXAMPLE 4.34   Applying inorder traversal to the expression tree in Figure 4.53 yields the expression $2 - 3 * 4 + 4 + 8/2$. ■

Other uses of traversals can be found in [6] in the suggested readings at the end of the chapter.

EXERCISES 4.5

In Exercises 1–6, construct an expression tree for each expression.

1. $a * b + c$

2. $(4 + 2) * (6 - 8)$

3. $((a - b) / c) * (d + e / f)$

4. $(((6 - 3) * 2) + 7) / ((5 - 1) * 4 + 8)$

5. $a * (b * (c * (d * e + f) - g) + h) + j$

6. $(((4 * 2) / 3) - (6 - 7)) + (((8 - 9) * 8) / 5)$

In Exercises 7–12 find the indicated subtrees.

7. the left subtree of vertex A

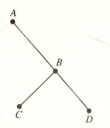

8. the right subtree of vertex A

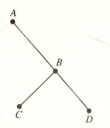

9. the left subtree of vertex C

10. the right subtree of vertex E

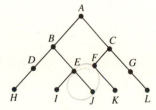

11. the left subtree of vertex E

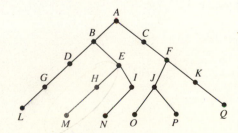

12. the right subtree of vertex D

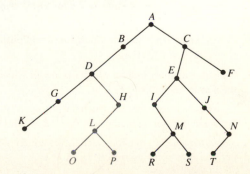

In Exercises 13–18 give the preorder listing of vertices for the binary trees in the indicated exercises.

13. Exercise 7　　　　**14.** Exercise 8　　　　**15.** Exercise 9

16. Exercise 10　　　　**17.** Exercise 11　　　　**18.** Exercise 12

In Exercises 19–24 give the postorder listing of vertices for the binary trees in the indicated exercises.

19. Exercise 7　　　　**20.** Exercise 8　　　　**21.** Exercise 9

22. Exercise 10　　　　**23.** Exercise 11　　　　**24.** Exercise 12

In Exercises 25–30 give the inorder listing of vertices for the binary trees in the indicated exercises.

25. Exercise 7　　　　**26.** Exercise 8　　　　**27.** Exercise 9

28. Exercise 10　　　　**29.** Exercise 11　　　　**30.** Exercise 12

In Exercises 31–36 find the Polish notation for the expressions in the indicated exercises.

31. Exercise 1　　　　**32.** Exercise 2　　　　**33.** Exercise 3

34. Exercise 4　　　　**35.** Exercise 5　　　　**36.** Exercise 6

In Exercises 37–42 find the reverse Polish notation for the expressions in the indicated exercises.

37. Exercise 1　　　　**38.** Exercise 2　　　　**39.** Exercise 3

40. Exercise 4　　　　**41.** Exercise 5　　　　**42.** Exercise 6

Evaluate the Polish notation expressions in Exercises 43–46.

43. $+ / 4 2 + 5 6$

44. $* + - + 4 3 6 2 8$

45. $+ * 4 / 6 2 - + 4 2 5$

46. $+ * + 3 4 - 1 2 - 3 / 4 2$

Evaluate the reverse Polish notation expressions in Exercises 47–50.

47. $4 5 - 7 * 2 3 + +$

48. $5 6 4 2 2 / + * -$

49. $2 3 + 4 6 - - 5 * 4 +$

50. $3 4 + 1 2 - * 4 2 / 3 - +$

51. Construct an expression tree for the Polish notation expression $* + B D - A C$.

52. Construct an expression tree for the Polish notation expression $* + B - D F + * A C E$.

53. Construct an expression tree for the reverse Polish notation expression $A C * B - D +$.

54. Construct an expression tree for the reverse Polish notation expression $E D - A + B C - F * +$.

55. Construct a binary tree for which the preorder listing of vertices is C, B, E, D, A and the inorder listing is B, E, C, A, D.

56. Construct a binary tree for which the preorder listing of vertices is E, C, A, D, B, F, G, H and the inorder listing is A, C, D, E, F, B, G, H.

57. Construct a binary tree for which the postorder listing of vertices is E, B, F, C, A, D and the inorder listing is E, B, D, F, A, C.

58. Construct a binary tree for which the postorder listing of vertices is D, H, F, B, G, C, A, E, and the inorder listing is D, F, H, E, B, A, G, C.

59. Construct a binary tree with 7 vertices for which the preorder listing is the same as the inorder listing.

60. Construct a binary tree with 8 vertices for which the postorder listing is the same as the inorder listing.

61. Construct a binary tree for which the preorder listing is the same as the postorder listing.

62. Construct two distinct binary trees which have 1, 2, 3 as their preorder listing of vertices.

63. Construct two distinct binary trees which have 1, 2, 3 as their postorder listing of vertices.

64. Verify for $n = 1, 2$, and 3 that the number of binary trees with n vertices is $\frac{(2n)!}{n!(n + 1)!}$. (Such numbers are called Catalan numbers.)

65. Draw a line around the outside of a binary tree, starting at the root, moving counterclockwise, and staying as close to the tree as possible. List a vertex the first time it is passed. This gives the preorder listing of the vertices. Illustrate this technique with the binary tree in Exercise 11.

66. Repeat the second part of Exercise 65 for the binary tree in Exercise 10.

67. Draw a line as described in Exercise 65. When must a vertex be listed to yield the postorder listing of vertices? Illustrate your statement with the binary tree in Exercise 11.

68. Repeat the second part of Exercise 67 for the binary tree in Exercise 10.

69. Draw a line as described in Exercise 65. When must a vertex be listed to yield the inorder listing of vertices? Illustrate your statement with the binary tree in Exercise 11.

70. Repeat the second part of Exercise 69 for the binary tree in Exercise 10.

71. Prove that if vertex X is a descendant of vertex Y in a binary tree, then Y precedes X in the preorder listing of vertices and X precedes Y in the postorder listing.

72. Prove that if the preorder and the inorder listings of vertices of a binary tree are given, then it is possible to reconstruct the binary tree.

73. The **Fibonacci trees** are defined recursively as follows: each of T_1 and T_2 is a vertex and for $n \geq 3$, T_n is the tree where the left subtree of the root is T_{n-1} and the right subtree is T_{n-2}. Find and prove a formula for the number of vertices in T_n.

4.6   Optimal Binary Trees and Binary Search Trees

In this section we present two applications, both of which require the construction of a binary tree to solve a problem. These can be studied in either order.

Optimal Binary Trees

To represent symbols, computers use strings of 0's and 1's called codewords. For example, in the ASCII (American Standard Code for Information Interchange) code, the letter A is represented by the codeword 01000001, B by 01000010, and C by 01000011. In this system each symbol is represented by some string of eight bits, where a bit is either a 0 or a 1. To translate a long string of 0's and 1's into its ASCII symbols we use the following procedure: Find the ASCII symbol represented by the first 8 bits, the ASCII symbol represented by the second 8 bits, etc. For example, 010000110100000101000010 is decoded as CAB.

For many purposes this kind of representation works well. However, there are situations, as in large-volume storage, where this is not an efficient method. In a fixed length representation, such as ASCII, every symbol is represented by a codeword of the same length. A more efficient approach is to use codewords of variable lengths, where the symbols used most often have shorter codewords than the symbols used less frequently. For example, in normal English usage the letters E, T, O, and A are used much more frequently than the letters Q, J, X, and Z. Is there a way to assign the shortest codewords to the most frequently used symbols? If messages use only these eight letters, a natural assignment to try is:

$$E: 0, \qquad T: 1, \qquad O: 01, \qquad A: 11,$$
$$Q: 00, \qquad J: 10, \qquad X: 101, \qquad Z: 011.$$

In this way we have assigned the shortest possible codewords to the most frequently used letters and longer codewords to the other letters. This appears to be a more efficient approach than assigning all these letters a codeword of the same fixed length, which would have to be three or more. (Why?)

But how can we decode a string of 0's and 1's? For example, how should the string 0110110 be decoded? Should we start by looking at only the first digit, or the first two, or the first three? Depending upon the number of digits used, the first letter could be E, O, or Z. We see that in order to use variable length codewords we need to select representations that permit unambiguous decoding.

A way to do this is to construct codewords so that no codeword is the first part of any other codeword. Such a set of codewords is said to have the **prefix property.** This property is not enjoyed by the above choice of codewords since the codeword for T is also the first part of the codeword for A. On the other hand, the set of codewords $S = \{000, 001, 01, 10, 11\}$ has the prefix property since no codeword appears as the first part of another codeword. The method to decode a string of 0's and 1's into codewords having the prefix property is to read one digit at a time until this string of digits becomes a codeword, then repeat this starting with the next digit, and continue until the decoding is done. For example, using the set of codewords S above, we would decode the string 001100100011 as 001,

10, 01, 000, 11. Thus, an efficient method of representation should use codewords such that (1) the codewords have the prefix property; and (2) the symbols used frequently have shorter codewords than those used less often.

Any binary tree can be used to construct a set of codewords with the prefix property by assigning 0 to each edge from a parent to its left child and 1 to each edge from a parent to its right child. Following the unique directed path from the root to a terminal vertex will give a string of 0's and 1's. The set of all strings formed in this way will be a set of codewords with the prefix property, because given any codeword, we can find the unique directed path to which it corresponds by working down from the root of the binary tree, going left or right according as each digit is 0 or 1. By definition we finish at a terminal vertex, and so this codeword cannot be the first part of another codeword.

EXAMPLE 4.35 For the binary tree in Figure 4.58(a) we assign 0's and 1's to its edges as shown in Figure 4.58(b). The directed paths from the root to all the terminal vertices then produce the codewords 000, 001, 01, 10, 11 as illustrated in Figure 4.59, where each codeword is written below the corresponding terminal vertex. ■

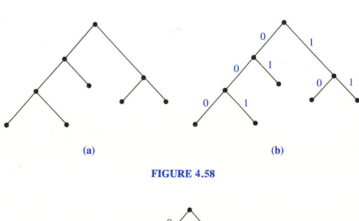

(a)   (b)

FIGURE 4.58

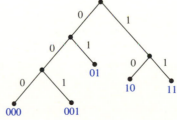

FIGURE 4.59

Thus, by using a binary tree we have found a way to produce codewords that have the prefix property. It remains to find a method for assigning shorter codewords to the more frequently used symbols. If we have only the 5 symbols in Figure 4.59, then we want to use the codewords 01, 10, and 11 for the three most frequently used symbols. Notice that these codewords correspond to the terminal vertices that are closest to the root. Thus, to obtain an efficient method for representing symbols by variable length codewords, we can use a binary tree and assign the most frequently used symbols to the terminal vertices that are closest to the root.

We will restrict our discussion to those binary trees for which every internal vertex has exactly 2 children. Suppose $w_1, w_2, \ldots, w_k$ are non-negative real numbers. A **binary tree for the weights** $w_1, w_2, \ldots, w_k$ is a binary tree with k terminal vertices labeled $w_1, w_2, \ldots, w_k$. A binary tree for the weights $w_1, w_2, \ldots, w_k$ has **weight** $d_1 w_1 + d_2 w_2 + \ldots + d_k w_k$, where d_i is the length of the directed path from the root to the vertex labeled w_i ($i = 1, \ldots, k$).

EXAMPLE 4.36 The binary tree in Figure 4.60(a) is a binary tree for the weights 2, 4, 5, 6 and has weight $3 \cdot 6 + 3 \cdot 5 + 2 \cdot 4 + 1 \cdot 2 = 43$. In Figure 4.60(b) is another binary tree for the weights 2, 4, 5, 6, but it has a smaller weight $2(2 + 4 + 5 + 6) = 34$, since the distance from the root to each terminal vertex is 2. ■

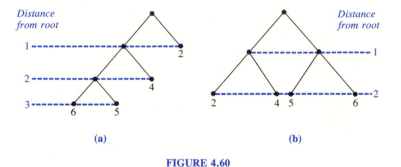

FIGURE 4.60

For the coding problem we want to find a binary tree of smallest possible weight in which the frequencies of the symbols to be encoded are the weights. A binary tree for the weights $w_1, w_2, \ldots, w_k$ is called an **optimal binary tree for the weights** $w_1, w_2, \ldots, w_k$ when its weight is as small as possible. Thus, the binary tree in Figure 4.60(a) is not an optimal tree for the weights 2, 4, 5, 6 since there is another binary tree with smaller weight, namely that in Figure 4.60(b).

The following algorithm due to David A. Huffman produces an optimal binary tree for the weights $w_1, w_2, \ldots, w_k$. The idea is to create a binary tree by using the two smallest weights, say w_1 and w_2, replace w_1 and w_2 by $w_1 + w_2$ in the list of weights, and then repeat the process.

Huffman's Optimal Binary Tree Algorithm

For non-negative real numbers $w_1, w_2, \ldots, w_k$, this algorithm constructs an optimal binary tree for the weights $w_1, w_2, \ldots, w_k$.

Step 1 (select smallest weights). If there are two or more weights in the list of weights, select the two smallest weights, say V and W. (Ties can be broken arbitrarily.) Otherwise, we are done.

Step 2 (make binary tree). Construct a binary tree with the root assigned the label $V + W$, its left child assigned the label V, and its right child assigned the label W. Include in this construction any binary trees that have the labels V or W assigned to a root. Replace V and W in the list of weights by $V + W$. Go to step 1.

EXAMPLE 4.37

FIGURE 4.61

For weights 2, 3, 4, 7, 8 we begin the construction of an optimal binary tree by choosing the weights 2 and 3 and forming the binary tree shown in Figure 4.61. (For convenience, we will refer to the vertices by their labels.) The new weights are 5*, 4, 7, 8, where an asterisk is placed on the 5 to remind us that 5* is the root of a binary tree. Next we choose 5* and 4 and form the binary tree in Figure 4.62(a) and then the one in Figure 4.62(b), where we have included the binary tree for which 5* is a root. The weights are now 9*, 7, 8. Repeating this process, we now choose 7 and 8 and form the binary tree shown in Figure 4.63. Note that now we have two binary trees as part of our construction process, the one in Figure

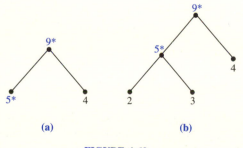

(a)   (b)

FIGURE 4.62

FIGURE 4.63

4.62(b) and the one in Figure 4.63. The weights are now 9*, 15*. Choosing both weights gives the binary tree in Figure 4.64(a), and when the other two binary trees are included, we have the one in Figure 4.64(b). From this we can then construct an optimal binary tree for the weights 2, 3, 4, 7, 8 as shown in Figure 4.65. The weight of this tree is $2(4 + 7 + 8) + 3(2 + 3) = 53$. ■

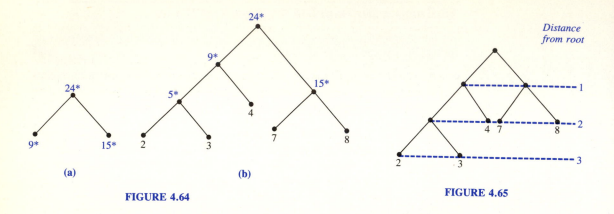

FIGURE 4.64

Distance from root

FIGURE 4.65

EXAMPLE 4.38   Using Huffman's optimal binary tree algorithm, we can construct an optimal binary tree for weights 2, 4, 5, 6 in the steps shown in Figure 4.66. This tree has weight $1 \cdot 6 + 2 \cdot 5 + 3(2 + 4) = 34$. Note that if in step (b) we had chosen weights 5 and 6 instead of 6* and 5 we would have obtained a different optimal binary tree, namely the one in Figure 4.60(b).

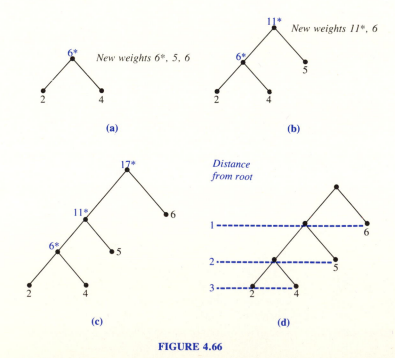

FIGURE 4.66

An analysis of this algorithm requires knowledge of sorting and inserting algorithms that we have not studied. Thus, we will state without proof (see suggested reading [3] at the end of the chapter) that Huffman's optimal binary tree algorithm is of order at most k^2, where k is the number of weights. A proof that the algorithm constructs an optimal binary tree is found in Exercises 45–47.

In order to find codewords with the prefix property such that the most frequently used symbols are assigned the shortest codewords, we construct an optimal binary tree with the stated frequencies of the symbols as its weights. Then by assigning 0's and 1's to the edges of this tree as described in Example 4.35, codewords can be efficiently assigned to the various symbols.

EXAMPLE 4.39   Suppose the characters E, T, A, Q, and Z have expected usage rates of 32, 28, 20, 4, and 1, respectively. In Figure 4.67 we see an optimal binary tree with weights 1, 4, 20, 28, 32 created by Huffman's optimal binary tree algorithm. Furthermore, each symbol has been placed in parentheses next to its usage rate. Then 0's and

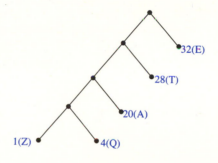

FIGURE 4.67

1's are assigned to the edges of the trees so that codewords with the prefix property are formed at the terminal vertices. (See Figure 4.68.) Thus, we see that E should be assigned the codeword 1, T should be assigned the codeword 01, A should be assigned 001, Q should be assigned 0001, and Z should be assigned 0000. ■

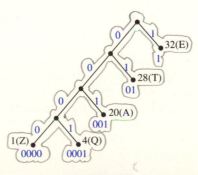

FIGURE 4.68

A binary tree describing the codewords can also be used to decode a string of 0's and 1's. To do so, we start with the digits in the string and follow the directed path from the root indicated by the 0's and 1's. When a terminal vertex is reached, the string is then decoded by the codeword at that vertex. Then this process is begun again at the root with the next digit. For example, to decode the string 00101 with the tree in Figure 4.68, we start at the root and go to the left child, then to the left child again, and then to the right child, which is a terminal vertex with the codeword 001 and symbol A. Then we go back to the root and decode the remaining bits 01, which correspond to the symbol T.

Another application of Huffman's optimal binary tree algorithm arises in regard to the merging of sorted lists. Suppose we have two sorted lists of numbers L_1 and L_2 that we want to merge together into one sorted list. Recall from Theorem 2.15 that if L_1 and L_2 have n_1 and n_2 numbers, respectively, then these two sorted lists can be merged into one sorted list with at most $n_1 + n_2 - 1$ comparisons. Now suppose we have 3 sorted lists L_1, L_2, and L_3 containing 150, 320, and 80 numbers, respectively. By merging only two lists at a time, how can we merge these 3 lists into one sorted list so that we minimize the number of comparisons needed? One way is to merge L_1 and L_2 into a third list with 470 numbers, which requires at most 469 comparisons. Then we merge this new list with L_3, which requires at most $470 + 80 - 1 = 549$ comparisons. Altogether this merge pattern requires at most $469 + 549 = 1018$ comparisons. This process can be represented by the binary tree in Figure 4.69(a), where the labels represent the sizes of the lists. A second merge pattern is to merge L_1 and L_3 first followed by merging this new list with L_2. This merge pattern requires at most $229 + 549 = 778$ comparisons and is represented in Figure 4.69(b). Finally, we can merge L_2

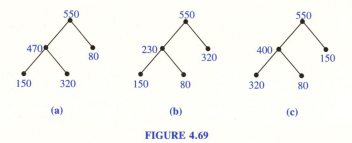

(a)   (b)   (c)

FIGURE 4.69

and L_3 first and then merge the result with L_1. This requires at most $399 + 549 = 948$ comparisons and is represented in Figure 4.69(c). Thus, we see that the second merge pattern requires the fewest comparisons. Furthermore, we observe that the optimal pattern of merging (the one requiring the fewest comparisons) occurs when the smallest lists are used first, since in this way fewer comparisons are made overall. In fact the number of times an item is sorted is the distance of its list from the root in the binary trees shown in Figure 4.69. Thus, we wish to minimize the weight of this tree, where the weight of a vertex is the number of items in the corresponding list. Our optimal merge pattern can be found by following the construction in Huffman's optimal binary tree algorithm.

EXAMPLE 4.40   In order to sort five sorted lists with 20, 30, 40, 60, and 80 numbers optimally, we begin by merging together the two sorted lists with the smallest number of items. These are the sorted lists with 20 and 30 numbers, giving a new sorted list of 50 numbers and using at most 49 comparisons. Now we consider the four sorted lists with 50, 40, 60, and 80 numbers and combine the lists with 50 and 40 items; this merging yields a new sorted list of 90 items using at most 89 comparisons. Next, from the three sorted lists with 90, 60, and 80 numbers, we merge the lists with 60 and 80 numbers to obtain a new sorted list of 140 items using at most 139 comparisons. Finally we combine the sorted lists with 90 and 140 numbers giving one sorted list with 230 numbers and using at most 229 comparisons. This optimal merge pattern uses at most 506 comparisons. A binary tree representation of this optimal merge pattern is given in Figure 4.70.   ▪

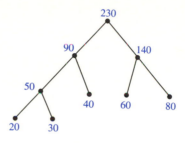

FIGURE 4.70

Binary Search Trees

Maintaining a large data set is a common problem for data processors. This consists not only of updating the data set by adding and deleting, but also of searching the data for a particular piece of information. Suppose, for instance, that the Acme Manufacturing Company maintains a list of its customers. When an order is received, the company must search this list to determine if the order is from an old or a new customer. If the order is from a new customer, then this customer's name must be added to the list. Moreover, when a customer goes out of business, that customer's name must be removed from the list.

One way to maintain such lists is to keep the data in the order in which they are received. For example, if Acme Manufacturing has ten customers named Romano, Cohen, Moore, Walters, Smith, Armstrong, Garcia, O'Brien, Young, and Tucker, they can keep these names in an array in the given order. This method enables items to be added to the list easily; for if Jones becomes a customer, this new name can be added to the end of the existing array. However, this method makes it very time consuming to determine if a particular name is in the list. Determining that Kennedy is not a customer, for instance, will require checking every name in the list. Of course the amount of checking required is minimal when Acme has only ten customers, but if Acme has a million customers, checking every name on the list is prohibitive.

Another approach is to keep the list in alphabetical order. For example, Acme Manufacturing's list of customer names can be stored as Armstrong, Cohen, Garcia, Moore, O'Brien, Romano, Smith, Tucker, Walters, and Young. With this method, it is easy to search the list for a particular name. (The procedure used in the proof of Theorem 2.14 can be adapted to give an efficient searching method.) However, adding or deleting from the list is more difficult because of the need to reposition the entries when an item is added or deleted. For example, if Acme gains a new customer named Baker, then we need to insert this name as the second entry in the list. This insertion requires repositioning every name in the original list except Armstrong's. Again this process is prohibitive if the list is very long.

Another approach is to store data at the vertices of a binary tree. For example, the list of Acme Manufacturing's customer names can be stored as in Figure 4.71.

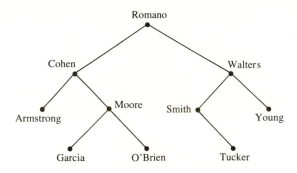

FIGURE 4.71

This binary tree is arranged so that if a vertex U belongs to the left subtree of vertex V, then U precedes V in alphabetical order; and if a vertex W lies in the right subtree of V, then W follows V in alphabetical order. Adding a new name to this tree is simple because we need only include one new vertex and edge in the tree, and searching the tree for a particular name requires no more than four comparisons if we search the tree properly.

In order to generalize the example above, suppose that we have a list of distinct numbers or words. We will use the symbol $\leq$ to denote the usual numerical or alphabetical (dictionary) order. For example, $7 \leq 9$ and ABGT $\leq$ ACE. A **binary search tree** for the list is a binary tree in which each vertex is labeled by an element of the list such that:

(1) No two vertices have the same label.
(2) If vertex U belongs to the left subtree of vertex V, then $U \leq V$.
(3) If vertex W belongs to the right subtree of vertex V, then $V \leq W$.

Thus, for each vertex V, all descendants of V in the left subtree of V precede V, and all descendants of V in the right subtree of V follow V.

EXAMPLE 4.41   One possible binary search tree for 1, 2, 4, 5, 6, 8, 9, 10 is given in Figure 4.72(a). Another possibility is shown in Figure 4.72(b). ■

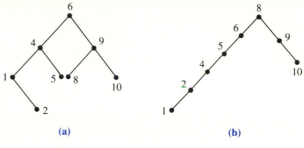

(a)   (b)

FIGURE 4.72

EXAMPLE 4.42   A binary search tree for IF, THEN, FOR, BEGIN, END, WHILE, TO is illustrated in Figure 4.73. ■

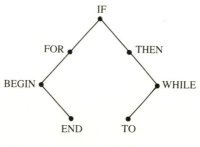

FIGURE 4.73

There is a systematic way to construct a binary search tree for a list. The basic idea is to put smaller elements as left children and larger elements as right children.

Binary Search Tree Construction Algorithm

This algorithm constructs a binary search tree with vertices labeled a_1, $a_2, \ldots , a_n$, where $a_1, a_2, \ldots , a_n$ are distinct. In the algorithm we refer to a vertex by its label.

Step 1 (start). Construct a root and label it a_1. If $n = 1$, we are done; otherwise, let $V = a_1$ and $k = 2$.

Step 2 (if smaller, go left). If $V \leq a_k$, go to step 3. Otherwise, we have $a_k \leq V$.
   (a) If V has no left child, construct a left child L for V and label L with a_k. If $k = n$, we are done; otherwise, increase k by 1, set $V = a_1$, and go to step 2.
   (b) Otherwise, if V has a left child L, set $V = L$, and go to step 2.

Step 3 (if larger, go right). We have $V \leq a_k$.
   (a) If V has no right child, construct a right child R for V and label R with a_k. If $k = n$, we are done; otherwise, increase k by 1, set $V = a_1$, and go to step 2.
   (b) Otherwise, if V has a right child R, set $V = R$, and go to step 2.

The construction described in the binary search tree construction algorithm does produce a binary tree. Furthermore, labels for the left descendants (those on the left side) are smaller than the label for the parent, and labels for the right descendants are larger. Thus, the algorithm does yield a binary search tree.

EXAMPLE 4.43   Using the binary search tree construction algorithm on the list 5, 9, 8, 1, 2, 4, 10, 6 results in the construction shown in Figure 4.74. ■

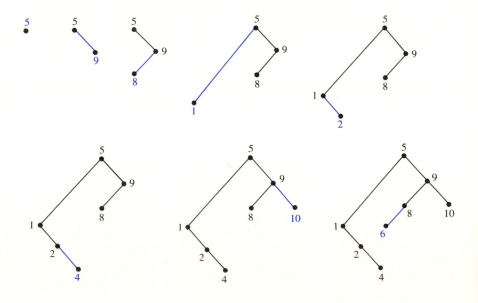

FIGURE 4.74

EXAMPLE 4.44   For the list of words in the sentence DISCRETE MATH IS FUN BUT HARD, the algorithm yields the binary search tree in Figure 4.75.   ■

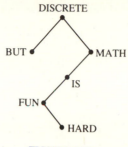

FIGURE 4.75

If an additional item is to be added to the binary search tree, then we can simply use the binary search tree construction algorithm one more time with that item. For example, to add SOMETIMES to the end of the list of words in the sentence DISCRETE MATH IS FUN BUT HARD, we would repeat the algorithm using the word SOMETIMES with the binary search tree in Figure 4.75 to obtain the one in Figure 4.76. This procedure for adding an item to a binary search tree is an efficient one.

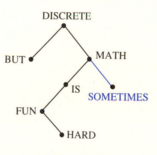

FIGURE 4.76

To determine if an item is in a binary search tree (or equivalently, a list), we follow closely the process used to construct a binary search tree. Specifically, we compare the item with the root and go left if it is smaller and go right if it is larger. This process is repeated until we either match some item in the tree or find that the item is not in the tree. This procedure is formalized in the following algorithm.

Binary Search Tree Search Algorithm

This algorithm will examine a binary search tree to decide if a given element *a* is in the tree.

Step 1 (start). Let V be the root of the binary search tree.

Step 2 (compare). If $a = V$, then a is in the tree; stop. Otherwise go to step 3.

Step 3 (if smaller, go left). If $V \leq a$, then go to step 4. Otherwise, $a \leq V$.
  (a) If there is no left child of V, then a is not in the tree; stop.
  (b) Otherwise V has a left child L; let $V = L$ and go to step 2.

Step 4 (if larger, go right). We have $V \leq a$.
  (a) If there is no right child of V, then a is not in the tree; stop.
  (b) Otherwise V has a right child R; let $V = R$ and go to step 2.

EXAMPLE 4.45   Let us use this algorithm to search for 7 in the binary search tree of Figure 4.74. We begin by comparing 7 to the root of the tree. Since it is larger, we go to the right child 9 and make another comparison. This time 7 is smaller, and so we go to the left child 8. Comparing 7 to 8, we go to the left child 6. Now comparing 7 to 6, we go to the right child of 6. Since there is none, 7 is not in the binary search tree. We also note that if we wanted to add 7 to this binary search tree at this time, it would become the right child of 6.   ■■

EXAMPLE 4.46   To search for the word FUN in the tree of Figure 4.75, we begin with the root DISCRETE. The first comparison takes us to the right child MATH. From there another comparison takes us to the left child IS. Again we go to the left child FUN. This comparison results in a match. Hence, we find that FUN is in the tree.   ■■

Applying the inorder traversal to the binary search tree in Figure 4.74 yields the inorder listing 1, 2, 4, 5, 6, 8, 9, 10, which is the usual numerical order for these numbers. Similarly, for the binary search tree in Figure 4.73 the inorder traversal gives the listing BEGIN, END, FOR, IF, THEN, TO, WHILE, which is the alphabetical order for these words. In general, when the inorder traversal is applied to a binary search tree, the resulting listing is the usual ordering of the elements. From this the smallest and largest elements of the tree can be found. Thus, by going left as far as possible in a binary search tree, the last vertex reached is the smallest element in the tree. Similarly by going right as much as possible, the last vertex reached is the largest element in the tree.

Deletions of items from a binary search tree can also be done efficiently. The details are left as exercises.

The construction of a binary search tree depends upon the order in which the items appear in the list. In other words, a different order for the items can produce a different binary search tree. For example, for the list 10, 9, 8, 6, 5, 4, 2, 1, which is the same set of items as in Example 4.43, the algorithm produces the binary search tree shown in Figure 4.77. It is easy to see that this tree offers no advantage over numerical order in storing the numbers since a search may require a comparison with every item in the tree. However, there is an extensive literature on the construction of binary trees that make for efficient searching. For example, binary search trees can be constructed so that the more frequently accessed items are closer to the root than those items that are not. Interested readers should consult suggested readings [6] and [8] at the end of the chapter.

FIGURE 4.77

EXERCISES 4.6

In Exercises 1–4 determine if the given sets of codewords have the prefix property.

1. {0, 100, 101, 11, 1011}   **2.** {00, 11, 010, 100, 011}

3. {00, 101, 111, 10001, 1010}   **4.** {00, 110, 101, 01}

5. Can there be a set of 6 codewords with the prefix property that contains 0, 10, and 11?

6. Can there be a set of 6 codewords with the prefix property that contains 10, 00, and 110?

7. Determine values for a, b, and c so that the set {00, 01, 101, a10, bc1} of 5 codewords has the prefix property.

8. Determine values for a, b, c, and d so that the set {00, 0a0, 0bc, d0, 110, 111} of 6 codewords has the prefix property.

For the values of n given in Exercises 9–14, draw a binary tree in which each vertex has 0 or 2 children that generates as in Example 4.35 a set of n codewords with the prefix property. Label the vertices with the codewords.

9. $n = 2$

10. $n = 3$

11. $n = 4$

12. $n = 7$

13. $n = 8$

14. $n = 9$

In Exercises 15–18 draw a binary tree that generates as in Example 4.35 the given codewords at the terminal vertices.

15. 1, 00, 011, 0100, 0101

16. 101, 00, 11, 011, 100, 010

17. 1111, 0, 1110, 110, 10

18. 1100, 000, 1111, 1101, 0010, 10, 0011

19. Decode the message 111010100000010110 with the assignment A: 010, B: 111, M: 000, N: 110, and T: 10.

20. Decode the message 110010111011110 with the assignment O: 0, B: 10, R: 110, I: 1110, N: 11110, and T: 11111.

21. Decode the message 100101110101001011 with the assignment A: 111, E: 0, N: 1010, O: 1011, and T: 100.

22. Decode the message 00111100010000111 with the assignment B: 1100, D: 111, E: 1101, J: 0011, N: 0000, O: 01, S: 0010, and T: 0001.

In Exercises 23–26 decode the messages using the given binary tree.

23. message: 01110111

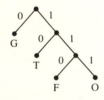

24. message: 11001001011000

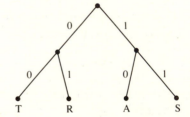

25. message: 0111111011101001100

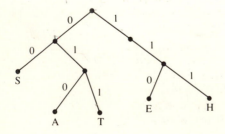

26. message: 000010010011000011101001

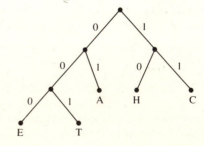

Locate a copy of the ASCII code in a computer programming book and use it to decode the messages given in Exercises 27–30.

27. 010001000100111101000111

28. 010010000100111101001101010000101

29. 01010001010101010100100101000010101010100

30. 010010000100010101001100010101010000

In Exercises 31–34 construct an optimal binary tree for the given weights. In the construction select a weight with an asterisk in preference to an equal weight without an asterisk, and use as the left child the smaller weight or the weight with an asterisk if two weights are equal.

31. 2, 4, 6, 8, 10

32. 4, 6, 8, 14, 15

33. 1, 4, 9, 16, 25, 36

34. 10, 12, 13, 16, 17, 17

In Exercises 35–38 determine the maximum number of comparisons needed to merge sorted lists with the given numbers of items into one sorted list.

35. 4 lists with 20, 30, 40, and 50 items

36. 5 lists with 15, 25, 35, 40, and 50 items

37. 6 lists with 20, 40, 60, 70, 80, and 120 items

38. 7 lists with 10, 30, 40, 50, 50, 50, and 70 items

In Exercises 39–42 in the construction of an optimal binary tree select a weight with an asterisk in preference to an equal weight without an asterisk, and use as the left child the smaller weight or the weight with an asterisk if two weights are equal.

39. The National Security Agency is helping American diplomats in foreign countries send coded messages back to the State Department in Washington, D.C. These messages are to be sent using the characters R, I, H, V with an expected usage rate of 40, 35, 20, 5, respectively, per 100 characters. Find an assignment of codewords that minimizes the number of bits needed to send a message.

40. Tom and Susan are exchanging love letters during class. In order to prevent others from reading these sweet words of romance, the messages are coded using only the characters T, A, I, L, P, and J with an expected usage rate of 34, 27, 21, 10, 6, 2, respectively, per 100 characters. Find an assignment of codewords that minimizes the amount of time (and hence, the number of bits) needed to send a message.

41. NASA is receiving information from one of its space probes. This information is in the form of numbers which represent pictures. (Each number corresponds to a shade of white, black, or gray.) The numbers used are 1, 2, 3, 4, 5, 6, 7, 8, 9, and 10 with expected usage rates of 125, 100, 75, 40, 60, 180, 20, 120, 150, and 130 per 1000 colored dots. Find an assignment of codewords for these numbers that minimizes the number of bits needed for the storage of this information.

42. The Gregory Computer Company has received a contract to store nursing data for all the hospitals in the Bloomington, Illinois, area. Even though the storage of this data will be on hard disks, the high volume of data makes it important that the data be stored efficiently. An analysis of sample data shows that only certain symbols are used: rn, @, c, s, po, os, od, tid, qod. Furthermore, the analysis shows a usage rate of 7, 12, 4, 9, 10, 8, 2, 18, 30, respectively, per 100 symbols. Find an assignment of codewords that minimizes the number of bits needed to store this data.

43. Prove that there exists a binary tree with n terminal vertices in which each vertex has 0 or 2 children.

44. In a binary tree in which each vertex has 0 or 2 children, prove that the number of terminal vertices is one more than the number of internal vertices.

Exercises 45–47 provide a proof that Huffman's optimal binary tree algorithm does generate an optimal binary tree. Suppose $w_1, w_2, \ldots, w_k$ are nonnegative real numbers and we assume $w_1 \leq w_2 \leq \ldots \leq w_k$ for convenience.

45. Prove that if T is an optimal binary tree for the weights $w_1, w_2, \ldots, w_k$, and if $w_i < w_j$, then the distance from the root to w_i is greater than or equal to the distance from the root to w_j.

46. Prove that there is an optimal binary tree for the weights $w_1, w_2, \ldots, w_k$ where w_1 and w_2 are children of the same parent.

47. Prove that if T is an optimal binary tree for the weights $w_1 + w_2, w_3, \ldots, w_k$, then the tree obtained by replacing the terminal vertex $w_1 + w_2$ by a binary tree with two children w_1 and w_2 is an optimal binary tree for the weights $w_1, w_2, \ldots, w_k$.

In Exercises 48–53 construct a binary search tree for the items in the order given.

48. The accounting department in the Busby Insurance Company has 8 divisions with 11, 15, 8, 3, 6, 14, 19, and 10 staff members in them. Construct a binary search tree for the number of staff in these units.

49. Reserved words in Pascal include LABEL, SET, OR, BEGIN, THEN, END, GOTO, DO, PACKED, and ELSE. Construct a binary search tree for these reserved words.

50. Predefined identifiers in Apple Pascal include ORD, CHR, WRITE, SEEK, PRED, EOF, WRITELN, BOOLEAN, PAGE, GET, TRUE, COPY, PUT, and ABS. Construct a binary search tree for these predefined identifiers.

51. The mathematics department has 13 faculty members with 14, 17, 3, 6, 15, 1, 20, 2, 5, 10, 18, 7, and 16 years of teaching experience. Construct a binary search tree for the years of teaching experience by the faculty.

52. In a survey of 15 mathematics departments it was found that there were 18, 9, 27, 20, 30, 15, 4, 13, 25, 31, 2, 19, 7, 5, and 28 faculty members. Construct a binary search tree for the sizes of the faculty.

53. ASCII code is used to represent more than just the alphabet. It is also used to represent the symbols), :, %, $-$, #, <, @, ?, $, (, !, and &. The corresponding ASCII codewords can be interpreted as binary numbers (with decimal values 41, 58, 37, 45, 35, 60, 64, 63, 36, 40, 33, and 38, respectively) and, hence, can be used to provide an ordering of these symbols. Construct a binary search tree for these symbols.

54. Construct a binary search tree for the letters of the alphabet so that at most 5 comparisons are needed to locate any specified letter.

55. In the binary search tree of Exercise 49 draw the directed path required to show that FILE is not in the tree. Then indicate where FILE would be added to the tree.

56. In the binary search tree of Exercise 48 draw the directed path required to show that 16 is not in the tree. Then indicate where 16 would be added to the tree.

57. In the binary search tree of Exercise 51 draw the directed path required to show that 4 is not in the tree. Then indicate where 4 would be added to the tree.

58. In the binary search tree of Exercise 50 draw the directed path required to show that POS is not in the tree. Then indicate where POS would be added to the tree.

59. In the binary search tree of Exercise 53 draw the directed path required to show that
$>$ (with decimal number 62) is not in the tree. Then indicate where $>$ would be
added to the tree.

60. In the binary search tree of Exercise 52 draw the directed path required to show that 8
is not in the tree. Then indicate where 8 would be added to the tree.

61. Suppose that the vertices of a binary tree are assigned distinct elements from a list of
either numbers or words with the property: If L is the left child of a vertex V, then
$L \leq V$ and if R is the right child of a vertex V, then $V \leq R$. Must the binary tree with
this assignment be a binary search tree for the list?

*Deletion of a terminal vertex V from a binary search tree is accomplished as follows:
Delete the vertex V and the edge on V and its parent.*

62. Draw the binary search tree obtained by deleting 6 from the binary search tree in
Exercise 48.

63. Repeat Exercise 62 for PACKED in the binary search tree in Exercise 49.

*When a binary search tree has a root R with only one child, deletion of the root is
accomplished as follows: Delete R and the edge on R and its child.*

64. Draw the binary search tree obtained by deleting the root from the following binary
search tree.

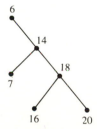

65. Repeat Exercise 64 for the binary search tree below.

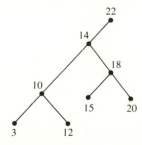

*Suppose that V is a vertex in a binary search tree such that V is not the root and V has only
one child C. Deletion of V from the binary search tree is accomplished as follows: Delete
V and the edge on V and C, and replace the tree formed by V and its descendants by the
tree formed by C and its descendants.*

66. Draw the binary search tree obtained by deleting $<$ from the binary search tree in
Exercise 53.

67. Repeat Exercise 66 for the vertex 3 in the binary search tree of Exercise 48.

In a binary search tree, deletion of a vertex V with 2 children is accomplished as follows: Find the largest item L in the left subtree of V. If L has no left child, delete L and the edge on L and its parent and replace V by L. If L has a left child C, delete L and the edge on L and C, replace the tree formed by L and its descendants by the tree formed by C and its descendants, and then replace V by L.

68. Draw the tree obtained by deleting 9 from the binary search tree in Exercise 52.

69. Draw the tree obtained by deleting 3 from the binary search tree in Exercise 51.

70. Draw the tree obtained by deleting ORD from the binary search tree in Exercise 50.

71. Draw the tree obtained by deleting WRITE from the binary search tree in Exercise 50.

72. Draw the tree obtained by deleting 27 from the binary search tree in Exercise 52.

73. Draw the tree obtained by deleting 4 from the binary search tree in Exercise 52.

74. Prove that the inorder listing of the vertices in a binary search tree gives the natural order for the elements in the tree.

Suggested Readings

1. Aho, Alfred V. John E. Hopcroft, and Jeffrey D. Ullman. *Data Structures and Algorithms*. Reading, MA: Addison-Wesley, 1983.
2. Bogart, Kenneth P. *Introductory Combinatorics*. Marsfield, MA: Pitman, 1983.
3. Horowitz, Ellis and Sartaj Sahni. *Fundamentals of Computer Algorithms*. Rockville, MD.: Computer Science Press, 1978.
4. Horowitz, Ellis and Sartaj Sahni. *Fundamentals of Data Structures in Pascal*. Rockville, MD.: Computer Science Press, 1984.
5. Hu, T. C. *Combinatorial Algorithms*. Reading, MA: Addison-Wesley, 1982.
6. Knuth, Donald E. *The Art of Computer Programming, vol. 1: Fundamental Algorithms*. 2nd ed. Reading, MA: Addison-Wesley, 1973.
7. Liu, C. L. *Elements of Discrete Mathematics*. 2nd ed. New York: McGraw-Hill, 1985.
8. Reingold, Edward, Jurg Nievergelt, and Narsingh Deo. *Combinatorial Algorithms*. Englewood Cliffs, NJ: Prentice-Hall, 1977.
9. Stubbs, Daniel and Neil W. Webre. *Data Structures with Abstract Data Types and Pascal*. Monterey, CA.: Brooks/Cole, 1985.
10. Tarjan, Robert Endrdre. *Data Structures and Network Algorithms*. Philadelphia, PA.: SIAM, 1983.

5

MATCHING

Many combinatorial problems involve matching items, subject to certain restrictions. An example is the problem of assigning airline pilots to flights introduced in Section 1.2. Another example is the assignment of pairs of participants at a conference to rooms so that roommates have the same smoking preference and sex. Sometimes an optimal matching may be desired. For example, a basketball coach must assign a player to guard each player on the opposing team in such a way as to minimize the opponents' total score. Such problems will be treated in this chapter.

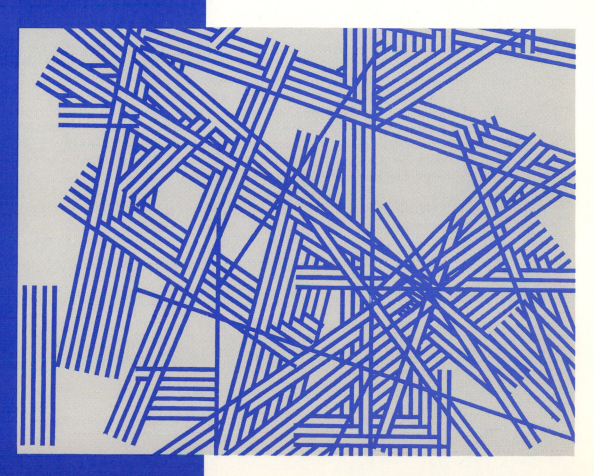

5.1   Systems of Distinct Representatives

The same matching problem may be viewed in various ways. As an example, let us consider the summer schedule of classes of the English department at a small college. There is a demand for 6 courses. To keep things simple we will call these Course 1, Course 2, . . . , Course 6. Certain professors are available to teach each course, as given in the following table.

Course	Professors
1	Abel, Crittenden, Forcade
2	Crittenden, Donohue, Edge, Gilmore
3	Abel, Crittenden
4	Abel, Forcade
5	Banks, Edge, Gilmore
6	Crittenden, Forcade

For brevity we will denote the professors by A, B, C, D, E, F, and G, according to their initials. In order to distribute the summer teaching jobs as fairly as possible, it is decided that no professor should teach more than one course. The question is whether all 6 courses can be taught, subject to this restriction. If not, what is the maximum number of courses that can be taught?

This is a problem of exactly the same sort as that of assigning airline pilots in Section 1.2. With only 6 courses and 7 professors we could probably find the answer by considering all possible matchings. One systematic way of doing this is the following. Let P_1 denote the set of professors available to teach Course 1, P_2 the set of professors available to teach Course 2, etc. Thus,

$$P_1 = \{A, C, F\},$$
$$P_2 = \{C, D, E, G\},$$
$$P_3 = \{A, C\},$$
$$P_4 = \{A, F\},$$
$$P_5 = \{B, E, G\},$$
$$P_6 = \{C, F\}.$$

If we forget for the moment the restriction that no professor teach more than one course, then a possible assignment of a professor to each course consists of a 6–tuple $(x_1, x_2, x_3, x_4, x_5, x_6)$, where $x_1 \in P_1$, $x_2 \in P_2$, etc. This is an element of the Cartesian product

$$P_1 \times P_2 \times P_3 \times P_4 \times P_5 \times P_6,$$

which has $3 \cdot 4 \cdot 2 \cdot 2 \cdot 3 \cdot 2 = 288$ elements. We need to know whether any of these 288 6–tuples has all its entries distinct (so that no professor teaches more than one course). Checking this without the help of a computer would be possible but extremely tedious. As in the case of the pilot assignment problem, however, such crude methods of searching for a solution quickly get beyond the capability of even a computer as the number of items to be matched gets larger. For example, if there were 30 courses and 3 professors available for each, then the Cartesian product would contain 3^{30} elements, and it would take a computer checking one million of these per second more than 6 years to go through them all.

There is a name for the sort of sequence of distinct elements, one from each of a given sequence of sets, that we are seeking in this example. Let $S_1, S_2, \ldots, S_n$ be a finite sequence of sets, not necessarily distinct. By a **system of distinct representatives** for $S_1, S_2, \ldots, S_n$ we mean a sequence $x_1, x_2, \ldots, x_n$ such that $x_i \in S_i$ for $i = 1$ to n, and such that the elements x_i are all distinct.

EXAMPLE 5.1   Find all systems of distinct representatives for the sets $S_1 = \{1, 2, 3\}$, $S_2 = \{1, 3\}$, $S_3 = \{1, 3\}$, $S_4 = \{3, 4, 5\}$.

Notice that the elements chosen from S_2 and S_3 must be 1 and 3 in some order. There are four systems of distinct representatives:

$$2, 1, 3, 4$$
$$2, 3, 1, 4$$
$$2, 1, 3, 5$$
$$2, 3, 1, 5.$$

EXAMPLE 5.2   Find all systems of distinct representatives for the sets $S_1 = \{2, 3\}$, $S_2 = \{2, 3, 4, 5\}$, $S_3 = \{2, 3\}$, $S_4 = \{3\}$.

There are none. For if x_1, x_2, x_3, x_4 was a system of distinct representatives, then x_1, x_3, and x_4 would be 3 distinct elements of $S_1 \cup S_3 \cup S_4 = \{2, 3\}$, which is impossible.

EXAMPLE 5.3   How many systems of distinct representatives does the sequence S, S, S, S have, where $S = \{1, 2, 3, 4\}$?

In this case a system of distinct representatives is simply a permutation of the integers $1, 2, 3, 4$. By Theorem 1.1 there are exactly $4! = 24$ of these.

Hall's Theorem

Now we return to our problem of assigning a professor to each summer English course. We are looking for a system of distinct representatives for the sequence

$$P_1 = \{A, C, F\},$$
$$P_2 = \{C, D, E, G\},$$
$$P_3 = \{A, C\},$$
$$P_4 = \{A, F\},$$
$$P_5 = \{B, E, G\},$$
$$P_6 = \{C, F\}.$$

The problem seems small enough that we might expect to find the solution, if there is one, by simply trying different combinations. Yet perhaps the best we can come up with is to cover 5 of the six courses. For example, we might assign the first 5 courses as in the list below.

Course 1 to Abel

Course 2 to Donohue

Course 3 to Crittenden

Course 4 to Forcade

Course 5 to Banks

We might suspect that it is not possible to do better than this, but not be certain. We would like a way to convince ourselves that no assignment of all 6 courses is possible without going through all 288 possibilities.

There is a way, and the key to it is to be found in Example 5.2. If we could discover some collection of sets chosen from P_1 through P_6, the union of which contained fewer elements than the number of sets in the collection, then we would know that a system of distinct representatives was impossible. Since this is a somewhat abstract idea, we will exhibit such a collection to make the argument more concrete. How such a collection might be found will be covered in a later section of this chapter.

The collection we have in mind is P_1, P_3, P_4, and P_6. Notice that

$$P_1 \cup P_3 \cup P_4 \cup P_6 = \{A, C, F\},$$

and the argument is the same as in Example 5.2. Suppose we had a system of distinct representatives $x_1, x_2, \ldots, x_6$. Then x_1, x_3, x_4, and x_6 would comprise 4 distinct elements lying in the union of the sets P_1, P_3, P_4, and P_6. But this is impossible because this union only contains 3 elements. There are only 3 professors (Abel, Crittenden, and Forcade) available to teach 4 of the courses, and so an assignment where no professor teaches more than one course cannot be made.

We have found a general principle which could be stated as follows. Suppose $S_1, S_2, \ldots, S_n$ is a finite sequence of sets, and suppose I is a subset of $\{1, 2, \ldots, n\}$ such that the union of the sets S_i for $i \in I$ contains fewer elements than the set I does. Then $S_1, \ldots, S_n$ has no system of distinct representatives. In our example (taking $S_i = P_i$ for $i = 1$ to 6), the set I is $\{1, 3, 4, 6\}$.

Finding such a set I enables us to be sure that no system of distinct representatives exists. The person responsible for assigning summer courses in our example will have to have the same professor teach two courses if all 6 courses are to be given. Professors with no summer employment may object that this is unfair, but the scheduler can use the set I to demonstrate to them that there is no way to cover all the courses otherwise.

If a sequence of sets has no system of distinct representatives, is there always some set I as above that can be used to demonstrate this fact in a compact way? The answer is yes, but the proof is somewhat complicated. This is the content of a famous theorem due to Philip Hall.

THEOREM 5.1   *Hall's Theorem*   The sequence of finite sets $S_1, S_2, \ldots, S_n$ has a system of distinct representatives if and only if whenever I is a subset of $\{1, 2, \ldots, n\}$, then the union of the sets S_i for $i \in I$ contains at least as many elements as the set I does.

The "only if" part of this theorem amounts to the principle we have already discovered. The "if" part will be proved in Section 5.4.

EXAMPLE 5.4   We will use Hall's theorem to show that the sequence

$$S_1 = \{A, C, E\},$$

$$S_2 = \{A, B\},$$

$$S_3 = \{B, E\}$$

has a system of distinct representatives. The subsets I of $\{1, 2, 3\}$ and the corresponding unions of sets S_i are given below.

I	Union of sets S_i, $i \in I$
$\varnothing$	$\varnothing$
$\{1\}$	$\{A, C, E\}$
$\{2\}$	$\{A, B\}$
$\{3\}$	$\{B, E\}$
$\{1, 2\}$	$\{A, B, C, E\}$
$\{1, 3\}$	$\{A, B, C, E\}$
$\{2, 3\}$	$\{A, B, E\}$
$\{1, 2, 3\}$	$\{A, B, C, E\}$

Since every set on the right has at least as many elements as the corresponding set on the left, the sequence has a system of distinct representatives. Of course, it is easy in this case to find one by inspection, for example A, B, E. ■

Applying Hall's theorem to our course scheduling example would involve examining the $2^6 = 64$ subsets of $\{1, 2, 3, 4, 5, 6\}$ and computing the corresponding union of sets P_i for each. (Of course, we would find that no system of distinct representatives exists.) Although this may seem better than our previous method, which entailed looking at 288 possible assignments, it is still not practical for finding whether a system of distinct representatives exists, since if there are n sets S_i, then there are 2^n sets I, and 2^n increases very quickly with n. Also, although the theorem tells us when a system of distinct representatives exists, it does not tell how to find one. Efficient methods for finding optimal matchings will be developed later in this chapter.

Readers interested in extensions of Hall's theorem should consult [7] in the suggested readings at the end of this chapter.

EXERCISES 5.1

In Exercises 1–6 tell how many systems of distinct representatives the given sequence of sets has.

1. $\{1, 2\}, \{2, 3\}, \{1, 3\}$

2. $\{1, 4\}, \{2\}, \{2, 3\}, \{1, 2, 3\}$

3. $\{1, 2, 3\}, \{1, 2, 3\}, \{1, 2, 3\}$

4. $\{1, 2, 3, 4, 5\}, \{1, 2, 3, 4, 5\}$

5. $\{1, 2, 5\}, \{2, 1\}, \{3, 4\}, \{1, 5\}, \{1, 2, 5\}, \{2, 4, 5\}$

6. $\{1, 2, 3\}, \{4, 5\}, \{6, 7\}$

In Exercises 7–10 a sequence of sets $S_1, \ldots, S_n$ is given. For each subset I of $\{1, 2, \ldots, n\}$ compute the union of the corresponding sets S_i, and determine from these unions whether the sequence has a system of distinct representatives or not.

7. $\{1, 2, 4\}, \{2, 4\}, \{2, 3\}, \{1, 2, 3\}$

8. $\{1, 2, 5\}, \{5, 1\}, \{1, 2\}, \{2, 5\}$

9. $\{1\}, \{1, 2\}, \{1, 2, 3\}, \{1, 3\}$

10. $\{4.5, 5, 7\}, \{9/2, 6, 7\}, \{\varnothing\}$

In Exercises 11–16 a sequence of sets $S_1, \ldots, S_n$ is given. Find a subset I of $\{1, 2, \ldots, n\}$ such that the union of the corresponding sets S_i has fewer elements than I does.

11. $\{1, 2\}, \{2, 3\}, \varnothing$

12. $\{1\}, \{1, 2\}, \{2, 3\}, \{2\}$

13. $\{1, 2, 3\}, \{1, 2, 4\}, \{1, 3, 4\}, \{1, 2, 3, 4\}, \{2, 3, 4\}$

14. $\{1, 2\}, \{2, 3\}, \{5\}, \{1, 3\}, \{4, 5\}, \{4, 5\}$

15. $\{2, 5, 7\}, \{1, 3, 4, 5\}, \{5, 7\}, \{2, 7\}, \{1, 3, 6\}, \{2, 5\}$

16. $\{1, 2\}, \{2, 4, 5, 7\}, \{1, 2, 3, 5, 6\}, \{1, 4, 7\}, \{2, 5, 7\}, \{1, 4, 5, 7\}, \{2, 4, 7\}$

17. Let $S_i = \{1, 2, \ldots, n\}$ for $i = 1, 2, \ldots, n$. How many systems of distinct representatives does the sequence $S_1, \ldots, S_n$ have?

18. Let $S_i = \{1, 2, \ldots, k\}$ for $i = 1, 2, \ldots, n$, where $n \leq k$. How many systems of distinct representatives does $S_1, \ldots, S_n$ have?

19. Let $S_i = \{1, 2, \ldots, k\}$ for $i = 1, 2, \ldots, n$, where $k < n$. How many systems of distinct representatives does $S_1, \ldots, S_n$ have?

20. Show that if the nonempty set S_i has k_i elements for $i = 1, 2, \ldots, n$, then the sequence $S_1, S_2, \ldots, S_n$ has exactly $k_1 k_2 \cdots k_n$ systems of distinct representatives if and only if the sets S_i are disjoint.

21. Mr. Jones brought home 6 differently flavored jelly beans for his 6 children. However, when he got home he found out that each child only likes certain flavors. Amy will only eat chocolate, banana, or vanilla, while Burt only likes chocolate and banana. Chris will only eat banana, strawberry, and peach, and Dan will only accept banana and vanilla. Edsel only likes chocolate and vanilla, and Frank only will eat chocolate, peach, and mint. Show that not every child will get a jellybean he or she likes.

22. Five girls go into a library to get a book. Jennifer only wants to read *The Velvet Room* or *Daydreamer*. Lisa only wants *Summer of the Monkeys* or *The Velvet Room*. Beth and Kim each only want *Jelly Belly* or *Don't Hurt Laurie!*, while Kara wants either one of the latter two books or else *Daydreamer*. If the library only has one copy of each book, can each girl take out a book she wants?

23. Show that if the union of the sets $S_1, \ldots, S_n$ contains more than n elements, and if the sequence $S_1, \ldots, S_n$ has a system of distinct representatives, then it has more than one.

24. Let S be a set with m elements, and let $S_i = S$ for $i = 1, 2, \ldots, n$. Show that the number of systems of distinct representatives of $S_1, \ldots, S_n$ is the same as the number of one-to-one functions from $\{1, 2, \ldots, n\}$ into $\{1, 2, \ldots, m\}$.

25. In the example in Section 1.2 there are 7 cities and a set of pilots who want to fly to each city. Either find a system of distinct representatives for this sequence of sets or else prove that none exists.

26. A sequence of sets $S_1, S_2, \ldots, S_n$ is said to satisfy **Hall's condition** if whenever $I \subseteq \{1, 2, \ldots, n\}$, then the number of elements in the union of the sets S_i, $i \in I$, is at least $|I|$. The "if" part of Hall's theorem amounts to the statement that any sequence satisfying Hall's condition has a system of distinct representatives. Prove this by using the strong induction principle on n. To prove the inductive step, consider two cases: (a) if I is a nonempty subset of $\{1, 2, \ldots, k + 1\}$ with fewer than $k + 1$ elements, then the union of the sets S_i for $i \in I$ has at least one more element than I does; (b) for some nonempty subset I of $\{1, \ldots, k + 1\}$ with fewer than $k + 1$ elements, the union of the sets S_i for $i \in I$ has the same number of elements as I.

5.2   Matchings in Graphs

There is a symmetry in matching problems that is hidden when they are formulated in terms of sets as in Section 5.1. For example, when we were trying to match a professor with each English course, we associated with each of the 6 courses a set, the set of professors who could teach that course. But we could just as well have turned the problem around and considered for each professor the set of courses he or she can teach. This symmetry is displayed better if we draw a graph as we did in Figure 1.10 for the airline pilot problem. We will let the courses and professors be the vertices of the graph, and put an edge between a course and a professor whenever the professor can teach the course. The result is shown in Figure 5.1.

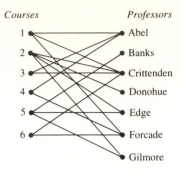

Courses   *Professors*

FIGURE 5.1

The graph we get is of a special form, since no edge joins a course to a course, or a professor to a professor. We say a graph with vertex set V and edge set E is **bipartite** in case V can be written as the union of two disjoint sets V_1 and V_2 such that each edge joins an element of V_1 with an element of V_2. The graph of Figure 5.1 is bipartite since we could take V_1 to be the set of courses and V_2 to be the set of professors.

EXAMPLE 5.5   The graph shown in Figure 5.2 is bipartite (even though it may not look it) because every edge goes between an odd-numbered vertex and an even-numbered one. Thus, we could take $V_1 = \{1, 3, 5, 7\}$ and $V_2 = \{2, 4, 6, 8\}$.   ■

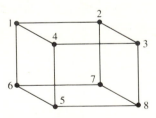

FIGURE 5.2

EXAMPLE 5.6   The graph shown in Figure 5.3 is not bipartite, as we can see by considering the vertices 1, 3, and 4. If, for example, 1 is in V_1, then 3 must be in V_2. But then 4 can be in neither of these sets.   ■

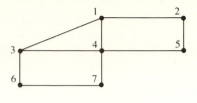

FIGURE 5.3

In our course assignment problem we wanted to pair up courses and professors. In terms of the graph representing the problem, this means that we want to choose a subset, say M, of the set of edges. No course can be taught by two professors, nor can a professor teach more than one course. This means that no vertex of the graph can be incident to more than one edge of M. In this application we would like M to contain as many edges as possible. These considerations motivate the following definitions.

Let G be a graph with vertices V and edges E. By a **matching** of G we mean a subset M of E such that no vertex in V is incident with more than one edge in M. By a **maximal matching** of G we mean a matching of G so that no other matching of G contains more edges.

EXAMPLE 5.7 The colored edges in Figure 5.4(a) form a matching of the bipartite graph pictured, since no two of them are incident with the same vertex. This matching of 3 edges is not maximal, however, since Figure 5.4(b) shows another matching with 4 edges. Note that even though the first matching is not maximal, no edge could be added to it and still have a matching. A maximal matching need not be unique. Figure 5.4(c) shows another maximal matching of the graph. ■

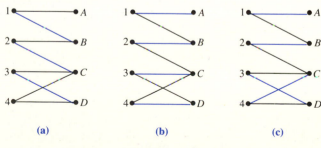

(a)   (b)   (c)

FIGURE 5.4

Our definition of a matching of a graph did not specify that the graph be bipartite. Finding a maximal matching is easier in the case of a bipartite graph, however, and many applications give rise to bipartite graphs. The following example gives a case when a maximal matching of a nonbipartite graph is desired.

EXAMPLE 5.8   A group of United Nations peace-keeping soldiers are to be divided into 2-person teams. It is important that the two members of a team speak the same language. The following table shows the languages spoken by the 7 soldiers available.

Soldier	Languages
1	French, German, English
2	Spanish, French
3	German, Korean
4	Greek, German, Russian, Arabic
5	Spanish, Russian
6	Chinese, Korean, Japanese
7	Greek, Chinese

If we make a graph, putting an edge between two soldiers whenever they speak a language in common, the result is exactly the graph of Figure 5.3, which we saw was not bipartite. One matching is pictured in Figure 5.5. It is clearly maximal since only one soldier is unmatched. ■

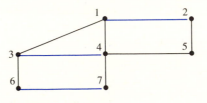

FIGURE 5.5

The Matrix of a Bipartite Graph

A convenient way to represent a bipartite graph is by a matrix of 0's and 1's, with the rows corresponding to the elements of V_1 and the columns to the elements of V_2. We put a 1 in the matrix whenever the vertices corresponding to the row and column are joined by an edge and a 0 otherwise. For example, the matrix of the graph of Figure 5.4 is

$$
\begin{array}{c c}
 & \begin{array}{cccc} A & B & C & D \end{array} \\
\begin{array}{c} 1 \\ 2 \\ 3 \\ 4 \end{array} &
\begin{bmatrix}
1 & 1 & 0 & 0 \\
0 & 1 & 1 & 0 \\
0 & 0 & 1 & 1 \\
0 & 0 & 1 & 1
\end{bmatrix}
\end{array}
$$

Of course, this matrix is uniquely determined only if we specify some order for the vertices in V_1 and V_2. Recall that a matching of a graph is a subset of its edges, and each edge corresponds to a 1 in the matrix. Two edges incident on the same vertex correspond to 1's in the same row or column of the matrix, depending on whether the vertex is in V_1 or V_2. Thus, a matching of a bipartite graph corresponds to some set of 1's in the matrix of the graph, no two of which are in the same row or column. Matrices have their own terminology, however.

By a **line** of a matrix we mean either a row or a column. Let A be a matrix. We say that a set of entries of A is **independent** if no two of them are in the same line. An independent set of 1's in A is a **maximal independent set** of 1's if no independent set of 1's in A contains more elements.

We will mark the 1's in a particular independent set with stars. The reader should check that the stars in the following three matrices mark independent sets corresponding to the three matchings shown in Figure 5.4.

$$\begin{bmatrix} 1 & 1^* & 0 & 0 \\ 0 & 1 & 1^* & 0 \\ 0 & 0 & 1 & 1^* \\ 0 & 0 & 1 & 1 \end{bmatrix} \quad \begin{bmatrix} 1^* & 1 & 0 & 0 \\ 0 & 1^* & 1 & 0 \\ 0 & 0 & 1^* & 1 \\ 0 & 0 & 1 & 1^* \end{bmatrix} \quad \begin{bmatrix} 1^* & 1 & 0 & 0 \\ 0 & 1^* & 1 & 0 \\ 0 & 0 & 1 & 1^* \\ 0 & 0 & 1^* & 1 \end{bmatrix}$$

For example, since one of the edges in the matching shown in Figure 5.4(a) is $\{1, B\}$, a star is placed on the 1 in row 1 and column B of the first matrix.

Although the language is different, finding a maximal matching in a bipartite graph and a maximal independent set of 1's in a matrix of 0's and 1's are really the same problem, and we will use whichever form is more convenient. Graphs are sometimes more accessible to the intuition, while matrices may be better for computational purposes.

Coverings

By a **covering** C of a graph we mean a set of vertices such that every edge is incident to at least one vertex in C. We say C is a **minimal covering** if no covering of the graph has fewer vertices. For example, the set $\{2, 3, 4, 5, 6\}$ may be seen to be a covering of the graph shown in Figure 5.5. This covering is not minimal, however, since the covering $\{1, 3, 5, 7\}$ has fewer elements.

EXAMPLE 5.9   Figure 5.6 represents the streets and intersections of the downtown area of a small city. A company wishes to place hot dog stands at certain intersections in such a way that no one in the downtown area will be more than one block from a stand. It would like to do this with as few stands as possible.

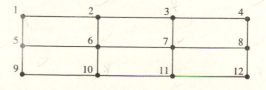

FIGURE 5.6

If we interpret Figure 5.6 as a graph with vertices at the intersections, then our problem is exactly one of finding a minimal covering. One covering is the set of vertices $\{1, 3, 6, 8, 9, 11\}$. We will see later that this covering is minimal. ▪

The next theorem gives a relation between the matchings and the coverings of a graph.

THEOREM 5.2   Suppose a graph has a matching M of m edges and a covering C of c vertices. Then $m \leq c$. Furthermore, if $m = c$, then M is maximal and C is minimal.

Proof.   By the definition of a covering every edge of the graph, and in particular every edge in M, is incident on some vertex in C. If the edge e is in M, let $v(e)$ be a vertex in C incident to e. Notice that if e_1 and e_2 are distinct edges in M, then $v(e_1)$ and $v(e_2)$ are also distinct, since by definition two edges in a matching cannot share a vertex. Thus, there are at least as many vertices in C as edges in M and so $m \leq c$.

Now suppose $m = c$. If M were not maximal there would be a matching with more than c edges, contradicting the first part of the theorem. Likewise, if C were not minimal there would be a covering with fewer than m vertices, leading to the same contradiction. ▪

In light of the second part of this theorem we can show the covering given in Example 5.9 is minimal by exhibiting a matching with the same number of elements, namely 6. One is given in Figure 5.7; of course, the theorem also implies that it is maximal.

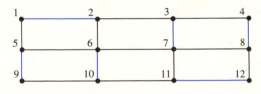

FIGURE 5.7

In the case of a bipartite graph we can make a translation of Theorem 5.2 into matrix language. The vertices of the graph correspond to lines of its matrix, and an edge is attached to a vertex when the 1 corresponding to the edge is in the line corresponding to the vertex. Thus, we define a **covering** of the 1's of a matrix of 0's and 1's to be a set of lines containing all the 1's of the matrix. Such a covering is **minimal** if there is no covering with fewer lines.

THEOREM 5.3 If a matrix of 0's and 1's has an independent set of m 1's and a covering of c lines, then $m \leq c$. If $m = c$, then the independent set is maximal and the covering is minimal.

EXAMPLE 5.10 The Scientific Matchmaking Service has as clients 5 men, Bob, Bill, Ron, Sam, and Ed, and 5 women, Cara, Dolly, Liz, Tammy, and Nan. The company believes that two people are not compatible if their first names do not contain a common letter. On the basis of this rule the company constructs the following matrix, in which a 1 means that the man and woman corresponding to the row and column are compatible.

$$
\begin{array}{c}
 \\
\text{Bob} \\
\text{Bill} \\
\text{Ron} \\
\text{Sam} \\
\text{Ed}
\end{array}
\begin{array}{ccccc}
\text{Cara} & \text{Dolly} & \text{Liz} & \text{Tammy} & \text{Nan} \\
\left[\begin{array}{ccccc}
0 & 1 & 0 & 0 & 0 \\
0 & 1 & 1 & 0 & 0 \\
1 & 1 & 0 & 0 & 1 \\
1 & 0 & 0 & 1 & 1 \\
0 & 1 & 0 & 0 & 0
\end{array}\right]
\end{array}
$$

The company would like to match as many clients as possible; that is, it wants a maximal independent set of 1's. Since all the 1's lie in just 4 lines, namely the 3rd and 4th rows and 2nd and 3rd columns, it is realized that no independent set of 1's can have more than 4 elements. An independent set with 4 elements does exist, however, and one is shown below.

$$
\begin{bmatrix}
0 & 1^* & 0 & 0 & 0 \\
0 & 1 & 1^* & 0 & 0 \\
1 & 1 & 0 & 0 & 1^* \\
1^* & 0 & 0 & 1 & 1 \\
0 & 1 & 0 & 0 & 0
\end{bmatrix}
$$ ■

EXERCISES 5.2

In Exercises 1–6 tell whether the graph is bipartite, and if so give a set V_1.

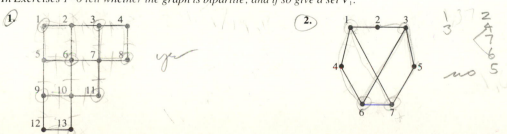

3.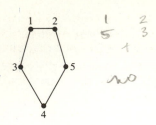

4.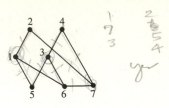

5.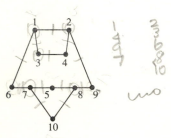

6.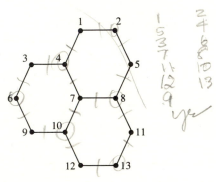

7. Give a maximal matching for each graph in Exercises 1, 2, and 3.

8. Give a maximal matching for each graph in Exercises 4, 5, and 6.

9. Give a minimal covering for each graph in Exercises 1, 2, and 3.

10. Give a minimal covering for each graph in Exercises 4, 5, and 6.

In Exercises 11–16 the graph given is bipartite with $V_1 = \{1, 3, 5, \ldots\}$ and $V_2 = \{2, 4, 6, \ldots\}$. Give the matrix of each graph. (Take the vertices in increasing order.)

11.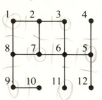

12.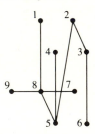

13.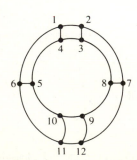

14.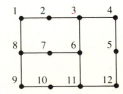

15.

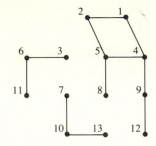

16.

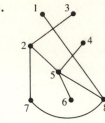

17. Find a maximal independent set of 1's for the matrices in Exercises 11, 12, and 13.

18. Find a maximal independent set of 1's for the matrices in Exercises 14, 15, and 16.

19. Find a minimal cover for the matrices of Exercises 11, 12, and 13.

20. Find a minimal cover for the matrices of Exercises 14, 15, and 16.

In Exercises 21 and 22 construct a bipartite graph and the corresponding matrix modeling the situation described. Indicate a maximal matching in the graph and the corresponding maximal independent set of 1's in the matrix.

21. Four airplane passengers want to read a magazine, but only 5 are available. Of these, Mr. Brown will only read *Time, Newsweek,* or *Fortune,* Ms. Garvey will only read *Newsweek* or *Organic Gardening,* Miss Rollo will only read *Organic Gardening* or *Time,* and Mrs. Onishi will only read *Fortune* or *Sunset.*

22. The Glumby family is going to Europe and each member is to choose one country that he or she knows the language of to study beforehand. Mr. Glumby knows Russian and French, Mrs. Glumby knows only Russian, Sally knows French, German, and Spanish, and Tim knows only French.

In Exercises 23 and 24 construct a graph modeling the situation described, and find a maximal matching for it.

23. The church sewing circle wants to break into 2–person groups to make altar cloths. The two people in a group should own the same brand of sewing machine. Ann has a Necchi, Betty has a Necchi and a Singer, Cora has a Necchi, a Singer, and a White, Debby has a Singer, a White, and a Brother, Eunice has a White and a Brother, and Frieda has a Brother.

24. The Weight Whittler Club wants to break up into 2–person support groups. The weights of the two men in a group should differ by no more than 20 pounds. Arthur weighs 185, Bob 250, Carl 215, Dan 210, Edgar 260, and Frank 205.

In Exercises 25 and 26 model the situation described with a bipartite graph, and construct the corresponding matrix. Find a minimal covering for the graph and indicate the corresponding lines of the matrix.

25. The police department has a policy of putting an experienced officer together with a rookie in a squad car. The experienced officers are Anderson, Bates, Coony, and Dotson, and the rookies are Wilson, Xavier, Yood, and Zorn. Anderson always works with Wilson or Xavier, Bates with Xavier, Yood, or Zorn, Coony with Wilson, and Dotson with Wilson or Xavier. The captain would like to call as small a number of officers as possible to tell each team its schedule for next month.

26. In mixed doubles Jimmy always plays with Crissy, Martina, or Pam, John always plays with Tracy, Ivan always plays with Crissy or Tracy, and Boris always plays with Crissy. The tournament director wants to tell each possible mixed-doubles pair its rating with as few phone calls as possible.

27. Show that the matrix of a bipartite graph is a submatrix of the adjacency matrix of that graph, as defined in Section 3.1. A **submatrix** of a matrix A is a matrix formed by removing some rows or columns (or both) from A.

28. Find a graph in which a maximal matching has fewer edges than a minimal covering has vertices.

29. Show that if a graph contains a cycle with an odd number of edges, then it is not bipartite.

30. Show that if a graph contains no cycle with an odd number of edges, then it is bipartite.

5.3   A Maximal Matching Algorithm

So far our examples have been small enough so that we could find a maximal matching by trial and error. For larger graphs, however, a better technique is needed; and, as was indicated in Section 5.1, simple exhaustion of all possibilities soon becomes impractical, even with a computer. There is an efficient algorithm for finding a maximal matching in a graph. For simplicity we consider the algorithm only for the case of a bipartite graph. To make explaining the algorithm easier, we will present it as a method of finding a maximal independent set of 1's in a matrix of 0's and 1's. As we saw in the previous section, this is equivalent to the problem of finding a maximal matching in a bipartite graph.

We will give an example of the use of the algorithm before we state it in a more formal way later in this section. We start with some independent set of 1's. This set could be found by inspection and could even be the empty set! Starting with a larger independent set will speed up finding a maximal such set, however. The algorithm will either tell us that the independent set of 1's we have is maximal, or else produce an independent set containing one more 1. We continue to apply the algorithm until a maximal independent set is reached.

For our example we will use the matrix

$$
\begin{array}{c c}
 & \begin{array}{cccc} A & B & C & D \end{array} \\
\begin{array}{c} 1 \\ 2 \\ 3 \\ 4 \end{array} &
\left[\begin{array}{cccc}
1^* & 0 & 1 & 1 \\
0 & 1^* & 0 & 0 \\
1 & 1 & 0 & 0 \\
0 & 1 & 0 & 0
\end{array} \right] ,
\end{array}
$$

in which an independent set of 1's has been indicated. Notice that if any 1 is added to this set it will no longer be independent. Our algorithm will involve performing two operations on some of the lines of this matrix, operations which we will call *labeling* and *scanning*. Once a line has been labeled it will never be labeled again in one application of the algorithm, and the same is true for scanning. A line must be labeled before it can be scanned. We begin by labeling, with the letter "*L*", all columns containing no starred 1's. (If there are no such columns our set of starred 1's is already maximal.) In our example this produces the following matrix.

$$
\begin{array}{c}
 \\
1 \\
2 \\
3 \\
4
\end{array}
\begin{array}{cccc}
A & B & C & D \\
\left[\begin{array}{cccc}
1^* & 0 & 1 & 1 \\
0 & 1^* & 0 & 0 \\
1 & 1 & 0 & 0 \\
0 & 1 & 0 & 0
\end{array}\right] \\
 & & L & L
\end{array}
$$

Now we scan each labeled column for unstarred 1's. In column *C* we find an unstarred 1 in the first row, so we label that row with a *C* to indicate that the unstarred 1 was found in column *C*. Then we put the letter "*S*" under column *C* to indicate that it has been scanned. The matrix now looks as follows.

$$
\begin{array}{c}
 \\
1 \\
2 \\
3 \\
4
\end{array}
\begin{array}{cccc}
A & B & C & D \\
\left[\begin{array}{cccc}
1^* & 0 & 1 & 1 \\
0 & 1^* & 0 & 0 \\
1 & 1 & 0 & 0 \\
0 & 1 & 0 & 0
\end{array}\right] \quad C \\
 & & LS & L
\end{array}
$$

When we scan column *D*, we also find an unstarred 1 in row 1. Since this row has already been labeled, we put the *S* under column *D* to indicate that it also has now been scanned.

Since all labeled columns have been scanned, we now turn our attention to the rows. Only row 1 has been labeled; so we scan it, now looking for starred 1's. There is one in column *A*; so we label this column with a 1 (the row scanned), and put an *S* after row 1 to show that it has been scanned.

$$
\begin{array}{c}
 \\
1 \\
2 \\
3 \\
4
\end{array}
\begin{array}{cccc}
A & B & C & D \\
\left[\begin{array}{cccc}
1^* & 0 & 1 & 1 \\
0 & 1^* & 0 & 0 \\
1 & 1 & 0 & 0 \\
0 & 1 & 0 & 0
\end{array}\right] \quad CS \\
1 & & LS & LS
\end{array}
$$

Since all labeled rows have been scanned, we go back to scanning columns. Column *A* is labeled but not scanned; so we scan it for unstarred 1's. There is one in row 3; so we label that row with an *A*, since we found it when scanning column *A*.

$$
\begin{array}{c}
 \\
1 \\
2 \\
3 \\
4
\end{array}
\begin{array}{cccc}
A & B & C & D \\
\left[\begin{array}{cccc}
1^* & 0 & 1 & 1 \\
0 & 1^* & 0 & 0 \\
1 & 1 & 0 & 0 \\
0 & 1 & 0 & 0
\end{array}\right] \quad CS \\
1S & & LS & LS
\end{array}
$$

We are now at a turning point in the algorithm. When we scan the labeled row 3 we find no starred 1, and so we mark this row with an exclamation point. This indicates that we will be able to improve on the independent set of 1's we started with. The labels on the lines of the matrix tell us exactly how to do this. Row 3 is labeled with an A, so we put a circle around the 1 in column A (and row 3). This column is labeled with a 1, so we put a circle around the starred 1 in row 1 (and column A). Row 1 is labeled with a C; so we put a circle around the 1 in column C (and row 1). Column C is labeled with the letter L so we stop drawing circles at this point. Our matrix now looks as follows.

$$
\begin{array}{c}
 \\
1 \\
2 \\
3 \\
4 \\
 \\
\end{array}
\begin{array}{cccc}
A & B & C & D \\
\end{array}
$$

$$
\begin{array}{c}
1 \\
2 \\
3 \\
4
\end{array}
\left[
\begin{array}{cccc}
①^* & 0 & ① & 1 \\
0 & 1^* & 0 & 0 \\
① & 1 & 0 & 0 \\
0 & 1 & 0 & 0
\end{array}
\right]
\begin{array}{c}
CS \\
\\
A! \\
\\
\end{array}
$$

$$
\quad\;\; 1S \qquad\;\; LS \quad LS
$$

At this point we find a larger independent set of 1's by reversing the stars on the circled 1's, that is, by adding a star to any circled 1 without a star, and removing the star from any circled 1 with a star. The result is an independent set of 1's with 3 elements instead of 2.

$$
\begin{array}{c}
1 \\
2 \\
3 \\
4
\end{array}
\begin{array}{cccc}
A & B & C & D \\
\end{array}
\left[
\begin{array}{cccc}
1 & 0 & 1^* & 1 \\
0 & 1^* & 0 & 0 \\
1^* & 1 & 0 & 0 \\
0 & 1 & 0 & 0
\end{array}
\right]
$$

It is instructive to see what we have done in this example in terms of graphs. Figure 5.8(a) shows the bipartite graph corresponding to our matrix, with the matching of our original set of two 1's indicated. Our matrix operations ended

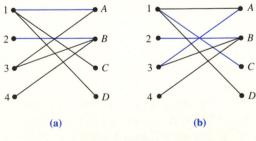

(a)   (b)

FIGURE 5.8

when we labeled a row, row 3, which contained no starred 1. Row 3 was labeled when we scanned column A. The 1 in row 3 and column A corresponds to the edge from 3 to A in our graph, and since row 3 contains no starred 1's this means vertex

3 has no edges of the matching attached to it. Likewise, column A was labeled when we scanned row 1. The 1 in row 1 and column A is starred and corresponds to the edge from 1 to A which is in our matching. Finally, row 1 was labeled when we scanned column C, from the unstarred 1 in row 1 and column C. Thus, the edge from 1 to C is not in the matching. Column C was labeled at the start of the algorithm, which means there are no starred 1's in it. Thus, no edge of the matching goes to vertex C.

We have found a simple path between vertex 3 and vertex C, namely 3, {3, A}, A, {A, 1}, 1, {1, C}, C, whose edges are alternately in and not in the matching. By its construction it includes an odd number of edges (it starts in V_1 and ends in V_2) and its first and last edge are not in the matching. Reversing the stars in our matrix corresponds to reversing which edges of this path are in the matching, as is laid out in Figure 5.9. The result is shown in Figure 5.8(b).

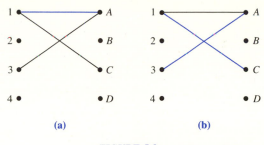

(a)   (b)

FIGURE 5.9

Applying the Algorithm to a Maximal Independent Set

Now we will apply the algorithm to the matrix with our new set of 3 starred 1's. Since only column D has no starred 1's, we start by labeling that column.

$$
\begin{array}{c}
\begin{array}{cccc} A & B & C & D \end{array} \\
\begin{array}{c} 1 \\ 2 \\ 3 \\ 4 \end{array}
\left[\begin{array}{cccc}
1 & 0 & 1^* & 1 \\
0 & 1^* & 0 & 0 \\
1^* & 1 & 0 & 0 \\
0 & 1 & 0 & 0
\end{array} \right] \\
L
\end{array}
$$

Scanning this column for unstarred 1's leads us to label row 1.

$$
\begin{array}{c}
\begin{array}{cccc} A & B & C & D \end{array} \\
\begin{array}{c} 1 \\ 2 \\ 3 \\ 4 \end{array}
\left[\begin{array}{cccc}
1 & 0 & 1^* & 1 \\
0 & 1^* & 0 & 0 \\
1^* & 1 & 0 & 0 \\
0 & 1 & 0 & 0
\end{array} \right] \begin{array}{c} D \\ \\ \\ \end{array} \\
LS
\end{array}
$$

Scanning row 1 for starred 1's leads us to label column C.

$$\begin{array}{c} \\ 1 \\ 2 \\ 3 \\ 4 \end{array} \begin{array}{cccc} A & B & C & D \\ \left[\begin{array}{cccc} 1 & 0 & 1* & 1 \\ 0 & 1* & 0 & 0 \\ 1* & 1 & 0 & 0 \\ 0 & 1 & 0 & 0 \end{array}\right] \end{array} \begin{array}{l} DS \\ \\ \\ \\ \end{array}$$

$$\qquad\qquad\qquad\qquad 1 \quad LS$$

This is the most important point in the present application of the algorithm. When we scan column C there is nothing to label, and we have the following matrix.

$$\begin{array}{c} \\ 1 \\ 2 \\ 3 \\ 4 \end{array} \begin{array}{cccc} A & B & C & D \\ \left[\begin{array}{cccc} 1 & 0 & 1* & 1 \\ 0 & 1* & 0 & 0 \\ 1* & 1 & 0 & 0 \\ 0 & 1 & 0 & 0 \end{array}\right] \end{array} \begin{array}{l} DS \\ \\ \\ \\ \end{array}$$

$$\qquad\qquad\qquad\qquad 1S \quad LS$$

All lines that are labeled have also been scanned, and there is nothing else we can do. This indicates that the independent set of 1's we started with is maximal.

Maximal Independent Set Algorithm

Given an independent set of 1's in a matrix of 0's and 1's, this algorithm either indicates that this set is maximal, or else finds a larger such set.

Step 1 (start). Label with an "L" all columns containing no starred 1.

Step 2 (scanning columns). If all labeled columns have been scanned, then go to step 4. Otherwise, for each column that is labeled but not scanned, look at any unstarred 1's in that column. If such a 1 is in an unlabeled row, then label that row with the name of the column being scanned. Put the letter "S" under each column after it has been scanned.

Step 3 (scanning rows). If all labeled rows have been scanned, then go to step 4. If some labeled but unscanned row contains no starred 1, then go to step 5. Otherwise, for each row that is labeled but not scanned, look for the starred 1 in that row. Label the column containing the starred 1 with the name of the row being scanned. Put the letter "S" after each row when it has been scanned. Go to step 2.

Step 4 (no improvement). Stop. The given independent set is maximal.

Step 5 (backtracking). A labeled row contains no starred 1. Circle the 1 in this row and in the column that the row is labeled with. Circle the starred 1 in this column and the row that this column is labeled with. Continue in this way until a 1 is circled in a column labeled with the letter "L."

Step 6 (larger independent set). Reverse the stars on all circled 1's. This gives an independent set of 1's with one more element than the original set.

This algorithm is due to Ford and Fulkerson, and can be found in suggested reading [3] at the end of this chapter. We will prove that it does what it says it does in the next section. Of course, by changing the language the algorithm could just as well be applied to a graph, but there are complications if it is not bipartite. A modification of the algorithm that applies to arbitrary graphs can be found in suggested reading [2].

Let us examine the complexity of this algorithm. In our analysis we will use the word "operation" in a somewhat vague way to indicate looking at some entry or row or column label of a matrix and perhaps taking some simple action such as applying or changing a symbol.

Suppose a matrix of 0's and 1's has m rows and n columns. Step 1 involves looking at all mn entries in the matrix, which we count as mn operations. After this the algorithm alternates between steps 2 and 3, both of which involve scanning. In order to scan one of the n columns we need to look at the m entries in that column, so all column scanning will take at most mn operations. Likewise, row scanning will take at most nm operations.

If we go to step 4 we are done, so we analyze steps 5 and 6. Backtracking will take at most $m + n$ operations, since each 1 we circle can be associated with a distinct row or column. Actually we could combine step 6 into step 5 with no additional work, reversing the stars as we backtracked. Thus, one application of the algorithm will take at most $3mn + m + n$ operations. To build up to a maximal independent set of 1's the algorithm will have to be repeated at most min$\{m,n\}$ times, even if we start with the empty set as our first independent set of 1's. Thus, the complexity of the algorithm for finding a maximal independent set of ones in an m by n matrix is of order no more than $(3mn + m + n)$min$\{m,n\}$. For the case $m = n = 30$, fewer than 90,000 operations would be necessary, and a fast computer could do the problem in less than one second.

Assigning Courses

As another example of the use of the algorithm we will go back to the example of assigning English professors to courses of Section 5.1. The matrix of the graph shown in Figure 5.1 is the following.

$$
\begin{array}{c}
 \\
1 \\
2 \\
3 \\
4 \\
5 \\
6
\end{array}
\begin{array}{c}
\begin{array}{ccccccc}
A & B & C & D & E & F & G
\end{array} \\
\begin{bmatrix}
1^* & 0 & 1 & 0 & 0 & 1 & 0 \\
0 & 0 & 1^* & 1 & 1 & 0 & 1 \\
1 & 0 & 1 & 0 & 0 & 0 & 0 \\
1 & 0 & 0 & 0 & 0 & 1^* & 0 \\
0 & 1^* & 0 & 0 & 1 & 0 & 1 \\
0 & 0 & 1 & 0 & 0 & 1 & 0
\end{bmatrix}
\end{array}
$$

The independent set of 1's shown was chosen by taking the first available 1 in the first row, second row, etc., subject to the condition that we not choose two 1's in the same column. We will show what our matrix looks like after each step in the algorithm.

	A	B	C	D	E	F	G	
1	1*	0	1	0	0	1	0	
2	0	0	1*	1	1	0	1	
3	1	0	1	0	0	0	0	
4	1	0	0	0	0	1*	0	
5	0	1*	0	0	1	0	1	
6	0	0	1	0	0	1	0	
	L	L				L		

After Step 1

	A	B	C	D	E	F	G	
1	1*	0	1	0	0	1	0	
2	0	0	1*	1	1	0	1	D
3	1	0	1	0	0	0	0	
4	1	0	0	0	0	1*	0	
5	0	1*	0	0	1	0	1	E
6	0	0	1	0	0	1	0	
			LS	LS		LS		

After Step 2

	A	B	C	D	E	F	G	
1	1*	0	1	0	0	1	0	
2	0	0	1*	1	1	0	1	DS
3	1	0	1	0	0	0	0	
4	1	0	0	0	0	1*	0	
5	0	1*	0	0	1	0	1	ES
6	0	0	1	0	0	1	0	
	5	2	LS	LS		LS		

After Step 3

	A	B	C	D	E	F	G	
1	1*	0	1	0	0	1	0	C
2	0	0	1*	1	1	0	1	DS
3	1	0	1	0	0	0	0	C
4	1	0	0	0	0	1*	0	
5	0	1*	0	0	1	0	1	ES
6	0	0	1	0	0	1	0	C
	5S	2S	LS	LS		LS		

After Step 2

	A	B	C	D	E	F	G	
1	1*	0	1	0	0	1	0	CS
2	0	0	(1*)	(1)	1	0	1	DS
3	1	0	(1)	0	0	0	0	C!
4	1	0	0	0	0	1*	0	
5	0	1*	0	0	1	0	1	ES
6	0	0	1	0	0	1	0	C
	1	5S	2S	LS	LS	LS		

After Steps 3 and 5

	A	B	C	D	E	F	G	
1	1*	0	1	0	0	1	0	
2	0	0	1	1*	1	0	1	
3	1	0	1*	0	0	0	0	
4	1	0	0	0	0	1*	0	
5	0	1*	0	0	1	0	1	
6	0	0	1	0	0	1	0	

After Step 6

If the algorithm is now applied with this new independent set of 1's, it tells us that the set is maximal, and we end up with the following configuration.

	A	B	C	D	E	F	G	
1	1*	0	1	0	0	1	0	
2	0	0	1	1*	1	0	1	ES
3	1	0	1*	0	0	0	0	
4	1	0	0	0	0	1*	0	
5	0	1*	0	0	1	0	1	ES
6	0	0	1	0	0	1	0	
	5S		2S	LS		LS		

It should be noted that in steps 2 and 3 the order in which the labeled but not scanned lines are chosen may affect what larger independent set of 1's the algorithm produces. In computing answers for the examples in this section we always chose the rows from top to bottom and the columns from left to right.

EXERCISES 5.3

Throughout these exercises, when applying the maximal independent set algorithm, choose rows from top to bottom and columns from left to right.

In Exercises 1 and 2 one stage in the application of the maximal independent set algorithm is shown. Apply step 2 or 3, as appropriate.

1.
$$\begin{array}{c} & A & B & C & D \\ 1 & \begin{bmatrix} 0 & 1* & 0 & 1 \\ 1* & 0 & 0 & 1 \\ 1 & 1 & 0 & 0 \\ 1 & 0 & 0 & 0 \end{bmatrix} & \begin{matrix} D \\ D \\ \\ \\ \end{matrix} \\ 2 \\ 3 \\ 4 \\ & & & LS & LS \end{array}$$

2.
$$\begin{array}{c} & A & B & C & D & E \\ 1 & \begin{bmatrix} 1* & 0 & 1 & 0 & 1 \\ 0 & 1* & 0 & 1 & 1 \\ 1 & 1 & 0 & 0 & 0 \\ 0 & 1 & 0 & 1* & 0 \end{bmatrix} & \begin{matrix} CS \\ ES \\ \\ \\ \end{matrix} \\ 2 \\ 3 \\ 4 \\ & 1 & 2 & LS & & LS \end{array}$$

In Exercises 3 and 4 the matrix is ready for step 5. What entries should be circled?

3.
$$\begin{array}{c} & A & B & C & D \\ 1 & \begin{bmatrix} 0 & 1* & 1 & 0 \\ 1* & 0 & 0 & 1 \\ 1 & 1 & 0 & 0 \\ 0 & 0 & 1* & 1 \end{bmatrix} & \begin{matrix} C \\ DS \\ A! \\ DS \end{matrix} \\ 2 \\ 3 \\ 4 \\ & 2S & & 4S & LS \end{array}$$

4.
$$\begin{array}{c} & A & B & C & D & E \\ 1 & \begin{bmatrix} 0 & 0 & 1* & 0 & 1 \\ 1* & 0 & 1 & 0 & 0 \\ 0 & 1* & 0 & 1 & 0 \\ 1 & 0 & 1 & 0 & 0 \end{bmatrix} & \begin{matrix} ES \\ C \\ DS \\ C! \end{matrix} \\ 2 \\ 3 \\ 4 \\ & & 3S & 1S & LS & LS \end{array}$$

In Exercises 5–10 a matrix is given with an independent set of 1's. Use the maximal independent set algorithm until it ends in step 4.

5.
$$\begin{bmatrix} 0 & 1* & 0 & 1 \\ 1* & 1 & 0 & 0 \\ 0 & 0 & 1* & 1 \\ 1 & 1 & 1 & 0 \end{bmatrix}$$

6.
$$\begin{bmatrix} 1* & 0 & 0 & 1 \\ 1 & 0 & 1* & 0 \\ 1 & 1* & 0 & 0 \\ 0 & 1 & 1 & 0 \end{bmatrix}$$

7.
$$\begin{bmatrix} 1* & 1 & 0 & 1 & 1 \\ 1 & 0 & 0 & 0 & 1* \\ 0 & 1* & 0 & 1 & 0 \\ 1 & 1 & 0 & 0 & 1 \end{bmatrix}$$

8.
$$\begin{bmatrix} 0 & 0 & 1* & 1 & 0 \\ 0 & 1* & 1 & 0 & 0 \\ 1* & 1 & 0 & 0 & 0 \\ 0 & 1 & 1 & 0 & 0 \end{bmatrix}$$

9.
$$\begin{bmatrix} 1* & 1 & 1 & 1 & 1 \\ 1 & 0 & 0 & 0 & 0 \\ 0 & 1* & 0 & 0 & 0 \\ 1 & 1 & 0 & 0 & 0 \\ 1 & 0 & 1* & 0 & 1 \end{bmatrix}$$

10.
$$\begin{bmatrix} 0 & 1* & 0 & 1 & 0 \\ 0 & 0 & 1* & 1 & 0 \\ 0 & 1 & 1 & 0 & 0 \\ 0 & 1 & 1 & 0 & 0 \\ 1* & 0 & 1 & 1 & 1 \end{bmatrix}$$

In Exercises 11–16 a bipartite graph is given with a matching. Convert it to a matrix and find a maximal matching by using the maximal independent set algorithm, starting with the corresponding independent set of 1's.

11.

12.

13.

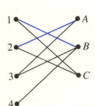

14.

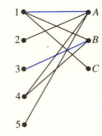

15.

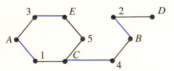

16.

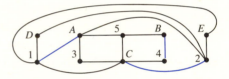

In Exercises 17–22 a sequence of sets is given with distinct elements in some of the sets starred. Convert to a matrix and use the maximal independent set algorithm to find a system of distinct representatives, if possible. Start with the corresponding independent set of 1's.

17. $\{B^*\}, \{A^*, B, C, D\}, \{A, B\}, \{B, D^*\}$

18. $\{C^*, D\}, \{A^*, B\}, \{A, D^*\}, \{A, C, D\}$

19. $\{W^*\}, \{Y^*, Z\}, \{W, Y\}, \{W, X^*, Y, Z\}$

20. $\{1^*, 3, 5\}, \{1, 4^*\}, \{2^*, 3, 5\}, \{1, 2, 4\}$

21. $\{carrot^*, egg\}, \{apple^*, banana, date, fennel\}, \{apple, carrot, egg^*\},$
$\{apple, carrot, egg\}$

22. $\{5^*, 13\}, \{1^*, 6, 9\}, \{1, 5\}, \{1, 6^*, 13\}$

23. Five ships, the Arabella, Constantine, Drury, Egmont, and Fungo, arrive at five loading docks. For technical reasons each dock can only accept certain ships. Dock 1 can only accept the Constantine or Drury. Likewise, Dock 2 can only accept the Egmont or Fungo, Dock 3 the Constantine, Egmont, or Fungo, Dock 4 the Arabella, Drury, or Fungo, and Dock 5 the Arabella, Constantine, or Egmont. The harbormaster sends the Constantine to Dock 1, the Egmont to Dock 2, the Fungo to Dock 3, and the Arabella to Dock 4. Use the maximal independent set algorithm to improve on this if possible.

24. A radio station wants to play an hour of rock music, followed by an hour each of classical, polka, and disco. Six disk jockeys are available, but each has his or her scruples. Only Barb, Cal, Dot, and Flo are willing to play rock. Likewise, only

Andy, Barb, Esther, and Flo will play classical, only Barb, Dot, and Flo will play polkas, and only Andy, Barb, and Dot will play disco. No disk jockey is allowed to work more than one hour per day. At present the station manager plans to use Barb, Andy, and Dot for the first 3 hours, but has no one left for the disco hour. Use the maximal independent set algorithm to find a better matching.

5.4   Applications of the Algorithm

In this section we will prove that our maximal independent set algorithm actually does what it claims to do. At the same time we will find that the algorithm actually leads to more insight about the relation between independent sets, coverings, and systems of distinct representatives. We start with a sequence of short lemmas concerning the results of applying the algorithm.

LEMMA 1

If the algorithm leads to step 5, then it produces an independent set of 1's with more elements than the original one.

Proof.   What goes on in the backtracking process was indicated in the previous section. Schematically, a pattern such as shown in Figure 5.10 emerges. Accord-

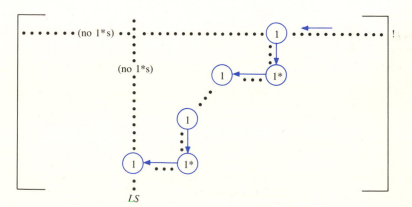

FIGURE 5.10

ing to the way the algorithm works, the circled symbols form an alternating sequence of unstarred and starred 1's, beginning and ending with an unstarred 1. Reversing the stars on these clearly increases the number of starred 1's by one. The new set is still independent since if a 1 given a star was in a line with any starred 1, the latter 1's star is removed in step 6. ∎

Now we prove some lemmas about what our matrix looks like if step 4 is reached and the algorithm indicates that our set of starred 1's is already maximal. Recall that in this case each line that has been labeled has also been scanned.

LEMMA 2   If step 4 is reached, then the labeled rows and unlabeled columns form a covering.

Proof.   If not, then some 1 is at the same time in an unlabeled row and labeled column. If this 1 is starred, then its column can only have been labeled when its row was scanned, contradicting the fact that the row is unlabeled. But if the 1 is unstarred, then when its column was scanned its row would have been labeled, another contradiction.   ▬▬

LEMMA 3   If step 4 is reached, then each labeled row and unlabeled column contains a starred 1.

Proof.   Each unlabeled column contains a starred 1 since columns that do not are labeled at step 1. On the other hand if a labeled row contained no starred 1, we would go to step 5 instead of step 4.   ▬▬

LEMMA 4   If step 4 is reached, then no starred 1 is in both a labeled row and unlabeled column.

Proof.   If a starred 1 is in a labeled row, then its column is labeled when the row is scanned.   ▬▬

THEOREM 5.4   The maximal independent set algorithm increases the number of elements when applied to an independent set that is not maximal. When applied to an independent set that is maximal, it tells us so.

Proof.   The flow of the algorithm is shown in Figure 5.11.

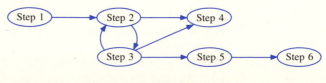

FIGURE 5.11

In steps 2 and 3 columns and rows are scanned. Since a matrix has only a finite number of lines, the algorithm eventually gets to step 4 or step 5. If it gets to step 5, then Lemma 1 tells us that it constructs an independent set with more elements than the one with which we started.

It remains to show that if the algorithm gets to step 4, then the independent set we started with is actually maximal. According to Lemma 2 the labeled rows and unlabeled columns form a covering. But Lemmas 3 and 4 say that the lines in this covering are in one-to-one correspondence with our independent set. Thus, the covering and the independent set contain the same number of elements. Then by Theorem 5.3 the covering is minimal and the independent set is maximal, which is what we want to prove. ■

König's Theorem

The argument just given amounts to a proof of a famous theorem of graph theory, first stated in 1931 by D. König, who pioneered the area. We state it in both its matrix and bipartite graph forms.

THEOREM 5.5   *König's Theorem*   In a matrix of 0's and 1's a maximal independent set of 1's contains the same number of elements as a minimal covering. Equivalently, in a bipartite graph a maximal matching contains the same number of elements as a minimal covering.

The maximal independent set algorithm gives us a construction of a minimal covering, namely the labeled rows and unlabeled columns.

EXAMPLE 5.11   We will use the algorithm to find a minimal covering for the graph shown in Figure 5.12.

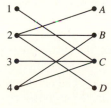

FIGURE 5.12

We convert the graph to the matrix that follows, and by inspection find the independent set shown.

$$
\begin{array}{c c c c c}
 & A & B & C & D \\
1 & \begin{bmatrix} 0 & 0 & 1^* & 0 \\ 1^* & 1 & 0 & 1 \\ 0 & 0 & 1 & 0 \\ 0 & 1^* & 1 & 0 \end{bmatrix} \\
2 \\
3 \\
4
\end{array}
$$

Applying the algorithm yields the following.

$$
\begin{array}{c c c c c}
 & A & B & C & D \\
1 & \begin{bmatrix} 0 & 0 & 1^* & 0 \\ 1^* & 1 & 0 & 1 \\ 0 & 0 & 1 & 0 \\ 0 & 1^* & 1 & 0 \end{bmatrix} & DS \\
2 \\
3 \\
4 \\
 & 2S & & LS
\end{array}
$$

We see that the matching we found by inspection was maximal, since we have reached step 4. A minimal covering consists of the labeled rows and unlabeled columns, namely row 2 and columns B and C. Thus, vertices 2, B, and C form a minimal covering for the original graph. ▆▆

Note that König's theorem only applies to bipartite graphs, even though matchings and coverings have been defined for arbitrary graphs. The reader should check that a maximal matching for the nonbipartite graph shown in Figure 5.13 contains 2 edges, while a minimal covering contains 3 vertices.

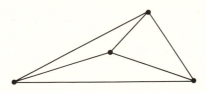

FIGURE 5.13

A Proof of Hall's Theorem

We can also use our conclusions about the algorithm to complete the proof of Hall's theorem, which was stated in Section 5.1. Recall that it remains to show that if $S_1, S_2, \ldots, S_n$ is a sequence of sets not having a system of distinct representatives, then there exists a subset I of $\{1, 2, \ldots, n\}$ such that the union of the sets S_i for $i \in I$ has fewer than $|I|$ elements.

Let the union of all the sets S_i for $i = 1, 2, \ldots, n$ be $\{t_1, t_2, \ldots, t_m\}$, where the t's are distinct. We construct a matrix of 0's and 1's with rows corresponding to the sets S_i and columns corresponding to the elements t_j. In particular, the entry in row i and column j is to be 1 if $t_j \in S_i$ and 0 otherwise. (We have already constructed such matrices; for example, the sequence $P_1, \ldots, P_6$ of sets of professors who teach certain courses mentioned in Section 5.1 leads to the matrix of the second example in Section 5.3.)

We use our algorithm as many times as necessary to confirm that we have a maximal independent set of 1's in this matrix. Let r_L, r_U, c_L, and c_U denote the number of labeled rows, unlabeled rows, labeled columns, and unlabeled columns in this matrix after the last application of the algorithm, respectively. Therefore, $r_L + r_U = n$ and $c_L + c_U = m$. By Lemmas 2, 3, and 4 our maximal independent set has $r_L + c_U$ elements.

If our maximal independent set had n elements, it would correspond to a system of distinct representatives, so we may assume that

$$r_L + c_U < n = r_L + r_U.$$

Thus, $c_U < r_U$.

We claim that the union of the r_U sets corresponding to the unlabeled rows contains fewer than r_U elements. For each 1 in an unlabeled row must be in an unlabeled column by Lemma 2. Thus, the union of the corresponding sets contains at most c_U elements, and we know that $c_U < r_U$. We can take I to be the numbers of the unlabeled rows. This completes the proof of Hall's theorem.

Notice that we have an actual construction of the set I using the algorithm. For example, the problem of the courses and professors led to the matrix

$$
\begin{array}{c}
 \\
1 \\
2 \\
3 \\
4 \\
5 \\
6
\end{array}
\begin{array}{cccccccc}
A & B & C & D & E & F & G \\
\begin{bmatrix} 1^* & 0 & 1 & 0 & 0 & 1 & 0 \\ 0 & 0 & 1 & 1^* & 1 & 0 & 1 \\ 1 & 0 & 1^* & 0 & 0 & 0 & 0 \\ 1 & 0 & 0 & 0 & 0 & 1^* & 0 \\ 0 & 1^* & 0 & 0 & 1 & 0 & 1 \\ 0 & 0 & 1 & 0 & 0 & 1 & 0 \end{bmatrix}
\end{array}
\begin{array}{c}
 \\
 \\
ES \\
 \\
 \\
ES \\

\end{array}
$$

$$5S 2S \;\; LS LS$$

in the previous section. The unlabeled rows are the four rows 1, 3, 4, and 6. As we saw in Section 5.1 the union of the corresponding sets contains fewer than four elements.

EXAMPLE 5.12

At a business meeting six speakers are to be scheduled at 9, 10, 11, 1, 2, and 3 o'clock. Mr. Brown can only talk before noon. Ms. Krull can only speak at 9 or 2. Ms. Zeno cannot speak at 9, 11, or 2. Mr. Toomey cannot speak until 2. Mrs. Abernathy cannot speak between 10 and 3. Mr. Ng cannot speak from 10 until 2. The scheduler cannot seem to fit everyone in. Show that it is impossible to do so, and give a way the scheduler can convince the speakers of this fact.

We construct the following matrix, where the rows correspond to speakers and the columns to times.

$$
\begin{array}{c}
 \\
B \\
K \\
Z \\
T \\
A \\
N
\end{array}
\begin{array}{cccccc}
9 & 10 & 11 & 1 & 2 & 3 \\
\begin{bmatrix} 1^* & 1 & 1 & 0 & 0 & 0 \\ 1 & 0 & 0 & 0 & 1^* & 0 \\ 0 & 1^* & 0 & 1 & 0 & 1 \\ 0 & 0 & 0 & 0 & 1 & 1^* \\ 1 & 0 & 0 & 0 & 0 & 1 \\ 1 & 0 & 0 & 0 & 1 & 1 \end{bmatrix}
\end{array}
$$

The independent set indicated was found by inspection. Applying the algorithm produces first

$$
\begin{array}{c}
 & \begin{array}{cccccc} 9 & 10 & 11 & 1 & 2 & 3 \end{array} & \\
\begin{array}{c} B \\ K \\ Z \\ T \\ A \\ N \end{array} &
\left[\begin{array}{cccccc}
① * & 1 & ① & 0 & 0 & 0 \\
1 & 0 & 0 & 0 & 1* & 0 \\
0 & 1* & 0 & 1 & 0 & 1 \\
0 & 0 & 0 & 0 & 1 & 1* \\
① & 0 & 0 & 0 & 0 & 1 \\
1 & 0 & 0 & 0 & 1 & 1
\end{array}\right] &
\begin{array}{c} 11S \\ 9 \\ 1S \\ \\ 9! \\ 9 \end{array} \\
 & \begin{array}{cccccc} BS & ZS & LS & & LS & \end{array} &
\end{array}
$$

and then

$$
\begin{array}{c}
 & \begin{array}{cccccc} 9 & 10 & 11 & 1 & 2 & 3 \end{array} & \\
\begin{array}{c} B \\ K \\ Z \\ T \\ A \\ N \end{array} &
\left[\begin{array}{cccccc}
1 & 1 & 1* & 0 & 0 & 0 \\
1 & 0 & 0 & 0 & 1* & 0 \\
0 & 1* & 0 & 1 & 0 & 1 \\
0 & 0 & 0 & 0 & 1 & 1* \\
1* & 0 & 0 & 0 & 0 & 1 \\
1 & 0 & 0 & 0 & 1 & 1
\end{array}\right] &
\begin{array}{c} 10S \\ \\ 1S \\ \\ \\ \end{array} \\
 & \begin{array}{cccccc} & ZS & BS & LS & & \end{array} &
\end{array}
$$

Notice that the rows corresponding to Krull, Toomey, Abernathy, and Ng are unlabeled. These four speakers all want to speak at 9, 2, or 3, which shows that all their restrictions cannot be accommodated. ■

The Bottleneck Problem

A foreman has 4 jobs that need doing and 5 workers to whom he could assign them. The time in hours each worker would need to do each job is shown in the following table.

	Job 1	Job 2	Job 3	Job 4
Worker 1	3	7	5	8
Worker 2	6	3	2	3
Worker 3	3	5	8	6
Worker 4	5	8	6	4
Worker 5	6	5	7	3

He needs all 4 jobs finished as soon as possible, and so is interested in making the maximum job time for the 4 workers chosen as small as possible.

Only one worker can do a job in 2 hours, so that it is obviously impossible to get all 4 jobs done that fast. Three hours is more reasonable. Let us make a matrix of 0's and 1's, putting a 1 in each position corresponding to a job time of 3 hours or less.

$$\begin{bmatrix} 1 & 0 & 0 & 0 \\ 0 & 1 & 1 & 1 \\ 1 & 0 & 0 & 0 \\ 0 & 0 & 0 & 0 \\ 0 & 0 & 0 & 1 \end{bmatrix}$$

We would like an independent set of 1's having 4 elements to correspond to the 4 jobs. Unfortunately, no such set exists. Since all the 1's lie in 3 lines (row 2 and columns 1 and 4), this fact is implied by König's theorem. Doing the jobs will take at least 4 hours, and so we add 1's to our matrix corresponding to the 4's in the original matrix.

$$\begin{bmatrix} 1 & 0 & 0 & 0 \\ 0 & 1 & 1 & 1 \\ 1 & 0 & 0 & 0 \\ 0 & 0 & 0 & 1 \\ 0 & 0 & 0 & 1 \end{bmatrix}$$

The same reasoning shows that still no independent set of four 1's exists, so we add 1's corresponding to the 5's in the original matrix.

$$\begin{bmatrix} 1^* & 0 & 1 & 0 \\ 0 & 1^* & 1 & 1 \\ 1 & 1 & 0 & 0 \\ 1 & 0 & 0 & 1^* \\ 0 & 1 & 0 & 1 \end{bmatrix}$$

The starred independent set was found by inspection. By applying the algorithm we find the larger set that follows.

$$\begin{bmatrix} 1 & 0 & 1^* & 0 \\ 0 & 1^* & 1 & 1 \\ 1^* & 1 & 0 & 0 \\ 1 & 0 & 0 & 1^* \\ 0 & 1 & 0 & 1 \end{bmatrix}$$

Thus, the shortest time all the jobs can be completed in is 5 hours.

Problems such as this are called **bottleneck** problems, since we are interested in making the job time of the slowest worker as small as possible. In other circumstances we might be interested instead in minimizing the total time to do all the jobs. Such problems will be treated in the next section.

EXERCISES 5.4

In Exercises 1–4 a matrix of 0's and 1's is given with an independent set indicated. Use the maximal independent set algorithm to find a minimal covering.

1.
$$\begin{bmatrix} 0 & 1^* & 0 & 1 \\ 1^* & 1 & 1 & 0 \\ 0 & 1 & 0 & 1^* \\ 0 & 0 & 0 & 1 \end{bmatrix}$$

2.
$$\begin{bmatrix} 1^* & 0 & 0 & 0 \\ 0 & 1^* & 0 & 1 \\ 1 & 0 & 0 & 0 \\ 1 & 0 & 1^* & 1 \end{bmatrix}$$

3.
$$\begin{bmatrix} 1^* & 0 & 1 & 0 & 0 \\ 0 & 0 & 1^* & 1 & 0 \\ 1 & 1^* & 0 & 1 & 1 \\ 1 & 0 & 1 & 1^* & 0 \\ 1 & 0 & 0 & 1 & 0 \end{bmatrix}$$

4.
$$\begin{bmatrix} 0 & 1^* & 0 & 0 & 0 \\ 0 & 0 & 1^* & 0 & 0 \\ 0 & 1 & 1 & 0 & 0 \\ 0 & 0 & 1 & 0 & 0 \\ 1^* & 1 & 0 & 1 & 1 \end{bmatrix}$$

In Exercises 5–8 a bipartite graph is given with a matching indicated. Use the maximal independent set algorithm to find a minimal covering.

5.
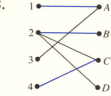

6.
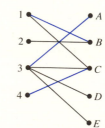

7.
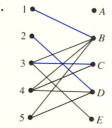

8.
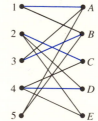

In Exercises 9–12 a sequence of sets $S_1, S_2, \ldots, S_n$ is given. Use the maximal independent set algorithm to find, if possible, a subset I of $\{1, 2, \ldots, n\}$ such that the union of the sets S_i for $i \in I$ has fewer elements than I.

9. $\{2, 4, 5\}, \{1, 3, 5\}, \{2, 3, 5\}, \{3, 4, 5\}, \{2, 3, 4\}$

10. $\{1, 2, 4\}, \{2, 3, 4, 5\}, \{2, 4, 6\}, \{1, 6\}, \{1, 4, 6\}, \{1, 2, 6\}$

11. $\{2, 7\}, \{1, 3, 6\}, \{5, 7\}, \{3, 4, 6\}, \{2, 5\}, \{2, 5, 7\}$

12. $\{1, 2\}, \{4, 6\}, \{0, 1, 3, 5, 6\}, \{1, 4, 7\}, \{2, 6\}, \{1, 4, 7\}, \{2, 6, 7\}$

13. A military commander must send a runner to each of 4 posts notifying them of a plan to attack. Because of differing terrain and skills, the time in hours for each runner to reach each post varies. Runner A takes 6 hours to get to Post 1, 5 hours to Post 2, 9 hours to Post 3, and 7 hours to Post 4. Runner *B* takes 4, 8, 7, and 8 hours to reach the four posts. Likewise, Runner *C* takes 5, 3, 9, and 8 hours, and Runner *D* takes 7, 6, 3, and 5 hours. The attack cannot begin until all posts have gotten the message. What is the shortest time until it can begin?

14. One step of a manufacturing process takes 5 operations which can be done simultaneously. These take different times in minutes on the 5 machines available, as given in the following table.

	M1	*M2*	*M3*	*M4*	*M5*
Operation 1	6	7	3	6	2
Operation 2	6	3	4	3	3
Operation 3	2	5	3	7	4
Operation 4	3	4	2	6	3
Operation 5	4	7	2	7	6

How fast can the entire step be accomplished?

15. The graph below shows a city map. Adjacent vertices are one block apart. It is desired that a policeman be stationed at some of the vertices so that no one is more than one block from a policeman. Use the algorithm to find the smallest number of policemen necessary to accomplish this, and tell where they should be positioned.

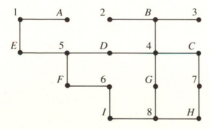

16. Show that if step 4 is reached in the maximal independent set algorithm, then the number of labeled columns equals the number of labeled rows plus the number of columns containing no starred 1.

5.5 The Hungarian Method

In the last section we considered a problem of assigning 4 jobs to 5 workers in such a way that all 4 jobs got done as soon as possible. Although this might be our goal in special circumstances, a more common aim is to minimize the total time necessary to do the 4 jobs. If each worker was paid the same hourly rate, for example, this would minimize the labor cost for the project.

For simplicity we will start with an example in which there are the same number of jobs and workers. The times in hours for each worker to do each job are given in the following table.

	Worker 1	Worker 2	Worker 3	Worker 4
Job 1	3	6	3	5
Job 2	7	3	5	8
Job 3	5	2	8	6
Job 4	8	3	6	4

An assignment of a worker to each job amounts to an independent set of 4 entries from the corresponding matrix, and we want the sum of the entries in that set to be as small as possible. For example, two possible assignments are indicated below.

$$\begin{bmatrix} 3^* & 6 & 3 & 5 \\ 7 & 3^* & 5 & 8 \\ 5 & 2 & 8^* & 6 \\ 8 & 3 & 6 & 4^* \end{bmatrix} \qquad \begin{bmatrix} 3 & 6 & 3 & 5^* \\ 7 & 3^* & 5 & 8 \\ 5^* & 2 & 8 & 6 \\ 8 & 3 & 6^* & 4 \end{bmatrix}$$

The first of these produces the sum $3 + 3 + 8 + 4 = 18$, and the second gives $5 + 3 + 5 + 6 = 19$; so the first independent set is better than the second for our purposes. Of course, other assignments might yield even smaller sums.

Suppose we subtract 3 from each entry in the first row of our matrix. The two assignments are shown below for the new matrix.

$$\begin{bmatrix} 0^* & 3 & 0 & 2 \\ 7 & 3^* & 5 & 8 \\ 5 & 2 & 8^* & 6 \\ 8 & 3 & 6 & 4^* \end{bmatrix} \qquad \begin{bmatrix} 0 & 3 & 0 & 2^* \\ 7 & 3^* & 5 & 8 \\ 5^* & 2 & 8 & 6 \\ 8 & 3 & 6^* & 4 \end{bmatrix}$$

Now the first set has the sum $0 + 3 + 8 + 4 = 15$ and the second has the sum $2 + 3 + 5 + 6 = 16$. The first assignment still has a sum 1 less than the second. The point is that although subtracting the same number from each entry of the first row changes the problem, it does not change which positions give the answer. Since any independent set of 4 entries will have exactly one of them in the first row, the sum of the entries in any such set will be decreased by 3 by our operation. Any assignment producing a minimal sum for the new matrix will also give a minimal sum for the original matrix. Furthermore, the same analysis applies to the other rows as well. In order to have entries as small as possible without introducing negative numbers, we will subtract from all entries in each row the smallest number in that row. This means subtracting 3 from the entries of the second row, 2 from those of the third row, and 3 from the entries of the fourth row.

$$\begin{bmatrix} 0 & 3 & 0 & 2 \\ 4 & 0 & 2 & 5 \\ 3 & 0 & 6 & 4 \\ 5 & 0 & 3 & 1 \end{bmatrix}$$

The same argument applies to columns, and so now we will subtract 1 from each entry of the fourth column.

$$\begin{bmatrix} 0 & 3 & 0 & 1 \\ 4 & 0 & 2 & 4 \\ 3 & 0 & 6 & 3 \\ 5 & 0 & 3 & 0 \end{bmatrix}$$

Finding a 4-entry independent set in this matrix will solve our original problem. Furthermore, now at least we might be able to recognize a solution. Suppose we could find an independent set of all 0's. This will clearly have minimal sum, since the matrix has no negative entries. Unfortunately, a maximal independent set of 0's has only 3 entries, as we can confirm with the maximal independent set algorithm (modified to find an independent set of 0's instead of 1's).

$$\begin{array}{c c} & \begin{array}{cccc} A & B & C & D \end{array} \\ \begin{array}{c} 1 \\ 2 \\ 3 \\ 4 \end{array} & \begin{bmatrix} 0^* & 3 & 0 & 1 \\ 4 & 0^* & 2 & 4 \\ 3 & 0 & 6 & 3 \\ 5 & 0 & 3 & 0^* \end{bmatrix} \; CS \\ & \begin{array}{cccc} 1S & & LS & \end{array} \end{array}$$

We have reached step 4 of that algorithm, and so the independent set of three 0's indicated is a maximal independent set of 0's. Now we will show how to change the matrix so as to have a better chance of finding an independent set of four 0's. Later we will show why the solution to the minimal sum problem has not been changed.

Since a maximal independent set of 0's has fewer than four entries, there is a minimal covering consisting of fewer than four lines. In fact, by what we discovered in the last section such a covering consists of the labeled rows and unlabeled columns of the above matrix. These lines are indicated below.

$$\begin{array}{c c} & \begin{array}{cccc} A & B & C & D \end{array} \\ \begin{array}{c} 1 \\ 2 \\ 3 \\ 4 \end{array} & \begin{bmatrix} 0^* & 3 & 0 & 1 \\ 4 & 0^* & 2 & 4 \\ 3 & 0 & 6 & 3 \\ 5 & 0 & 3 & 0^* \end{bmatrix} \; CS \\ & \begin{array}{cccc} 1S & & LS & \end{array} \end{array}$$

Look at the entries not in any line of this covering. (By the definition of a covering they are all positive.) The smallest of these is 2. Now we change our matrix as follows:

(1) Subtract 2 from each entry not in a line of the cover.
(2) Add 2 to each entry in both a row and column of the cover.
(3) Leave unchanged entries in exactly one line of the cover.

The resulting matrix is shown below.

$$\begin{bmatrix} 0 & 5 & 0 & 3 \\ 2 & 0 & 0 & 4 \\ 1 & 0 & 4 & 3 \\ 3 & 0 & 1 & 0 \end{bmatrix}$$

Now we can find an independent set of four 0's, and pick out the corresponding set in the original matrix, as shown.

$$
\begin{bmatrix}
0^* & 5 & 0 & 3 \\
2 & 0 & 0^* & 4 \\
1 & 0^* & 4 & 3 \\
3 & 0 & 1 & 0^*
\end{bmatrix}
\qquad
\begin{bmatrix}
3^* & 6 & 3 & 5 \\
7 & 3 & 5^* & 8 \\
5 & 2^* & 8 & 6 \\
8 & 3 & 6 & 4^*
\end{bmatrix}
$$

The minimal sum for an independent set of four entries in the original matrix is $3 + 5 + 2 + 4 = 14$.

Of course, several questions need to be answered. One is whether the operation involving a minimal covering we just described is legitimate, that is, does not change the solution to the minimal sum problem. Another is whether this operation even does any good for the purpose of producing an independent set of four 0's, since although we subtract from some entries, we add to others. These questions will be answered after we state our method in a formal way.

Hungarian Algorithm

Starting with an n-by-n matrix with integer entries, this algorithm finds an independent set of n entries, the sum of which is minimal.

Step 1 (reduce matrix). Subtract from each entry of each row the smallest entry in that row. Do the same for columns.

Step 2 (check for independent set of n 0's). Find a maximal independent set of 0's. If this has n entries, then we are done, and the corresponding set of entries in the original matrix has a minimal sum.

Step 3 (use a minimal covering to adjust matrix). Find a minimal covering for the 0's of the matrix. Let k be the smallest matrix entry not in any line of the covering. Subtract k from each entry not in a line of the covering, and add k to each entry in both a row and column of the covering. Go to step 2.

Justification of the Hungarian Algorithm

First we will show why the operation of step 3 in the algorithm does not change which independent set is a solution. The reason is that this operation may be broken down into adding and subtracting numbers from rows and columns of the matrix, which we have already seen do not change which independent set is a solution. In particular, let k be the smallest (positive) entry not in any line of a covering. Let us subtract k from every entry of every row of the whole matrix, and then add k to every entry of every line of the covering, line by line. The net effect is exactly that of step 3. The number k is subtracted from each entry not in a line of the covering. If an entry is in a line of the covering exactly once, then it is not changed, since k is both subtracted from and added to it. Entries in both a row and column of the covering have k subtracted once but added twice, a net result of $+k$.

Now we address the question of whether step 3 does any good. It is conceivable that the algorithm could cycle through steps 2 and 3 forever without ever producing an independent set of n 0's. We will show that this cannot happen. After step 1 our matrix will contain only nonnegative integers as entries. We will show that the sum of all entries in the matrix will decrease whenever step 3 is performed. Obviously if this sum were 0 then all matrix entries would be 0 and an independent set of n 0's would exist. Thus, if the algorithm went on forever, the sums of all matrix entries would give an infinite decreasing sequence of positive integers, which is impossible.

Step 3 is performed only when no independent set of n 0's exists. Then a minimal covering will contain c rows and columns, where $c < n$. (This is a consequence of König's theorem.) Let us compute the effect of step 3 on the sum of all the entries of the matrix. As we have just seen step 3 amounts to subtracting k from each entry of the entire matrix, and then adding k to each entry of each line of the covering. Since there are n^2 entries in the matrix, the subtraction decreases the sum of all entries by kn^2. Likewise, since there are c lines in the covering, each containing n entries, the addition increases the sum of all the entries by kcn. The net amount added to the sum of all the entries is

$$-kn^2 + kcn = kn(-n + c).$$

But this quantity is negative because $c < n$, and so the net effect is to decrease the sum of all entries, as was claimed.

The reason this method is called "Hungarian," is to honor König, who was from Hungary, and upon whose theorem it is based. The algorithm is due to H. W. Kuhn.

Matrices That Are Not Square

Let us suppose that in our example a fifth worker becomes available, so that now our table becomes as follows.

	Worker 1	Worker 2	Worker 3	Worker 4	Worker 5
Job 1	3	6	3	5	3
Job 2	7	3	5	8	5
Job 3	5	2	8	6	2
Job 4	8	3	6	4	4

It is still a reasonable question to ask how to assign the 4 jobs in such a way as to make the sum of their times minimal, but our matrix is no longer square, and the algorithm only applies to square matrices. Of course, one worker is not going to get a job, and this simple idea provides a key to how to adapt the method. We introduce a fifth job, one requiring no time at all to do. This amounts to adding a row of 0's to the matrix, producing the square matrix on the left below.

$$\begin{bmatrix} 3 & 6 & 3 & 5 & 3 \\ 7 & 3 & 5 & 8 & 5 \\ 5 & 2 & 8 & 6 & 2 \\ 8 & 3 & 6 & 4 & 4 \\ 0 & 0 & 0 & 0 & 0 \end{bmatrix} \qquad \begin{bmatrix} 0 & 3 & 0 & 2 & 0 \\ 4 & 0 & 2 & 5 & 2 \\ 3 & 0 & 6 & 4 & 0 \\ 5 & 0 & 3 & 1 & 1 \\ 0 & 0 & 0 & 0 & 0 \end{bmatrix}$$

The second matrix above shows the result of applying step 1. Applying the maximal independent set algorithm to this matrix yields the matrix on the left below. The matrix on the right shows the result of applying step 3 (with $k = 1$) to it.

An independent set of five 0's is shown below for this matrix, along with the corresponding set for the original matrix.

$$\begin{bmatrix} 0^* & 4 & 0 & 2 & 1 \\ 3 & 0^* & 1 & 4 & 2 \\ 2 & 0 & 5 & 3 & 0^* \\ 4 & 0 & 2 & 0^* & 1 \\ 0 & 1 & 0^* & 0 & 1 \end{bmatrix} \qquad \begin{bmatrix} 3^* & 6 & 3 & 5 & 3 \\ 7 & 3^* & 5 & 8 & 5 \\ 5 & 2 & 8 & 6 & 2^* \\ 8 & 3 & 6 & 4^* & 4 \\ 0 & 0 & 0^* & 0 & 0 \end{bmatrix}$$

By using the fifth worker we can do all jobs in $3 + 3 + 4 + 2 = 12$ hours instead of the previous minimum of 14 hours.

Independent Sets with Maximal Sum

A sweater factory has 4 workers and 4 machines on which sweaters can be made. The number of sweaters a worker can make in a day depends on the machine he or she uses, as indicated in the following table.

	Machine 1	Machine 2	Machine 3	Machine 4
Worker 1	3	6	7	4
Worker 2	4	5	5	6
Worker 3	6	3	4	4
Worker 4	5	4	3	5

In this case what we are looking for is an independent set with 4 entries, the sum of which is *maximal*, instead of minimal. We reduce this to a problem we already know how to solve by multiplying the corresponding matrix by -1. The result is shown at the left below.

$$\begin{bmatrix} -3 & -6 & -7 & -4 \\ -4 & -5 & -5 & -6 \\ -6 & -3 & -4 & -4 \\ -5 & -4 & -3 & -5 \end{bmatrix} \qquad \begin{bmatrix} 4 & 1 & 0 & 3 \\ 2 & 1 & 1 & 0 \\ 0 & 3 & 2 & 2 \\ 0 & 1 & 2 & 0 \end{bmatrix}$$

Finding a maximal sum in the original matrix is equivalent to finding a minimal sum in this matrix. The negative entries cause no problems, since they disappear when we subtract the least entries of each row (here -7, -6, -6, and -5). The result is shown at the right above. Thus, a maximum sum problem may be solved by applying the Hungarian method to the negative of the original matrix. The reader should check that a maximum of 23 sweaters can be produced per day.

EXERCISES 5.5

In Exercises 1–8 find the smallest sum of an independent set of entries of the matrix with as many elements as the matrix has rows.

1.
$$\begin{bmatrix} 1 & 2 & 3 \\ 6 & 5 & 4 \\ 7 & 8 & 9 \end{bmatrix}$$

2.
$$\begin{bmatrix} 1 & 4 & 3 & 8 \\ 2 & 7 & 9 & 3 \\ 8 & 2 & 5 & 5 \\ 6 & 6 & 4 & 7 \end{bmatrix}$$

3.
$$\begin{bmatrix} 6 & 2 & 5 & 8 \\ 6 & 7 & 1 & 6 \\ 6 & 3 & 4 & 5 \\ 5 & 4 & 3 & 4 \end{bmatrix}$$

4.
$$\begin{bmatrix} 2 & 3 & 5 & 1 & 2 \\ 4 & 3 & 5 & 4 & 2 \\ 3 & 6 & 3 & 1 & 4 \\ 3 & 6 & 4 & 5 & 4 \\ 4 & 2 & 4 & 5 & 4 \end{bmatrix}$$

5.
$$\begin{bmatrix} 3 & 5 & 5 & 3 & 8 \\ 4 & 6 & 4 & 2 & 6 \\ 4 & 6 & 1 & 3 & 6 \\ 3 & 4 & 4 & 6 & 5 \\ 5 & 7 & 3 & 5 & 9 \end{bmatrix}$$

6.
$$\begin{bmatrix} 0 & 1 & 0 & -1 & 1 \\ 3 & 0 & 4 & 4 & 5 \\ 1 & 3 & 7 & 4 & 7 \\ -1 & -2 & 2 & 3 & 3 \\ 2 & 4 & 7 & 5 & 9 \end{bmatrix}$$

7.
$$\begin{bmatrix} 3 & 4 & 5 & 7 & 6 \\ 5 & 3 & 4 & 5 & 2 \\ 1 & 3 & 4 & 5 & 3 \\ 5 & 6 & 5 & 4 & 3 \end{bmatrix}$$

8.
$$\begin{bmatrix} 5 & 6 & 2 & 3 & 4 & 3 \\ 6 & 4 & 4 & 2 & 0 & 3 \\ 5 & 4 & 5 & 2 & 6 & 6 \\ 5 & 6 & 1 & 4 & 7 & 6 \end{bmatrix}$$

In Exercises 9–12 find the largest sum of an independent set of entries with as many elements as the matrix has rows.

9.
$$\begin{bmatrix} 5 & 4 & 2 & 3 \\ 3 & 1 & 4 & 3 \\ 1 & 1 & 1 & 3 \\ 5 & 3 & 6 & 3 \end{bmatrix}$$

10.
$$\begin{bmatrix} 5 & 4 & 3 & 4 \\ 5 & 3 & 1 & 7 \\ 7 & 5 & 2 & 10 \\ 2 & 4 & 2 & 7 \end{bmatrix}$$

11.
$$\begin{bmatrix} 6 & 5 & 3 & 1 & 4 \\ 2 & 5 & 3 & 7 & 8 \\ 8 & 3 & 7 & 5 & 4 \\ 7 & 1 & 5 & 3 & 8 \end{bmatrix}$$

12.
$$\begin{bmatrix} 6 & 7 & 3 & 8 & 9 \\ 4 & 7 & 5 & 6 & 2 \\ 2 & 5 & 8 & 6 & 9 \end{bmatrix}$$

13. A newspaper sports editor must send 4 of his reporters to 4 cities. From past experience he knows what expenses to expect from each reporter in each city. He can expect Addams to spend $700 in Los Angeles, $500 in New York, $200 in Las Vegas, and $400 in Chicago. Hart can be expected to spend $500, $500, $100, and

$600 in these cities, Young to spend $500, $300, $400, and $700, and Herriman to spend $400, $500, $600, and $500. How should the editor make the assignments to keep the total expenses minimal?

14. A supervisor has 5 salespeople who can be assigned to 5 different routes next month. Art can be expected to sell $9000 worth of goods on Route 1, $8000 on Route 2, $10,000 on Route 3, $7000 on Route 4, and $8000 on Route 5. Betty would sell $6000, $9000, $5000, $7000, and $4000 on these routes; Chester would sell $4000, $5000, $4000, $8000, and $2000; Denise would sell $4000, $7000, $5000, $4000, and $2000; and Ed would sell $5000, $5000, $7000, $9000, and $3000. What is the maximal total expected sales possible next month?

15. A foreman has 4 jobs and 5 workers he could assign them to. The time in hours each worker needs for each job is shown in the following table.

	Worker 1	Worker 2	Worker 3	Worker 4	Worker 5
Job 1	7	3	5	7	2
Job 2	6	1	4	2	6
Job 3	8	3	8	9	1
Job 4	7	2	1	5	6

After subtracting the minimum entries from the rows and columns of the corresponding matrix we have the matrix

$$\begin{bmatrix} 0^* & 1 & 3 & 4 & 0 \\ 0 & 0^* & 3 & 0 & 5 \\ 2 & 2 & 7 & 7 & 0^* \\ 1 & 1 & 0^* & 3 & 5 \end{bmatrix},$$

didn't use a square matrix

in which the stars indicate a maximal independent set of 0's. The corresponding job assignment will require a total of $7 + 1 + 1 + 1 = 10$ hours. But by assigning the jobs to workers 2, 4, 5, and 3 the total time could be reduced to $3 + 2 + 1 + 1 = 7$ hours. What is wrong?

Suggested Readings

1. Berge, C. *Graphs and Hypergraphs*. New York: American Elsevier, 1973.
2. Edmonds, J. "Paths, Trees, and Flowers." *Canad. J. Math.* 17, 1965, 449–467.
3. Ford, L. R. and D. R. Fulkerson. *Flows in Networks*. Princeton, NJ: Princeton University Press, 1962.
4. Hall, P. "On Representations of Subsets." *J. London. Math. Soc.* 10, 1935, 26–30.
5. Kuhn, H. W. "The Hungarian Method for the Assignment Problem." *Naval Res. Logist. Quart.* 2, 1955, 83–97.
6. Lawler, E. L. *Combinatorial Optimization: Networks and Matroids*. New York: Holt, Rinehart and Winston, 1976.
7. Mirsky, L. *Transversal Theory*. New York: Academic Press, 1971.
8. Roberts, Fred S. *Applied Combinatorics*. Englewood Cliffs, N. J.: Prentice-Hall, 1984.

6

NETWORK FLOWS

Many practical problems require the movement of some commodity from one location to another. For example, an oil company must move crude oil from the oil fields to its refinery, and a long-distance telephone company must move messages from one city to another. In both of these situations there is a limitation to the amount of the commodity that can be moved at one time. The volume of crude oil that the oil company can move, for instance, is limited by the capacity of the pipeline through which the oil must flow. And the number of telephone calls that the phone company can handle is limited by the capacity of its cable and its switching equipment. This type of problem, in which some commodity must be moved from one location to another subject to the restriction that certain capacities not be exceeded, is called a **network flow** problem. In this chapter we will be primarily concerned with solving such problems.

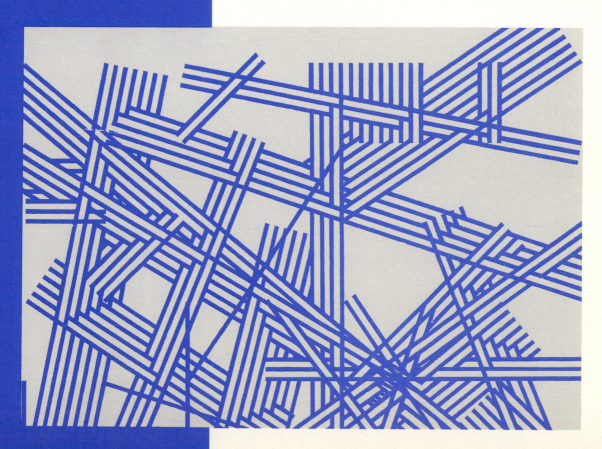

6.1   Flows and Cuts

In the example described above in which an oil company must ship crude oil from the oil fields to its refinery, there is one origin for the oil (the oil fields) and one destination (the refinery). However, there may be many different pipelines available through which the oil can be sent. Figure 6.1 shows this situation for an oil company with oil fields at Prudhoe Bay and a refinery in Seward, Alaska. (Here the pipeline capacities are given in thousands of barrels per day.) This figure showing the possible routes from the oil fields to the refinery is a special type of weighted directed graph.

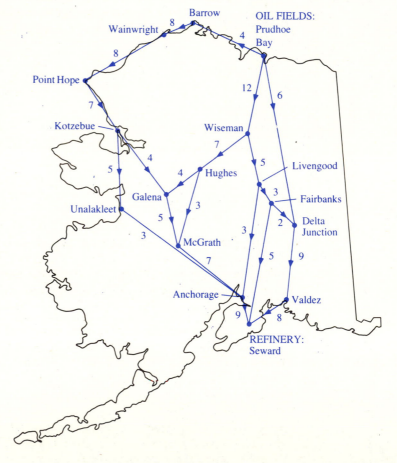

FIGURE 6.1

By a **transportation network,** or more simply a **network,** we mean a weighted directed graph satisfying the following three conditions.

(1) There is exactly one vertex having no incoming edges, i.e., exactly one vertex with indegree 0. This vertex is called the **source.**

(2) There is exactly one vertex having no outgoing edges, i.e., exactly one vertex with outdegree 0. This vertex is called the **sink.**

(3) The weight assigned to each edge is a nonnegative number.

In this context a directed edge of the network will be called an **arc,** and the weight of an arc will be called its **capacity.**

EXAMPLE 6.1   Figure 6.2 shows a weighted directed graph with five vertices and seven arcs. The seven arcs are: (A, B) with capacity 6, (A, C) with capacity 8, (A, D) with capacity 3, (B, C) with capacity 5, (B, D) with capacity 6, (C, E) with capacity 4, and

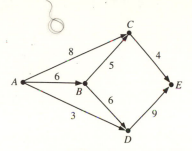

FIGURE 6.2

(D, E) with capacity 9. Clearly the capacity of each arc is a nonnegative number. Note that vertex A is the only vertex having no incoming arcs and vertex E is the only vertex having no outgoing arcs. Thus, the directed graph in Figure 6.2 is a transportation network with vertex A as its source and vertex E as its sink. ■

In a transportation network we consider a commodity flowing along arcs from the source to the sink. The amount carried by each arc must not exceed the capacity of the arc, and none of the commodity is lost along the way. Thus, at each vertex other than the source and the sink, the amount of the commodity that arrives must equal the amount of the commodity that leaves. We will formalize these ideas in the following definition.

Let A be the set of arcs in a transportation network N, and for each arc e in A let $C(e)$ denote the capacity of e. A **flow** in N is a function F that assigns to each arc e a number $F(e)$, called the **flow along arc e,** such that

(1) $0 \le F(e) \le C(e)$, and

(2) for each vertex v other than the source and sink, the total flow into v (the sum of the flows along all arcs ending at v) equals the total flow out of v (the sum of the flows along all arcs beginning at v).

Since the capacity of an arc is nonnegative, it is clear that the function F assigning the number 0 to each arc is always a flow in a transportation network. Consequently, every network has a flow.

EXAMPLE 6.2

For the transportation network in Figure 6.2, the function F defined by: $F(A, B) = 6, F(A, C) = 0, F(A, D) = 3, F(B, C) = 4, F(B, D) = 2, F(C, E) = 4$, and $F(D, E) = 5$ is a flow. This flow is shown in Figure 6.3, where the first number on each arc is its capacity and the second number is the flow along that

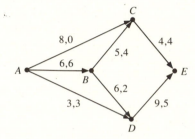

FIGURE 6.3

arc. Notice that each value of F is a nonnegative number that does not exceed the capacity of the corresponding arc. In addition, at vertices B, C, and D the total flow into the vertex equals the total flow out of the vertex. For instance, the total flow into vertex B is 6 along arc (A, B); and the total flow out of vertex B is also 6, 4 along arc (B, C) and 2 along arc (B, D). Likewise, the total flow into vertex D is 5, 3 along arc (A, D) and 2 along arc (B, D); and the total flow out of vertex D is also 5 along arc (D, E). ■

   In Figure 6.3 the total flow out of vertex A is 9, 6 along arc (A, B), 0 along arc (A, C), and 3 along arc (A, D). Notice that this number is the same as the total flow into vertex E, which is 4 along arc (C, E) and 5 along arc (D, E). This equality is a basic property of every flow.

THEOREM 6.1

For any flow in a transportation network, the total flow out of the source equals the total flow into the sink.

Proof.   Let $v_1, v_2, \ldots, v_n$ denote the vertices of the network, with v_1 being the source and v_n being the sink. Let F be a flow in this network, and for each k ($1 \le k \le n$) define I_k to be the total flow into v_k and O_k to be the total flow out of v_k. Finally, let S denote the sum of the flows along every arc in the network.
   For each arc $e = (v_j, v_k)$, $F(e)$ is included exactly once in the sum $I_1 + I_2 +$

$\ldots + I_n$ (in the term I_k) and exactly once in the sum $O_1 + O_2 + \ldots + O_n$ (in the term O_j). Hence, $I_1 + I_2 + \ldots + I_n = S$ and $O_1 + O_2 + \ldots + O_n = S$; so $O_1 + O_2 + \ldots + O_n = I_1 + I_2 + \ldots + I_n$. But for any vertex v_k other than the source and sink, $I_k = O_k$. Cancelling these common terms in the preceding equation gives $O_1 + O_n = I_1 + I_n$. Now $I_1 = 0$ because the source has no incoming arcs, and $O_n = 0$ because the sink has no outgoing arcs. Hence, we see that $O_1 = I_n$, that is, the total flow out of the source equals the total flow into the sink. ▉

If F is a flow in a transportation network, the common value of the total flow out of the source and the total flow into the sink is called the **value** of the flow F.

In the network shown in Figure 6.1 in which crude oil is to be shipped through pipelines, the oil company would be interested in knowing how much oil can be sent per day from the oil fields to the refinery. Likewise, in any transportation network it is important to know the amount of a commodity that can be shipped from the source to the sink without exceeding the capacities of the arcs. In other words, we would like to know the largest possible value of a flow in a transportation network. A flow having maximum value in a network is called a **maximal flow.**

In Section 6.2 we will present an algorithm for finding a maximal flow in a transportation network. In order to understand this algorithm better, we will first consider some of the ideas that are involved in finding a maximal flow. Suppose, for example, that we want to find a maximal flow in the transportation network shown in Figure 6.2. Because this network is so small, it will not be difficult to determine a maximal flow by a little experimentation. Our approach will be to find a sequence of flows with increasing values. We will begin by taking the flow to be zero along every arc. Thus, the current flow is as in Figure 6.4, where the numbers along each arc are the arc capacity and the current flow along the arc, respectively.

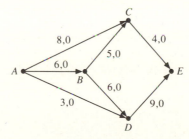

FIGURE 6.4

Now we will try to find a path from the source to the sink along which we can increase the present flow. Such a path is called a **flow-augmenting** path. In this case, since there is no arc along which the flow equals the capacity, any directed path from the source to the sink will suffice. Suppose that we choose the path A, C, E. By how much can we increase the flow along the arcs in this path?

Because the capacities of the arcs (A, C) and (C, E) in this path are 8 and 4, respectively, it is clear that we can increase the flows along these two arcs by 4 without exceeding their capacities. Recall that we are only changing the flow along arcs in our chosen path A, C, E and that the flow out of vertex C must equal the flow into C. Consequently, if we tried to increase the flow along arc (A, C) by more than 4, then the flow along arc (C, E) would also need to be increased by more than 4. But a flow along arc (C, E) that is greater than 4 would exceed the capacity of this arc. Hence, the largest amount by which we can increase the flow along the path A, C, E is 4. When we increase the flow in this manner, we obtain the flow shown in Figure 6.5.

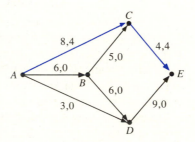

FIGURE 6.5

Now we will try to find another flow-augmenting path so that we can increase the present flow. Note that such a path cannot use arc (C, E) because the flow in this arc is already at its capacity. One acceptable path is A, D, E. For this path we can increase the flow by as much as 3 without exceeding the capacity of any arc. (Why?) If we increase the flows in arcs (A, D) and (D, E) by 3, we obtain the new flow shown in Figure 6.6.

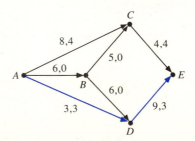

FIGURE 6.6

Again we will try to find a flow-augmenting path. Path A, B, D, E is such a path. For this path we can increase the flow by as much as 6 without exceeding the capacity of any arc. If we increase the flows in arcs (A, B), (B, D), and (D, E) by 6, we obtain the new flow shown in Figure 6.7.

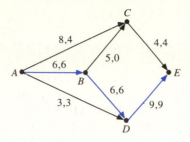

FIGURE 6.7

Is it possible to find another flow-augmenting path? Note that any path leading to the sink must use either arc (C, E) or arc (D, E) because these are the only arcs leading to the sink. But the flow along these arcs is already at the capacity of the arcs. Consequently, it is not possible to increase the flow in Figure 6.7 any further, and so this flow is a maximal flow. The value of this flow is 13, the common value of the flow out of the source and into the sink.

The argument used above to justify that there could be no flow having a value larger than 13 is an important one. As this argument suggests, the value of a maximal flow is limited by the capacities of certain sets of arcs. Recall once more the oil pipeline network in Figure 6.1. Suppose that after analyzing this network you have determined that the value of a maximal flow is 18 but that your colleagues at the oil company are questioning your calculation. They point out that it is possible to ship 22 thousand barrels per day out of Prudhoe Bay and 22 thousand barrels per day into Seward; so they believe that there should be a flow having the value 22. How can you convince them that there can be no flow having a value greater than 18?

Suppose that the vertices of the network are partitioned into two sets S and T such that the source belongs to S and the sink belongs to T. (Recall that this statement means that each vertex belongs to exactly one of the sets S or T.) Since every path from the source to the sink begins at a vertex in S and ends at a vertex in T, each such path must contain an arc that joins some vertex in S to some vertex in T. So if we can partition the vertices of the network into sets S and T in such a way that the total capacity of the arcs going from a vertex in S to a vertex in T is 18, we will have proved that there can be no flow with a value greater than 18.

It can be seen in Figure 6.8 that such a partition is obtained by taking

$$T = \{\text{Fairbanks, Delta Junction, Valdez, Seward}\}$$

and S to be the other cities in the figure. The black line in Figure 6.8 separates the cities in S (northwest of the line) from the cities in T (southeast of the line). Notice that the only arcs joining a city in S to a city in T are those from Anchorage to Seward (with capacity 9), Livengood to Fairbanks (with capacity 3), and Prudhoe Bay to Delta Junction (with capacity 6). These arcs have a total capacity of $9 + 3 + 6 = 18$, and so no flow from a vertex in S to a vertex in T can exceed this number.

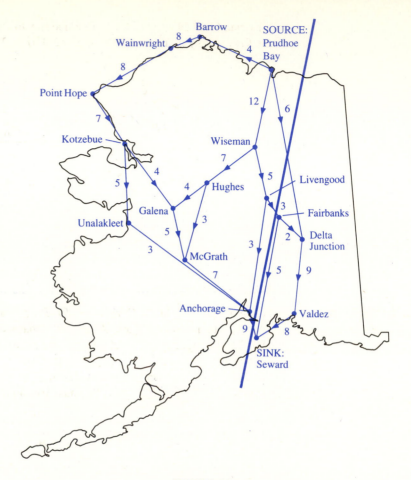

FIGURE 6.8

Generalizing from this example, we define a **cut** in a network to be a partition of its vertices into two sets S and T such that the source lies in S and the sink belongs to T. The sum of the capacities of all arcs leading from a vertex in S to a vertex in T is called the **capacity** of the cut. Note that in determining the capacity of the cut we consider only the capacity of arcs leading from a vertex in S to a vertex in T and not those leading from a vertex in T to a vertex in S.

EXAMPLE 6.3   In Figure 6.8 let

$$S = \{\text{Prudhoe Bay, Barrow, Wainwright, Point Hope, Kotzebue}\}$$

and T contain the cities not in S. Then $\{S, T\}$ is a cut because Prudhoe Bay is in S and Seward is in T. The arcs leading from a city in S to a city in T are Kotzebue to Unalakleet (with capacity 5), Kotzebue to Galena (with capacity 4), Prudhoe Bay to Wiseman (with capacity 12), and Prudhoe Bay to Delta Junction (with capacity 6). So the capacity of this cut is $5 + 4 + 12 + 6 = 27$. ■

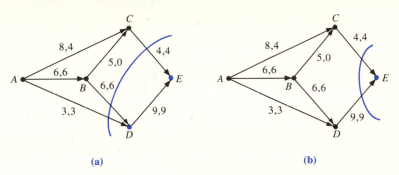

(a) (b)

FIGURE 6.9

EXAMPLE 6.4   In Figure 6.9(a), $S = \{A, B, C\}$ and $T = \{D, E\}$ form a cut. The arcs leading from a vertex in S to a vertex in T are (A, D) with capacity 3, (B, D) with capacity 6, and (C, E) with capacity 4. Therefore, the capacity of this cut is $3 + 6 + 4 = 13$.

The sets $S' = \{A, B, C, D\}$ and $T' = \{E\}$ also form a cut. See Figure 6.9(b). In this case the arcs leading from a vertex in S' to a vertex in T' are (C, E) with capacity 4 and (D, E) with capacity 9. Thus, this cut also has capacity 13. ■

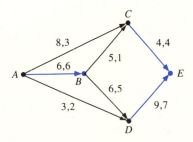

FIGURE 6.10

In Figure 6.10 let $\{S, T\}$ be the cut with $S = \{A, C, D\}$ and $T = \{B, E\}$. Let us consider the total flow (not the capacities) along the arcs joining vertices in S and T. Notice first that the total flow from S to T (that is, the total flow along arcs leading from a vertex in S to a vertex in T) is $6 + 4 + 7 = 17$, the sum of the flows along the blue arcs (A, B), (C, E), and (D, E), respectively. Likewise, the total flow from T to S is $1 + 5 = 6$, the sum of the flows along the black arcs (B, C) and (B, D). The difference between the total flow from S to T and the total flow from T to S is therefore $17 - 6 = 11$, which is the value of the flow shown in Figure 6.10. This equality is true in general, as the next theorem shows.

THEOREM 6.2   If F is a flow in a transportation network and $\{S, T\}$ is a cut, then the value of F equals the total flow along arcs leading from a vertex in S to a vertex in T minus the total flow along arcs leading from a vertex in T to a vertex in S.

Proof.   If X and Y are sets of vertices in the network, we will denote by $F(X, Y)$ the total flow along arcs leading from a vertex in X to a vertex in Y. With this notation the result to be proved can now be written as $a = F(S, T) - F(T, S)$, where a is the value of F. Note that if $Y_1 \cap Y_2 = \emptyset$, then clearly $F(X, Y_1 \cup Y_2) = F(X, Y_1) + F(X, Y_2)$; and likewise, if $X_1 \cap X_2 = \emptyset$, then $F(X_1 \cup X_2, Y) = F(X_1, Y) + F(X_2, Y)$.

By condition (2) in the definition of a flow, $F(\{v\}, S \cup T) - F(S \cup T, \{v\}) = 0$ if v is neither the source nor the sink, and $F(\{v\}, S \cup T) - F(S \cup T, \{v\}) = a$ if v is the source. Summing these equations for all v in S gives the equation $F(S, S \cup T) - F(S \cup T, S) = a$. Thus,

$$a = F(S, S \cup T) - F(S \cup T, S)$$
$$= [F(S, S) + F(S, T)] - [F(S, S) + F(T, S)]$$
$$= F(S, T) - F(T, S). \quad \blacksquare$$

Corollary.   If F is a flow in a transportation network and $\{S, T\}$ is a cut, then the value of F cannot exceed the capacity of $\{S, T\}$.

Proof.   Using the notation in the proof of Theorem 6.2, we have

$$a = F(S, T) - F(T, S) \leq F(S, T)$$

since $F(T, S) \geq 0$. But the flow along any arc leading from a vertex in S to a vertex in T cannot exceed the capacity of that arc. Thus, $F(S, T)$ cannot exceed the capacity of the cut $\{S, T\}$. It follows that the value of F cannot exceed the capacity of the cut $\{S, T\}$. $\quad \blacksquare$

The corollary to Theorem 6.2 is a useful result. It implies that the value of a maximal flow in a transportation network cannot exceed the capacity of *any* cut in the network. By using this fact we can easily obtain an upper bound on the value of a maximal flow. In Section 6.3 we will be able to strengthen this result by showing that every transportation network contains at least one cut with capacity equal to the value of a maximal flow. (Notice, for instance, that Example 6.4 presents two cuts with capacity equal to the value of the maximal flow in the network shown in Figure 6.7.) This fact will enable us to prove that a particular flow is a maximal flow as we did in analyzing the flow in Figure 6.7.

EXERCISES 6.1

In Exercises 1–6 tell whether the given weighted directed graph is a transportation net-work or not. If so, identify the source and sink. If not, tell why.

1.

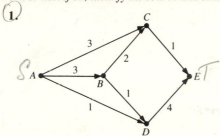

2.

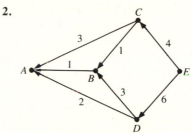

3.

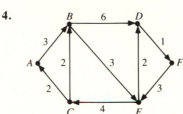

4.

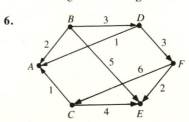

5.

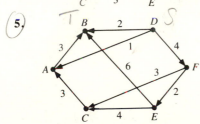

6.

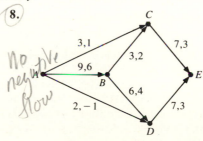

In Exercises 7–12 a transportation network is given. The first number along each arc gives the capacity of the arc. Tell whether the second set of numbers along the arcs is a flow for the network. If so, give the value of the flow. If not, tell why.

7.

8.

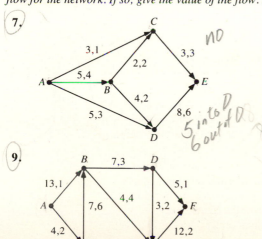

9.

10.

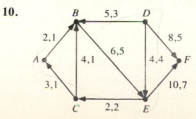

11.

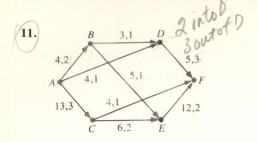

12.

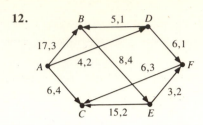

In Exercises 13–18 tell whether the given sets S and T form a cut for the network indicated. If so, give the capacity of the cut. If not, tell why.

13. $S = \{A, B\}$ and $T = \{D, E\}$ for the network in Exercise 7   *no*

14. $S = \{A, D\}$ and $T = \{B, C, E\}$ for the network in Exercise 8   *yes*

15. $S = \{A, D, E\}$ and $T = \{B, C, F\}$ for the network in Exercise 9

16. $S = \{A, B, C, D\}$ and $T = \{D, E, F\}$ for the network in Exercise 10

17. $S = \{A, D, E\}$ and $T = \{B, C, F\}$ for the network in Exercise 11   *yes*

18. $S = \{A, B, C\}$ and $T = \{D, E, F\}$ for the network in Exercise 12

In Exercises 19–24 find by inspection a flow satisfying the given conditions.

19. A flow of value 11 for the network in Exercise 7

20. A flow of value 13 for the network in Exercise 8

21. A flow of value 11 for the network in Exercise 9

22. A flow of value 17 for the network in Exercise 10

23. A flow of value 18 for the network in Exercise 11

24. A flow of value 18 for the network in Exercise 12

In Exercises 25–30 find by inspection a cut satisfying the given conditions.

25. A cut of capacity 11 for the network in Exercise 7

26. A cut of capacity 13 for the network in Exercise 8

27. A cut of capacity 11 for the network in Exercise 9

28. A cut of capacity 17 for the network in Exercise 10

29. A cut of capacity 18 for the network in Exercise 11

30. A cut of capacity 18 for the network in Exercise 12

31. A telephone call can be routed from Chicago to Atlanta along various lines. The line from Chicago to Indianapolis can carry 40 calls at the same time. Other lines and their capacities are: Chicago to St. Louis (30 calls), Chicago to Memphis (20 calls), Indianapolis to Memphis (15 calls), Indianapolis to Lexington (25 calls), St. Louis to Little Rock (20 calls), Little Rock to Memphis (15 calls), Little Rock to Atlanta (10 calls), Memphis to Atlanta (25 calls), and Lexington to Atlanta (15 calls). Draw a transportation network displaying this information.

32. A power generator at a dam is capable of sending 300 megawatts to substation 1, 200 megawatts to substation 2, and 250 megawatts to substation 3. In addition, substation 2 is capable of sending 100 megawatts to substation 1 and 70 megawatts to substation 3. Substation 1 can send at most 280 megawatts to the distribution center, and substation 3 can send at most 300 megawatts to the distribution center. Draw a transportation network displaying this information.

In Exercises 33–36 let $F(X, Y)$ be defined as in the proof of Theorem 6.2.

33. Find $F(X, Y)$ and $F(Y, X)$ if F is the flow in Exercise 10, $X = \{B, C, D\}$, and $Y = \{A, E, F\}$.

34. Find $F(X, Y)$ and $F(Y, X)$ if F is the flow in Exercise 12, $X = \{C, E, F\}$, and $Y = \{A, B, D\}$.

35. Give an example to show that $F(X, Y_1 \cup Y_2)$ may not equal $F(X, Y_1) + F(X, Y_2)$ if Y_1 and Y_2 are not disjoint.

36. Prove that $F(X, Y \cup Z) = F(X, Y) + F(X, Z) - F(X, Y \cap Z)$ for any sets of vertices X, Y, and Z.

6.2   A Maximal Flow Algorithm

In this section we will present an algorithm for finding the maximal flow in a transportation network. This algorithm is based on a procedure formulated by Ford and Fulkerson and utilizes a modification suggested by Edmonds and Karp. (See suggested readings [5] and [3] at the end of this chapter.) The essence of the algorithm is described in Section 6.1:

(1) Begin with any flow, for example the one having zero flow along every arc.
(2) Find a flow-augmenting path (a path from the source to the sink along which the present flow can be increased), and increase the flow along this path by as much as possible.
(3) Repeat step (2) until it is no longer possible to find a flow-augmenting path.

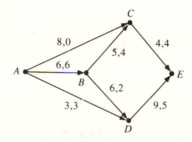

FIGURE 6.11

Some care must be taken in deciding if there is a flow-augmenting path. Consider, for example, the transportation network in Figure 6.11, where the numbers along each arc are the capacity of the arc and the present flow along the arc, respectively. This flow was obtained by sending 4 units of flow along path A, B, C, E; then 3 units along path A, D, E; and then 2 units along path A, B, D, E.

The value of the flow in Figure 6.11 is 9, and we know from the argument following Figure 6.7 that the value of a maximal flow in this network is 13. Consequently, we will look for a path from A to E along which the present flow can be increased. Clearly the only way to increase the flow out of A is to use arc

(*A*, *C*). But (*C*, *E*) is the only arc leading out of vertex *C*, and the present flow along this arc equals its capacity. Thus, we cannot increase the flow along arc (*C*, *E*). Therefore, there is no *directed* path from *A* to *E* along which the flow can be increased. But if we allowed flow from vertex *C* to vertex *B* along the arc (*B*, *C*), we could send 4 units of flow along path *A*, *C*, *B*, *D*, *E*. This additional 4 units would give us a maximal flow from *A* to *E*.

How can we justify sending 4 units from *C* to *B* when the arc is directed from *B* to *C*? Since there is already 4 units of flow along arc (*B*, *C*), sending 4 units of flow from *C* to *B* has the effect of cancelling the previous flow along (*B*, *C*). Thus, by sending 4 units of flow along path *A*, *C*, *B*, *D*, *E*, we obtain the maximal flow shown in Figure 6.7.

If we look at the network in Figure 6.11 more carefully, we can see that our first path *A*, *B*, *C*, *E* was not well chosen. For by using the arc (*C*, *E*) as part of this path, we prevent the later use of arc (*A*, *C*). (Note that because there is no arc except (*C*, *E*) leaving vertex *C*, any flow sent into vertex *C* along arc (*A*, *C*) must leave along arc (*C*, *E*).) Therefore, the use of arc (*B*, *C*) in the path *A*, *C*, *B*, *D*, *E* corrects the original poor choice of path *A*, *B*, *C*, *E*. Clearly our algorithm will need some method to correct a poor choice of path from source to sink made earlier. In the version of the algorithm stated below, this correction occurs in step 4(b).

This algorithm, like the maximal independent set algorithm in Section 5.3, is based on the labeling procedure devised by Ford and Fulkerson. In the algorithm, we perform two operations on vertices called *labeling* and *scanning*. Here again a vertex must be labeled before it can be scanned.

Maximal Flow Algorithm

For a transportation network in which arc (*x*, *y*) has capacity $C(x, y)$, this algorithm either indicates that the current flow *F* is a maximal flow or else replaces *F* by a flow with a larger value.

Step 1 (label the source). Label the source with the triple (source, $+$, ∞).

Step 2 (check for breakthrough). If the sink has been labeled, go to step 5.

Step 3 (check for a maximal flow). If every labeled vertex has been scanned, the present flow is a maximal flow; stop.

Step 4 (scan a vertex). Among all the vertices that have been labeled but not scanned, let *v* denote the one that was labeled first, and suppose that the label on *v* is (*u*, $\pm$, *a*). For each unlabeled vertex *w*, perform exactly one of the following three actions:
   (a) If (*v*, *w*) is an arc and $F(v, w) < C(v, w)$, assign to *w* the label (*v*, $+$, *b*), where *b* is the smaller of *a* and $C(v, w) - F(v, w)$.
   (b) If (*w*, *v*) is an arc and $F(w, v) > 0$, assign to *w* the label (*v*, $-$, *b*), where *b* is the smaller of *a* and $F(w, v)$.
   (c) If neither (a) nor (b) holds, do not label *w*.
   When each unlabeled vertex has been considered, vertex *v* has been scanned. Go to step 2.

Step 5 (breakthrough). Let *v* denote the sink.

Step 6 (increase the flow). If the label on *v* is $(u, +, a)$, replace $F(u, v)$ by $F(u, v) + a$; otherwise, if the label on *v* is $(u, -, a)$, replace $F(v, u)$ by $F(v, u) - a$. If *u* is the source, stop; otherwise, if *u* is not the source, let $v = u$ and repeat step 6.

If the present flow is not a maximal flow, this algorithm uses breadth-first search to find a shortest flow-augmenting path (that is, one with the fewest arcs). Each vertex *v* along this path is labeled in one of two ways: $(u, +, a)$ or $(u, -, a)$. The first entry of the label, *u*, signifies that vertex *u* precedes *v* on this path. The second entry of the label denotes that (u, v) is an arc or that (v, u) is an arc on this path, depending on whether the entry is + or −, respectively. And the third entry of the label, *a*, is a positive number indicating how much the present flow can be increased (if the second entry of the label is +) or decreased (if the second entry of the label is −) without violating the restrictions in condition (1) of the definition of a flow for any arc along the path from the source to *v*.

We will illustrate the use of the maximal flow algorithm by finding a maximal flow for the network discussed in Section 6.1. When we reach step 4 of the algorithm, we will examine the unlabeled vertices in alphabetical order. In order to begin the algorithm, we will take the flow to be 0 along every arc, as shown in Figure 6.12. (Again the two numbers written beside each arc are the capacity and the present flow along that arc.)

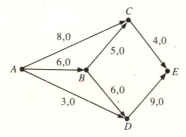

FIGURE 6.12

In step 1 we assign the label (source, +, ∞) to the source, vertex *A*. In step 2 we check to see if the sink (vertex *E*) has been labeled; since it has not been labeled, we move to step 3. At this stage vertex *A* has been labeled but not scanned; so we proceed to step 4. In step 4 we will scan vertex *A* by examining the unlabeled vertices (*B*, *C*, *D*, and *E*) to see if any of them can be assigned labels. Note that vertex *B* is unlabeled, (A, B) is an arc, and the flow along this arc (0) is less than the capacity (6). Thus, we can perform action (a) in step 4 on vertex *B*. Since 6 is the smaller of ∞ (the third entry in the label on *A*) and $6 - 0$, we label *B* with $(A, +, 6)$. Likewise, we can perform action (a) on the vertices *C* and *D*, which assigns them the labels $(A, +, 8)$ and $(A, +, 3)$, respectively. Because vertex *E* is not joined to vertex *A*, it cannot yet be given a label. This completes the scanning of vertex *A*. The current labels are shown in Figure 6.13.

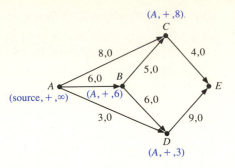

FIGURE 6.13

After scanning vertex A, we return to step 2. The sink is still unlabeled, so we proceed to step 3. There are 3 vertices that have been labeled but not scanned (namely, vertices B, C, and D), and so we move again to step 4. Of the three labeled vertices that have not been scanned, vertex B was the first one to be labeled, and so we scan vertex B. Because there are no unlabeled vertices joined by an arc to vertex B, no changes result from the scanning of vertex B. Therefore, we return to step 2 once more. As before, we proceed to step 3 and then to step 4. At this stage there are two labeled vertices that have not been scanned (namely, vertices C and D), and of these C was the first to be labeled. Thus, we scan vertex C in step 4. Since E is unlabeled, (C, E) is an arc, and the flow along this arc (0) is less than its capacity (4), we perform action (a). This action assigns to vertex E the label $(C, +, 4)$ because 4 is the smaller of 8 (the third entry in the label on C) and $4 - 0$ (the capacity minus the flow along arc (C, E)). Since there are no unlabeled vertices remaining, this completes the scanning of vertex C. The present labels are shown in Figure 6.14.

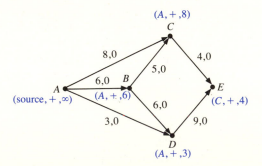

FIGURE 6.14

Having completed the scanning of vertex C, we return again to step 2. This time, however, the sink has been labeled, and so we proceed to steps 5 and 6. The fact that the sink has been labeled $(C, +, 4)$ tells us that the current flow can be increased by 4 along a path through vertex C. The vertex that precedes C in this path is the first entry in the label on C, which is $(A, +, 8)$. Thus, the path from the source to the sink along which the flow can be increased by 4 is A, C, E. When we increase the flow along the arcs in this path by 4, the resulting flow is as in Figure 6.15.

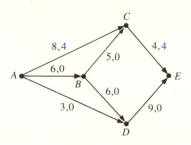

FIGURE 6.15

This finishes step 6, and so the first iteration of the algorithm has been completed. We now remove all the labels and repeat the algorithm with the flow in Figure 6.15. As before, we assign to vertex A the label (source, $+$, ∞) and then proceed to steps 2, 3, and 4. In step 4 we scan vertex A. This results in vertices B, C, and D receiving the respective labels $(A, +, 6)$, $(A, +, 4)$, and $(A, +, 3)$. After returning to step 2, we reach step 4 again. Since B is the unscanned vertex that was labeled first, we now scan vertex B. As in the first iteration of the algorithm, no changes result from the scanning of vertex B. So we return to step 2 and eventually reach step 4 again. This time vertex C is the unscanned vertex that was labeled first. But unlike the first iteration we cannot label vertex E because the flow along arc (C, E) is not less than the capacity of the arc. Consequently, no changes result from scanning vertex C. Once more we return to step 2 and eventually reach step 4. This time vertex D is the only labeled vertex that has not been scanned, and so we scan vertex D. Since (D, E) is an arc along which the flow is less than the capacity, we perform action (a). As a result of this action vertex E is labeled $(D, +, 3)$. This completes the scanning of vertex D, and so we return to step 2. But now the sink has been labeled; so we proceed to steps 5 and 6. (See Figure 6.16.)

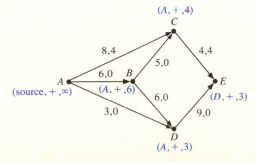

FIGURE 6.16

Because the label on the sink is $(D, +, 3)$, we can increase the current flow by 3 along a path through vertex D. To find the vertex that precedes D in this path, we examine the label on D, which is $(A, +, 3)$. Since the first entry of this label is A, we see that the path along which the flow can be increased by 3 is A, D, E. When we increase the flow along the arcs in this path by 3, we obtain the flow shown in Figure 6.17. This completes the second iteration of the algorithm.

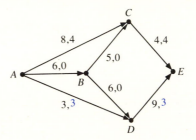

FIGURE 6.17

Again we discard all of the labels and perform another iteration of the algorithm. In this third iteration we reach steps 5 and 6 with the labels shown in Figure 6.18. From these labels we see that the flow can be increased by 6 along the path

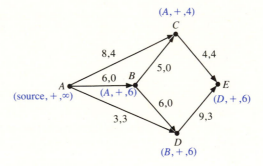

FIGURE 6.18

A, B, D, E. By increasing the flows along the arcs in this path by 6, we obtain the flow in Figure 6.19.

Again we discard all of the labels and perform another iteration of the algorithm. This time, however, we can label only vertex C when scanning vertex A. (See Figure 6.20.) Moreover, when vertex C is scanned, no changes occur. Consequently, in returning to step 2, we find that all of the labeled vertices have been scanned. Thus, step 3 assures us that the present flow (the one shown in Figure 6.19) is a maximal flow. When the algorithm ends, the set of labeled vertices $S = \{A, C\}$ and the set of unlabeled vertices $T = \{B, D, E\}$ form a cut. Notice that

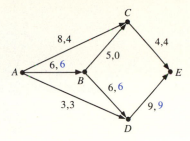

FIGURE 6.19

the capacity of this cut is $4 + 6 + 3 = 13$, which equals the value of the maximal flow in Figure 6.19. We will see in Section 6.3 that this is no coincidence: *When the maximal flow algorithm ends, the sets of labeled and unlabeled vertices always determine a cut with capacity equal to the value of a maximal flow.*

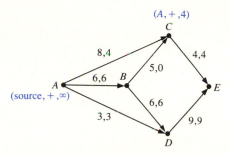

FIGURE 6.20

EXAMPLE 6.5   We will use the maximal flow algorithm to find a maximal flow for the network shown in Figure 6.21. When labeling vertices in step 4, we will consider them in alphabetical order.

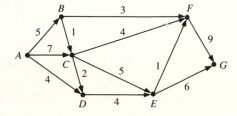

FIGURE 6.21

Iteration 1.   The labels assigned in iteration 1 are shown in Figure 6.22. Thus, we increase the flow by 3 along the path A, B, F, G.

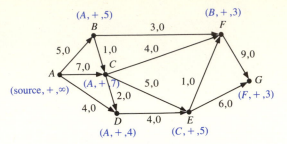

FIGURE 6.22

Iteration 2.   The labels assigned in iteration 2 are shown in Figure 6.23. Thus, we increase the flow by 5 along the path A, C, E, G.

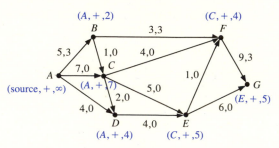

FIGURE 6.23

Iteration 3.   The labels assigned in iteration 3 are shown in Figure 6.24. Thus, we increase the flow by 2 along the path A, C, F, G.

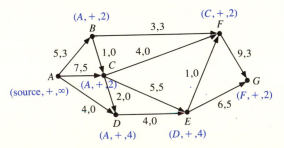

FIGURE 6.24

Iteration 4.   The labels assigned in iteration 4 are shown in Figure 6.25. Thus, we increase the flow by 1 along the path A, D, E, G.

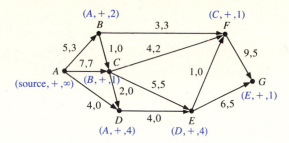

FIGURE 6.25

Iteration 5.   The labels assigned in iteration 5 are shown in Figure 6.26. Thus, we increase the flow by 1 along the path A, B, C, F, G.

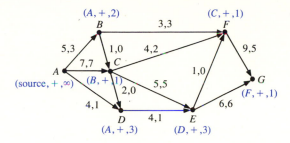

FIGURE 6.26

Iteration 6.   The labels assigned in iteration 6 are shown in Figure 6.27. Thus, we increase the flow by 1 along the path A, D, E, F, G.

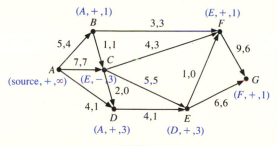

FIGURE 6.27

Iteration 7.   The labels assigned in iteration 7 are shown in Figure 6.28. Thus, we increase the flow by 1 along the path A, D, E, C, F, G. (Note that we are using arc (C, E) in the wrong direction to cancel 1 unit of the flow sent along this arc in iteration 2.)

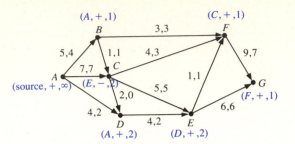

FIGURE 6.28

Iteration 8.   The labels assigned in iteration 8 are shown in Figure 6.29.

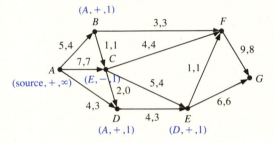

FIGURE 6.29

Since the sink is not labeled, the flow shown in Figure 6.29 is a maximal flow. The value of this maximal flow is 14. Note that the set $S = \{A, B, C, D, E\}$ of labeled vertices and the set $T = \{F, G\}$ of unlabeled vertices form a cut with capacity $3 + 4 + 1 + 6 = 14$. ■

Maximal flows need not be unique. For instance, the flow shown in Figure 6.30 is a maximal flow for the network in Example 6.5. This flow is different from the one shown in Figure 6.29.

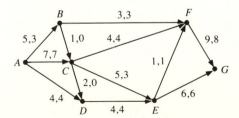

FIGURE 6.30

We conclude this section with a useful observation about the maximal flow algorithm. Any flow in which the flow along each arc is an integer is called an **integral flow.** Suppose that all of the arc capacities in a network are integers and we begin the maximal flow algorithm with an integral flow. In this case the third entry of each label, which is assigned during step 4 of the algorithm, is a minimum of integers. Consequently, *if all the arc capacities are integers and we begin with zero flow along every arc, the maximal flow algorithm produces an integral flow.*

EXERCISES 6.2

Throughout these exercises, if there is a choice of vertices to label when using the maximal flow algorithm, label the vertices in alphabetical order.

In Exercises 1–4 a network, a flow, and a flow-augmenting path are given. Determine the amount by which the flow can be increased along the given path.

1. Path: *A, B, D, E*

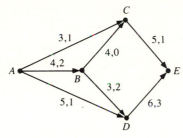

2. Path: *A, B, C, E*

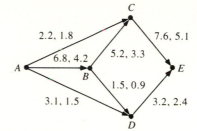

3. Path: *A, B, E, D, F*

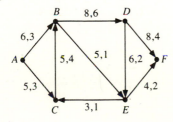

4. Path: *D, B, C, E, F*

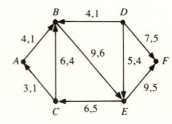

In Exercises 5–8 a network and flow are given. By performing the maximal flow algorithm on this network and flow, we obtain the labels shown in each network. Determine a flow having a larger value than the given flow by performing steps 5 and 6 of the maximal flow algorithm.

5.

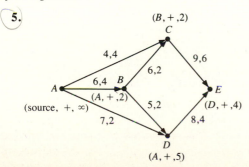

6.

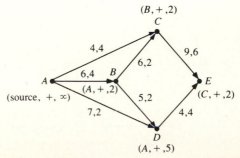

7.

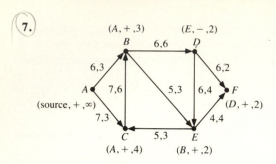

(A, +, 3)   (E, −, 2)

8.

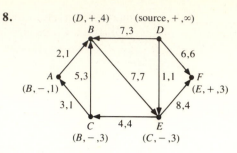

(D, +, 4)   (source, +, ∞)

In Exercises 9–16 a network and a flow are given. Use the maximal flow algorithm to show that the given flow is maximal or else to find a flow with a larger value. If the given flow is not maximal, name the flow-augmenting path and the amount by which the flow can be increased.

9.

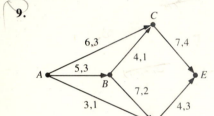

10.

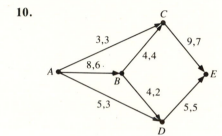

11.

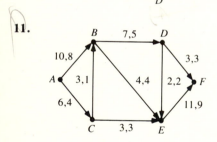

12.

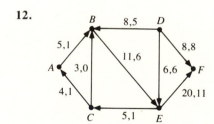

13.

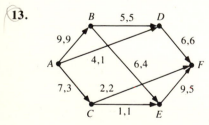

14.

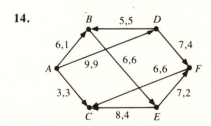

15.

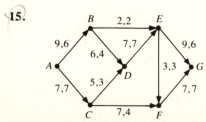

16.

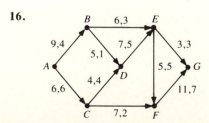

In Exercises 17–20 a transportation network and a flow are given. Use the maximal flow algorithm to find a maximal flow for each network.

17.

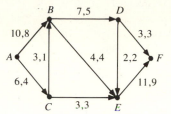

18.

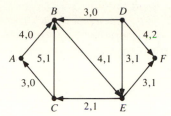

19.

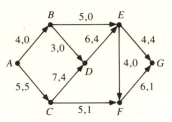

20.

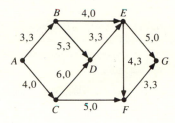

In Exercises 21–26 a transportation network is given. Find a maximal flow in each network by starting with the flow that is 0 along every arc and applying the maximal flow algorithm.

21.

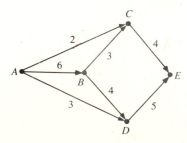

22.

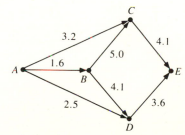

23.

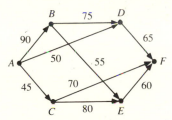

24.

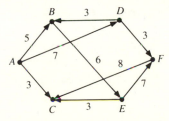

25.

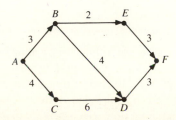

26.

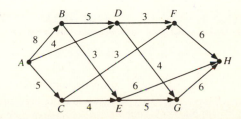

By a ***multisource transportation network*** *we mean a weighted directed graph that satis-fies the conditions in the definition of a transportation network except that instead of its containing only one vertex of indegree 0, there is a nonempty finite set S_0 of vertices with indegree 0. We say that F is a **flow** on such a network if:*

(i) $0 \leq F(e) \leq C(e)$ for each arc e, where C(e) is the capacity of arc e; and

(ii) for each vertex v other than the sink and the elements of S_0, the total flow into v equals the total flow out of v.

*The **value** of such a flow is the total flow into the sink, and a flow is called a **maximal flow** if its value is as large as possible.*

27. Show that in a multisource transportation network the value of a flow equals the total flow out of all the vertices in S_0.

28. Given a multisource transportation network N, create from N a transportation network N' by introducing a new vertex u and edges with infinite capacity from u to each element in S_0. Show that a maximal flow in N may be found by applying the maximal flow algorithm to N'.

By using Exercise 28, find a maximal flow in each of the multisource transportation networks given in Exercises 29–30.

29.

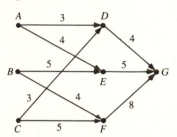

30.

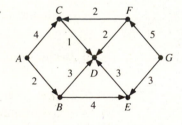

31. Generalize the concept of a transportation network to allow multiple sources and sinks; then define "flow," "value of a flow," and "maximal flow" for such a network. State and prove an analogue of Theorem 6.1 for such a network.

32. For a transportation network in which each arc has an integral capacity, give an example of a maximal flow such that the flow in some arcs is not an integer.

33. Consider a transportation network with source u and sink v in which each arc has capacity 1. Show that the value of a maximal flow equals the maximal number of directed paths from u to v that have no arcs in common.

34. Write a computer program to implement the maximal flow algorithm.

6.3   The Max-Flow Min-Cut Theorem

In this section we will show that the maximal flow algorithm described in Section 6.2 does what we claim; that is, it either confirms that the current flow is a maximal flow or else finds a flow having a larger value. We will also verify the observation that when the algorithm ends, the sets of labeled and unlabeled vertices determine a cut with capacity equal to the value of a maximal flow. Such cuts are of special interest because they are cuts having the smallest possible capacity, as we will see in Theorem 6.4.

A cut in a transportation network is called a **minimal cut** if no other cut has a smaller capacity. The theorem below provides a method for detecting maximal flows and minimal cuts.

THEOREM 6.3

In any transportation network, if F is a flow and $\{S, T\}$ is a cut such that the value of F equals the capacity of $\{S, T\}$, then F is a maximal flow and $\{S, T\}$ is a minimal cut.

Proof.   Let F be a flow having value c, and let $\{S, T\}$ be a cut having capacity c. Let F' be any other flow in this network, and let the value of F' be v. Applying the corollary to Theorem 6.2 to F' and $\{S, T\}$, we see that $v \leq c$. Hence, there is no flow in this network having a value greater than c, the value of F. It follows that F is a maximal flow.

Now let $\{S', T'\}$ be a cut in this network with capacity k. Applying the corollary to Theorem 6.2 to F and $\{S', T'\}$ gives $c \leq k$. Consequently, there is no cut in this network having a value less than c, the capacity of $\{S, T\}$. Therefore, $\{S, T\}$ is a minimal cut.   ▪

In order to justify the validity of the maximal flow algorithm we need to show that:

(1) The algorithm ends after a finite number of iterations.
(2) When the algorithm ends in step 3, the current flow is a maximal flow.
(3) When the algorithm ends in step 6, the original flow has been replaced by a flow with a larger value.

Statement 1 above will be considered in Theorem 6.5, and the proof of statement 3 will be left as an exercise (Exercise 24). The theorem below verifies statement 2 by showing that if an iteration of the maximal flow algorithm ends with the sink unlabeled, then the present flow is a maximal flow.

THEOREM 6.4

If, during an iteration of the maximal flow algorithm, the sink is not labeled, then the present flow is maximal. Moreover, the sets of labeled and unlabeled vertices form a minimal cut having capacity equal to the value of the present flow.

Proof.   Suppose that during some iteration of the maximal flow algorithm the sink is not labeled. Let F denote the current flow, $C(x, y)$ the capacity of arc (x, y), S the set of labeled vertices, and T the set of unlabeled vertices. Then the source is in S and the sink is in T; so $\{S, T\}$ is a cut.

Let (x, y) be an arc leading from a vertex x in S to a vertex y in T. Since x is in S, x has been labeled during this iteration of the maximal flow algorithm. If $F(x, y) < C(x, y)$, then when x was scanned, we would have labeled y in step 4(a) of the algorithm. But y is in T and, hence, is unlabeled; thus, we must have $F(x, y) = C(x, y)$.

Now let (y, x) be an arc leading from a vertex y in T to a vertex x in S. Since x is in S, x has been labeled during this iteration of the algorithm. When x is scanned, we would have labeled y in step 4(b) of the algorithm if $F(y, x) > 0$. But y is in T and so is unlabeled; thus, we must have $F(y, x) = 0$.

By Theorem 6.2 the value of F equals the total flow p along all arcs leading from a vertex in S to a vertex in T minus the total flow q along all arcs leading from a vertex in T to a vertex in S. But the two preceding paragraphs show that p equals the capacity of the cut $\{S, T\}$ and $q = 0$. Therefore, the value of F equals p, the capacity of the cut $\{S, T\}$. It then follows from Theorem 6.3 that F is a maximal flow and $\{S, T\}$ is a minimal cut. ■

Theorem 6.4 also proves our earlier assertion that when the maximal flow algorithm ends, the cut determined by the sets of labeled and unlabeled vertices is a minimal cut. Thus, for example, in Figure 6.20 we see that $S = \{A, C\}$ and $T = \{B, D, E\}$ form a minimal cut, and in Figure 6.29 we see that $S = \{A, B, C, D, E\}$ and $T = \{F, G\}$ form a minimal cut.

We have already seen that a network may have more than one maximal flow. Likewise, a network may have more than one minimal cut. In Figure 6.20, for instance, $\{A, B, C\}$ and $\{D, E\}$ is a different minimal cut from the one mentioned in the paragraph above.

EXAMPLE 6.6

A natural gas utility delivers gas to Little Rock from a source in Amarillo through the network of pipelines shown in Figure 6.31. In this diagram the first number beside each pipeline is the capacity of the pipeline and the second is the present flow, both measured in hundreds of millions of cubic feet per day. The utility has proposed raising its rates to pay for additional pipelines. Although the Arkansas Regulatory Commission agrees that more than the present 14.7 hundred million cubic feet of gas are needed in Little Rock each day, it is not convinced that additional pipelines need to be built. It questions the need for more pipelines because most of the pipelines operated by the utility are not being used to capacity, and some are not being used at all. How should the utility argue for new pipelines?

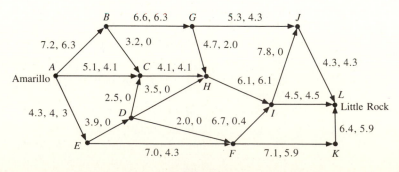

FIGURE 6.31

In order to justify its request for additional pipelines, the utility should apply the maximal flow algorithm to the network and flow in Figure 6.31. By doing so, it will find that only the vertices A, B, C, G, H, and J are labeled. Consequently, the flow in Figure 6.31 is a maximal flow, and

$$S = \{A, B, C, G, H, J\} \text{ and } T = \{D, E, F, I, K, L\}$$

form a minimal cut. The utility should, therefore, prepare a map as in Figure 6.32 with A, B, C, G, H, and J in the northwestern region and D, E, F, I, K, and L in the southeastern region. This map shows that only three pipelines (shown in color) carry gas from the northwestern region to the southeastern region, and each of these is being used to capacity. On this basis the utility can argue the need for more pipelines from the northwestern region to the southeastern region. ■

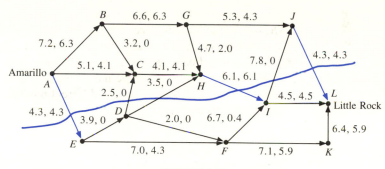

FIGURE 6.32

It is conceivable that the maximal flow algorithm may never end because no iteration occurs in which the sink cannot be labeled. Our next result, however, shows that this situation cannot occur if all the capacities in the network are rational numbers.

THEOREM 6.5   If all the capacities in a transportation network are rational numbers and we start with zero flow along each arc, then the maximal flow algorithm produces a maximal flow in a finite number of iterations.

Proof.   Suppose first that all the capacities in the network are integers. Let $\{S, T\}$ be the cut in which S consists of the source alone and T contains all the other vertices. Since all the arc capacities are integers, the capacity of the cut $\{S, T\}$ is an integer c.

Apply the maximal flow algorithm, beginning with zero flow along every arc. Now consider any iteration of the algorithm in which the sink is labeled. The label on the sink must be of the form $(u, +, a)$ or $(u, -, a)$, where $a > 0$. Moreover, because all the capacities are integers, a is a minimum of integers and hence is an integer. Therefore, $a \geq 1$, and each iteration of the maximal flow algorithm increases the value of the flow by at least 1. But by the corollary to Theorem 6.2, no flow in this network can have a value exceeding c. Hence, after at most $c + 1$ iterations the maximal flow algorithm must end with the sink unlabeled. But if the sink is unlabeled, then Theorem 6.4 guarantees that a maximal flow has been obtained.

Suppose now that all the capacities in the network are rational numbers. Find the least common denominator d of all the arc capacities, and consider the new network obtained by multiplying all of the original capacities by d. In this new network all the capacities are integers. Thus, applying the maximal flow algorithm to the new network must produce a maximal flow F in a finite number of steps by the argument above. But then this same sequence of steps will produce a maximal flow for the original network in which the flow along arc (x, y) is $F(x, y)/d$. (See Exercises 13–15.) ▬

Theorem 6.5 can be proved without the requirement that the capacities be rational numbers. More generally, Edmonds and Karp have shown that the maximal flow algorithm produces a maximal flow in no more than $\frac{1}{2}mn$ iterations, where m is the number of arcs and n is the number of vertices in the network. (See pages 117–119 of suggested reading [8] at the end of the chapter.) Note that in each iteration of the maximal flow algorithm we consider an arc (v, w) at most twice, once in the proper direction from v to w and once in the opposite direction from w to v. Thus, if we count the number of times that an arc is considered, the complexity of the maximal flow algorithm is at most $\frac{2m(mn)}{2} = m^2 n$. Since the number of arcs m cannot exceed $n(n - 1)$, it follows that the complexity of the maximal flow algorithm is at most $n^3(n - 1)^2$.

We will end this section by proving a famous theorem discovered independently by Ford and Fulkerson and Elias, Feinstein, and Shannon. (See suggested readings [6] and [4] at the end of the chapter.)

THEOREM 6.6   ***Max-Flow Min-Cut Theorem***   In any transportation network, the value of a maximal flow equals the capacity of a minimal cut.

Proof.   Let F be a maximal flow in a transportation network. Apply the maximal flow algorithm to this network with F as the current flow. Clearly the sink will not be labeled, for otherwise we would obtain a flow having a greater value than F, which is a maximal flow. But if the sink is not labeled, then Theorem 6.4 shows that the sets of labeled and unlabeled vertices form a minimal cut having capacity equal to the value of F. ▬

EXERCISES 6.3

In Exercises 1–4 give the capacity of the indicated cut for the network below.

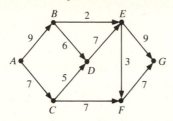

1. $S = \{A, C, F\}$ and $T = \{B, D, E, G\}$

2. $S = \{A, B, E\}$ and $T = \{C, D, F, G\}$

3. $S = \{A, D, E\}$ and $T = \{B, C, F, G\}$

4. $S = \{A, E, F\}$ and $T = \{B, C, D, G\}$

In Exercises 5–8 a network and a maximal flow are given. Find a minimal cut for the network by applying the maximal flow algorithm to this network and flow.

5.

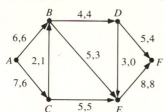

6.

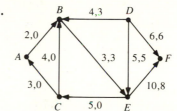

7.

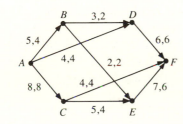

8.

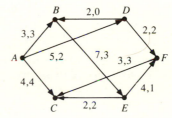

In Exercises 9–12 use the maximal flow algorithm to find a minimal cut.

9.

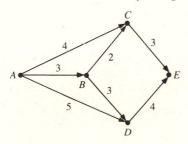

10.

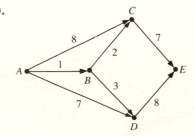

11.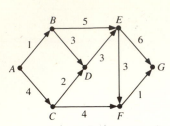

12.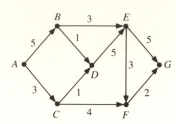

In Exercises 13–14 a network N with rational arc capacities is given. Let N′ be the network obtained from N by multiplying all the capacities in N by d, the least common denominator of the capacities. Apply the maximal flow algorithm to N′, and use the result to determine a maximal flow for the original network N.

13.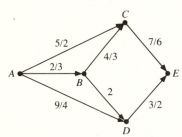

14.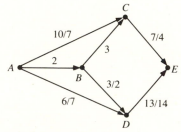

15. Let N be a transportation network and $d > 0$. Define $N′$ as in Exercises 13 and 14 to be the network with the same directed graph as N but with all the arc capacities of N multiplied by d.
(a) Show that $\{S, T\}$ is a minimal cut for $N′$ if and only if it is a minimal cut for N.
(b) Prove that if v and $v′$ are the values of maximal flows for N and $N′$, respectively, then $v′ = dv$.
(c) Show that F is a maximal flow for N if and only if $F′$ is a maximal flow for $N′$, where $F′$ is defined by $F′(x, y) = dF(x, y)$.

16. Suppose that D is a weighted directed graph having nonnegative weights (capacities) on each directed edge. Show that if any two distinct vertices of D are designated as the source and the sink, then the maximal flow algorithm will produce a maximal flow from the source to the sink. (Thus, the maximal flow algorithm can be used even if conditions 1 and 2 in the definition of a transportation network are not satisfied.)

17. How many cuts are there in a transportation network with n vertices?

18. Let D be a directed graph, and let s and t be distinct vertices in D. Make D into a network by giving each directed edge a capacity of 1. Show that the value of a maximal flow in this network equals the minimum number n of directed edges that must be removed from D so that there is no directed path from s to t. (*Hint:* Show that if $\{S, T\}$ is a minimal cut, then n equals the number of arcs from S to T.)

In Exercises 19 and 20 use the result of Exercise 18 to find a minimal set of directed edges whose removal leaves no directed path from s to t.

19.

19.

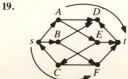

20.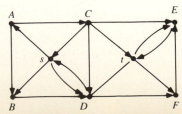

21. Consider an (undirected) graph G in which each edge $\{x, y\}$ is assigned a nonnegative number $C(x, y) = C(y, x)$ representing its capacity to transmit the flow of some substance *in either direction*. Suppose that we want to find the maximum possible flow between distinct vertices s and t of G, subject to the condition that, for any vertex x other than s and t, the total flow into x must equal the total flow out of x. Show that this problem can be solved with the maximal flow algorithm by replacing each edge $\{x, y\}$ of G by two directed edges (x, y) and (y, x) each having capacity $C(x, y)$.

For the graphs in Exercises 22–23, use the method described in Exercise 21 to find the maximal possible flow from s to t if the numbers on the edges represent the capacity of flow along the edge in either direction.

22. 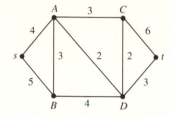   23.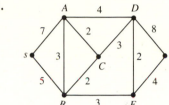

24. Prove that if the maximal flow algorithm ends in step 6, then the original flow has been replaced by a flow with a larger value.

6.4   Flows and Matchings

In this section we will relate network flows to the matchings studied in Section 5.2.

Recall from Sections 5.1 and 5.2 that a graph G is called **bipartite** if its vertex set V can be written as the union of two disjoint sets V_1 and V_2 in such a way that all the edges in G join a vertex in V_1 to a vertex in V_2. A **matching** of G is a subset M of the edges of G such that no vertex in V is incident with more than one edge in M. Furthermore, a matching of G with the property that no matching of G contains more edges is called a **maximal matching** of G.

From a bipartite graph G we can form a transportation network N as follows.

(1) The vertices of N will be the vertices of G together with two additional vertices s and t. These vertices s and t will be the source and the sink for N, respectively.

(2) The arcs in N will be of three types.
   (a) There will be an arc in N from s to every vertex in V_1.
   (b) There will be an arc in N from every vertex in V_2 to t.
   (c) If x is in V_1, y is in V_2, and $\{x, y\}$ is an edge in G, there will be an arc from x to y in N.

(3) All arcs in N will have capacity 1.

We will call N the **network associated with G.**

EXAMPLE 6.7   In the bipartite graph G in Figure 6.33, the vertex set $V = \{A, B, C, W, X, Y, Z\}$ is partitioned into the sets $V_1 = \{A, B, C\}$ and $V_2 = \{W, X, Y, Z\}$.

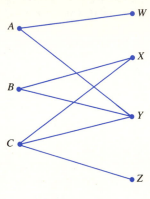

FIGURE 6.33

The network N associated with G is shown in Figure 6.34. Note that N contains a copy of G and two new vertices s and t, which are the source and the

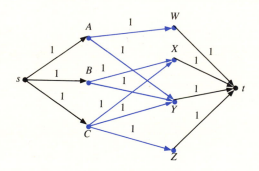

FIGURE 6.34

sink for N, respectively. The edges of G that join vertices in V_1 to vertices in V_2 become arcs in N of capacity 1 directed from the vertices in V_1 to the vertices in V_2. The other arcs in N are directed from the source s to each vertex in V_1 and from each vertex in V_2 to the sink t; these arcs also have capacity 1.   ■

Consider the bipartite graph in Figure 6.35. The network associated with this graph is shown in Figure 6.36.

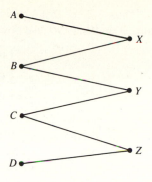

FIGURE 6.35

When the maximal flow algorithm is applied to the network in Figure 6.36, the zero flow can be increased by 1 unit along the path s, A, X, t, by 1 unit along

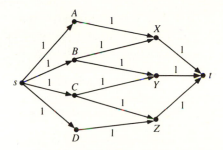

FIGURE 6.36

the path s, B, Y, t, and by 1 unit along the path s, C, Z, t. The resulting maximal flow is shown in Figure 6.37. Notice that this is an integral flow.

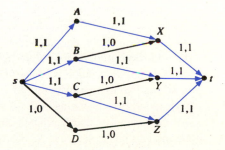

FIGURE 6.37

Thus, we see that a maximal flow in the network shown in Figure 6.36 has value 3, and one maximal flow is obtained by sending:

$$1 \text{ unit along } s, A, X, t;$$

$$1 \text{ unit along } s, B, Y, t; \text{ and}$$

$$1 \text{ unit along } s, C, Z, t.$$

If we disregard the source and sink in these three paths, we obtain the three arcs (A, X), (B, Y), and (C, Z). These arcs correspond to the edges $\{A, X\}$, $\{B, Y\}$, and $\{C, Z\}$ in Figure 6.35. Clearly these edges are a maximal matching of the bipartite graph in Figure 6.35, because in this graph the set V_2 contains only three vertices.

Thus, we have obtained a maximal matching of a bipartite graph by using the maximal flow algorithm on the network associated with the graph. The theorem below shows that this technique will always work.

THEOREM 6.7   Let G be a bipartite graph in which the set of vertices can be written as the union of disjoint sets V_1 and V_2. Let N be the network associated with G.

(a) Every integral flow in N corresponds to a matching of G, and every matching of G corresponds to an integral flow in N. This correspondence is such that two vertices x in V_1 and y in V_2 are matched in G if and only if there is 1 unit of flow in N along arc (x, y).

(b) A maximal flow in N corresponds to a maximal matching of G.

Proof.   (a) Let F be an integral flow in N, and let M be the set of edges $\{x, y\}$ in G for which x is in V_1, y is in V_2, and $F(x, y) = 1$. To prove that M is a matching of G, we must show that no vertex of G is incident with more than one edge in M. Let u be any vertex in G. Since G is the union of the disjoint sets V_1 and V_2, u belongs to exactly one of the sets V_1 or V_2.

Suppose without loss of generality that u belongs to V_1 and that u is incident with the edge $\{u, v\}$ in M. We will show that vertex u is not incident with any other edge in M. For suppose that $\{u, w\}$ is another edge in M. Then $F(u, v) = 1$ and $F(u, w) = 1$ by the definition of M. Thus, in N the total flow out of vertex u is at least 2. But in N the only arc entering u is (s, u), and this arc has capacity 1. So the total flow into vertex u does not equal the total flow out of vertex u, a fact which contradicts that F is a flow in N. Hence, u is incident with at most one edge in M, and so M is a matching of G. This proves that every integral flow in N corresponds to a matching of G.

Now suppose that M is a matching in G. Let N be the network associated with G, and let s be the source in N and t be the sink. For each arc in N define a function F by:

$$F(s, x) = 1 \text{ if } x \in V_1 \text{ and there exists } z \in V_2 \text{ such that } (x, z) \in M;$$

$$F(y, t) = 1 \text{ if } y \in V_2 \text{ and there exists } w \in V_1 \text{ such that } (w, y) \in M;$$

$$F(x, y) = 1 \text{ if } x \in V_1, y \in V_2, \text{ and } (x, y) \in M; \text{ and}$$

$$F(u, v) = 0 \text{ otherwise.}$$

Since each arc e in N has capacity 1 and $0 \leq F(e) \leq 1$, F satisfies condition (1) in the definition of a flow.

Now consider any vertex x of N other than s and t. Such a vertex is a vertex of G and, hence, belongs to either V_1 or V_2. Suppose without loss of generality that x belongs to V_1. By the definition of F, either $F(s, x) = 0$ or $F(s, x) = 1$. If $F(s, x) = 0$, then there exists no $z \in V_2$ such that $(x, z) \in M$; so the total flow into x and the total flow out of x are both 0. On the other hand, if $F(s, x) = 1$, then there exists $z \in V_2$ such that $(x, z) \in M$. Because M is a matching of G, z is unique. Thus, in this case the total flow into x and the total flow out of x are both 1. Therefore, in either case the total flow into x equals the total flow out of x, and so F satisfies condition (2) in the definition of a flow. It follows that F is a flow in N. This proves that every matching of G corresponds to an integral flow in N.

(b) Under the correspondence described in part (a) above, the total number of vertices in V_1 that are matched with vertices in V_2 is the value of the flow F. Thus, M is a maximal matching of G if and only if F is a maximal flow in N. ■

EXAMPLE 6.8   Recall the example from Section 5.1 in which an English department wishes to assign courses to professors, one course per professor. The list of professors available to teach the courses is given below.

Course	Professors
1	Abel, Crittenden, Forcade
2	Crittenden, Donohue, Edge, Gilmore
3	Abel, Crittenden
4	Abel, Forcade
5	Banks, Edge, Gilmore
6	Crittenden, Forcade

The English department would like to obtain a maximal matching so that it can offer the largest possible number of courses.

As in Section 5.2 we can represent this problem by a bipartite graph with vertex set $V = \{1, 2, 3, 4, 5, 6, A, B, C, D, E, F, G\}$, where we have denoted the professors by their initials. Here the set V can be partitioned as the union of the disjoint sets of courses and professors

$$V_1 = \{1, 2, 3, 4, 5, 6\} \qquad \text{and} \qquad V_2 = \{A, B, C, D, E, F, G\},$$

respectively. By drawing an edge between each professor and the courses he or she can teach, we obtain the graph shown in Figure 6.38. (This is the graph obtained previously in Figure 5.1.)

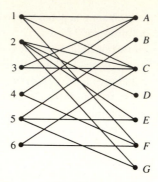

FIGURE 6.38

We will obtain a maximal matching for the graph in Figure 6.38 using the maximal flow algorithm. First, however, note that if we match each course with the first available professor who has not been assigned a course, we obtain the matching with edges $\{1, A\}, \{2, C\}, \{4, F\}, \{5, E\}$. The network associated with the graph in Figure 6.38 is shown in Figure 6.39. Here all arcs are directed from

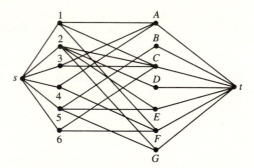

FIGURE 6.39

the left to the right and have capacity 1. The matching $\{1, A\}, \{2, C\}, \{4, F\}, \{5, E\}$ obtained above corresponds to the flow shown in Figure 6.40, where arcs having a flow of zero are shown in black and those with a flow of 1 are shown in blue.

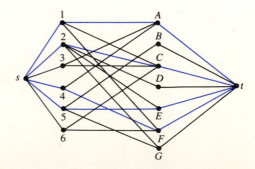

FIGURE 6.40

If we apply the maximal flow algorithm to the network and flow in Figure 6.40, we find that $s, 3, C, 2, D, t$ is a flow-augmenting path. Increasing the flow by 1 along this path gives the flow in Figure 6.41.

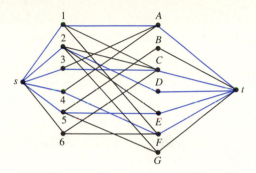

FIGURE 6.41

If another iteration of the maximal flow algorithm is performed on the flow in Figure 6.41, only vertices $s, 1, 3, 4, 6, A, C,$ and F will be labeled. Thus, the flow shown in Figure 6.41 is a maximal flow. By Theorem 6.7 this means that the corresponding matching $\{1, A\}, \{2, D\}, \{3, C\}, \{4, F\},$ and $\{5, E\}$ is a maximal matching for the bipartite graph in Figure 6.38. Hence, the English department can offer 5 of the 6 courses by assigning course 1 to Abel, course 2 to Donohue, course 3 to Crittenden, course 4 to Forcade, and course 5 to Edge. ■

EXERCISES 6.4

In Exercises 1–6 determine whether the given graph is bipartite or not. If it is, construct the network associated with the graph.

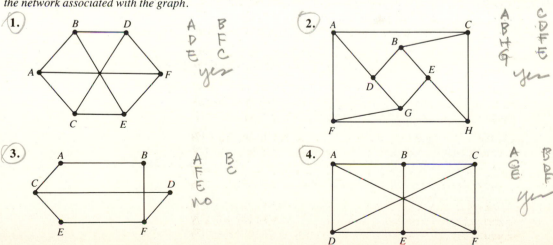

5.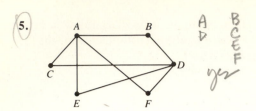

6.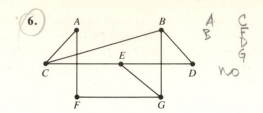

In Exercises 7–10 a bipartite graph is given with a matching indicated in color. Construct the network associated with the given graph, and use the maximal flow algorithm to determine whether this matching is maximal. If not, find a larger matching.

7.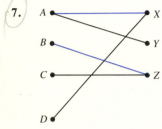

8.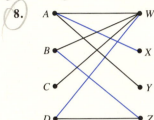

9.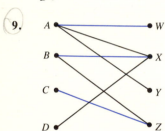

10.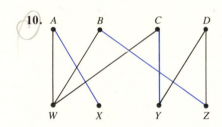

In Exercises 11–14 use the maximal flow algorithm to find a maximal matching for the given bipartite graph.

11.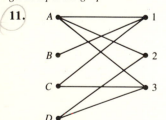

12.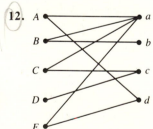

13.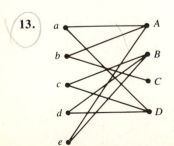

14.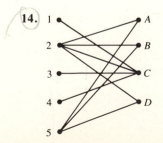

15. Four mixed couples are needed for a tennis team, and 5 men and 4 women are available. Alan will not play with Flo, Greta, or Helga; Bob will not play with Iris; Flo, Greta, and Helga will not play with Ed; Dan will not play with Helga or Iris; and Cal will only play with Iris. Can a team be put together under these conditions? If so, how?

16. Five actresses are needed for parts in a play that require Chinese, Danish, English, French, and German accents. Sally does English and French accents; Tess does Chinese, Danish, and German; Ursula does English and French; Vickie does all accents except English; and Wanda does all except Danish and German. Can the five roles be filled under these conditions. If so, how?

17. When the maximal flow algorithm is applied to the network and flow in Figure 6.41, only the vertices s, 1, 3, 4, 6, A, C, and F will be labeled. What is the significance of courses 1, 3, 4, and 6 and professors A, C, and F in the context of Example 6.8?

18. By a **network with vertex capacities** we mean a transportation network N along with a function K from its set of vertices to the nonnegative real numbers. In such a network a flow must satisfy the additional restriction that, for each vertex v, neither the total flow into v nor the total flow out of v can exceed $K(v)$. (Of course, these totals are the same if v is a vertex other than the source or sink.) Show that the value of a maximal flow in such a network equals the value of a maximal flow in the ordinary transportation network N^*, where N^* is formed as follows.
(a) For each vertex x in N, include two vertices x' and x'' in N^*.
(b) For each vertex x in N, include an arc (x', x'') in N^* with capacity $K(x)$.
(c) For each arc (x, y) in N, include an arc (x'', y') of the same capacity in N^*.
(Note that if in N the source is s and the sink is t, then in N^* the source is s' and the sink is t''.)

For the networks with vertex capacities in Exercises 19–21, construct the network N^* described in Exercise 18.

19. $K(A) = 9$, $K(B) = 8$, $K(C) = 9$, $K(D) = 7$, and $K(E) = 10$

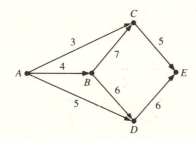

20. $K(A) = 8$, $K(B) = 4$, $K(C) = 7$, $K(D) = 7$, $K(E) = 6$, and $K(F) = 9$

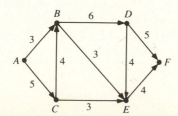

21. $K(A) = 16$, $K(B) = 9$, $K(C) = 6$, $K(D) = 5$, $K(E) = 8$, $K(F) = 7$, and $K(G) = 15$

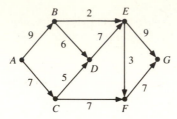

In Exercises 22 and 23 use the method of Exercise 18 to find a maximal flow for the network with vertex capacities in the indicated exercise.

22. Exercise 19   **23.** Exercise 20

24. Let s and t be distinct vertices in a directed graph D. Make D into a transportation network N with vertex capacities by letting s and t have infinite capacity, the other vertices have capacity 1, and each arc have capacity 1. Use Exercise 18 above and Exercise 33 of Section 6.2 to show that the value of a maximal flow for N equals the maximum number of directed paths from s to t that use no vertex other than s and t more than once.

Suggested Readings

1. Bondy, J. A. and U. S. R. Murty. *Graph Theory with Applications*. New York: North-Holland, 1976.
2. Burr, Stefan A., ed. *The Mathematics of Networks*. In *Proceedings of Symposia in Applied Mathematics,* vol. 26. Providence, RI: American Mathematical Society, 1982.
3. Edmonds, Jack and Richard M. Karp. "Theoretical Improvements in Algorithmic Efficiency for Network Flow Problems." *Journal of the Association for Computing Machinery,* vol. 19, no. 2 (April 1972): 248–264.
4. Elias, P., A. Feinstein, and C. E. Shannon. "Note on Maximum Flow Through a Network." *IRE Transactions on Information Theory, IT-2* (1956): 117–119.
5. Ford, Jr., L. R. and D. R. Fulkerson. *Flows in Networks*. Princeton, NJ: Princeton University Press, 1962.
6. Ford, Jr., L. R. and D. R. Fulkerson. "Maximal Flow through a Network." *Canadian Journal of Mathematics,* vol. 8, no. 3 (1956): 399–404.
7. Frank, Howard and Ivan T. Frisch. "Network Analysis." *Scientific American,* vol. 223, no. 1 (July 1970): 94–103.
8. Lawler, Eugene L. *Combinatorial Optimization: Networks and Matroids*. New York: Holt, Rinehart, and Winston, 1976.
9. Zadeh, Norman. "Theoretical Efficiency of the Edmonds-Karp Algorithm for Computing Maximal Flows." *Journal of the Association for Computing Machinery,* vol. 19, no. 1 (January 1972): 184–192.

7

COUNTING TECHNIQUES

As we saw in Sections 1.2 and 1.3, many combinatorial problems involve counting. Since the number of objects under consideration is often extremely large, it is desirable to be able to count a set of objects without having to list them all. In this chapter we will discuss several fundamental counting techniques that are frequently used in solving combinatorial problems. The reader should carefully review Sections 1.2 and 2.6, which contain several results to which we will refer.

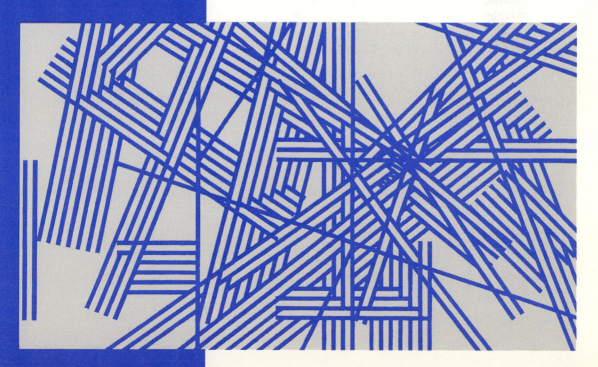

7.1   Pascal's Triangle and the Binomial Theorem

One of the most basic problems that arises in combinatorial analysis is to count the number of subsets of a given set that contain a specified number of elements. Since this number occurs frequently, we will introduce the notation $C(n, r)$† to denote the number of different subsets containing r elements that can be formed from a set of n elements. Because any set of n elements can be thought of as the set $A = \{1, 2, \ldots, n\}$, Theorem 2.16 shows that

$$C(n, r) = \frac{n!}{r!(n - r)!}. \tag{7.1}$$

EXAMPLE 7.1 By (7.1) we see that the number of subsets containing two elements that can be formed from the set of vowels $\{a, e, i, o, u\}$ is

$$C(5, 2) = \frac{5!}{2! \, 3!} = \frac{5 \cdot 4 \cdot 3!}{2 \cdot 1 \cdot 3!} = \frac{5 \cdot 4}{2 \cdot 1} = 10.$$

The ten subsets in question are easily seen to be $\{a, e\}$, $\{a, i\}$, $\{a, o\}$, $\{a, u\}$, $\{e, i\}$, $\{e, o\}$, $\{e, u\}$, $\{i, o\}$, $\{i, u\}$, and $\{o, u\}$. ▪

In the proof of Theorem 2.16 we saw that any subset S of A containing r elements ($r \geq 1$) was either:

(1) a subset of $\{1, 2, \ldots, n - 1\}$ containing r elements (if $n \notin S$), or
(2) the union of $\{n\}$ and a subset of $\{1, 2, \ldots, n - 1\}$ containing $r - 1$ elements (if $n \in S$).

Thus, $C(n, r)$, the number of subsets of A containing r elements, is the sum of $C(n - 1, r)$, the number of subsets of type (1), and $C(n - 1, r - 1)$, the number of subsets of type (2). We have proved the following result.

THEOREM 7.1 If $n > r \geq 1$, then $C(n, r) = C(n - 1, r) + C(n - 1, r - 1)$.

†Another common way of denoting this number is $\binom{n}{r}$.

EXAMPLE 7.2   It follows from Theorem 7.1 that $C(7, 3) = C(6, 3) + C(6, 2)$. We will verify this equation by evaluating $C(7, 3)$, $C(6, 3)$, and $C(6, 2)$ using (7.1). Now

$$C(6, 2) = \frac{6!}{2! \, 4!} = \frac{6 \cdot 5 \cdot 4!}{2 \cdot 1 \cdot 4!} = \frac{6 \cdot 5}{2 \cdot 1} = 15,$$

$$C(6, 3) = \frac{6!}{3! \, 3!} = \frac{6 \cdot 5 \cdot 4 \cdot 3!}{3 \cdot 2 \cdot 1 \cdot 3!} = \frac{6 \cdot 5 \cdot 4}{3 \cdot 2 \cdot 1} = 20, \text{ and}$$

$$C(7, 3) = \frac{7!}{3! \, 4!} = \frac{7 \cdot 6 \cdot 5 \cdot 4!}{3 \cdot 2 \cdot 1 \cdot 4!} = \frac{7 \cdot 6 \cdot 5}{3 \cdot 2 \cdot 1} = 35.$$

So $C(6, 3) + C(6, 2) = 20 + 15 = 35 = C(7, 3)$. ■

The triangular array below

$n = 0$				$C(0, 0)$				
$n = 1$			$C(1, 0)$		$C(1, 1)$			
$n = 2$		$C(2, 0)$		$C(2, 1)$		$C(2, 2)$		
$n = 3$	$C(3, 0)$		$C(3, 1)$		$C(3, 2)$		$C(3, 3)$	
$n = 4$	$C(4, 0)$	$C(4, 1)$		$C(4, 2)$		$C(4, 3)$		$C(4, 4)$

$$\vdots$$

is called **Pascal's triangle.** Although this array was first known to the Chinese, its name comes from the seventeenth century French mathematician Blaise Pascal (1623–1662), whose paper *Traite du triangle arithmetique* developed many of the triangle's properties.

Note that the rows of the triangle are numbered beginning with row $n = 0$ and that the entries $C(n, r)$ for a fixed value of r lie along a diagonal extending from the upper right to the lower left.

Let us consider the entries of Pascal's triangle in more detail. Since every set has exactly one subset containing 0 elements (namely the empty set), $C(n, 0) = 1$. Moreover, there is only one subset of n elements that can be formed from a set containing n elements (namely the entire set); so $C(n, n) = 1$. Therefore, the first and last numbers in every row of Pascal's triangle are 1's. In addition Theorem 7.1 states that every entry that is not first or last in its row is the sum of the two nearest entries in the row above. For example, $C(4, 2) = C(3, 1) + C(3, 2)$ and $C(4, 3) = C(3, 2) + C(3, 3)$. By repeatedly using these properties it is easy to evaluate the entries in Pascal's triangle. The resulting numbers are shown below.

```
              1
            1   1
          1   2   1
        1   3   3   1
      1   4   6   4   1
                ⋮
```

EXAMPLE 7.3   Continuing in the triangle above, we see that the numbers in the next row (the row $n = 5$) are $1, 1 + 4 = 5, 4 + 6 = 10, 6 + 4 = 10, 4 + 1 = 5$, and 1. ■

Pascal's triangle contains an important symmetry: Each row reads the same from left to right as it does from right to left. In terms of our notation this statement says that $C(n, r) = C(n, n - r)$ for any r satisfying $0 \le r \le n$. Although this property is easily verified by computing $C(n, r)$ and $C(n, n - r)$ using (7.1), it is also possible to prove this equality by a combinatorial argument, that is, by using the definition of these numbers. Recall that $C(n, r)$ represents the number of subsets containing r elements that can be formed from a set of n elements. Thus, the equality $C(n, r) = C(n, n - r)$ asserts that a set with n elements has the same number of subsets containing r elements as it has subsets containing $n - r$ elements. It is easy to see why this statement is true: If, from a set of n elements, we select a subset S of r elements, then we have also implicitly formed a subset containing $n - r$ elements, namely the subset consisting of all the elements not in S.

The numbers $C(n, r)$ are called **binomial coefficients** because they appear in the algebraic expansion of the binomial $(x + y)^n$. More specifically, in this expansion $C(n, r)$ is the coefficient of the term $x^{n-r}y^r$. Thus, the coefficients that occur in the expansion of $(x + y)^n$ are the numbers in row n of Pascal's triangle. For example,

$$(x + y)^3 = (x + y)(x + y)^2 = (x + y)(x^2 + 2xy + y^2)$$
$$= x(x^2 + 2xy + y^2) + y(x^2 + 2xy + y^2)$$
$$= (x^3 + 2x^2y + xy^2) + (x^2y + 2xy^2 + y^3)$$
$$= x^3 + 3x^2y + 3xy^2 + y^3.$$

Note that the coefficients (1, 3, 3, and 1) occurring in this expansion are the numbers in the $n = 3$ row of Pascal's triangle.

THEOREM 7.2   *The Binomial Theorem*   For every positive integer n,

$$(x + y)^n = C(n, 0)x^n + C(n, 1)x^{n-1}y + \ldots + C(n, n - 1)xy^{n-1} + C(n, n)y^n.$$

Proof.   In the expansion of

$$(x + y)^n = (x + y)(x + y) \ldots (x + y)$$

we choose either an x or a y from each of the n factors $x + y$. The term in the expansion involving $x^{n-r}y^r$ results from combining all the terms obtained by choosing x from $n - r$ factors and y from r factors. The number of such terms is, therefore, the number of ways to select a subset of r factors from which to choose y. (We will select x from each factor from which we do not choose y.) By Theorem 2.16 this number is $C(n, r)$. Hence, the coefficient of $x^{n-r}y^r$ in the expansion of $(x + y)^n$ is $C(n, r)$. ■

EXAMPLE 7.4   Using the coefficients from the $n = 4$ row of Pascal's triangle and the binomial theorem, we see that

$$(x + y)^4 = C(4, 0)x^4 + C(4, 1)x^3y + C(4, 2)x^2y^2 + C(4, 3)xy^3 + C(4, 4)y^4$$
$$= x^4 + 4x^3y + 6x^2y^2 + 4xy^3 + y^4. \quad \blacksquare$$

EXERCISES 7.1

Evaluate the numbers in Exercises 1–12.

1. $C(5, 3)$ **2.** $C(7, 2)$ **3.** $C(8, 3)$ **4.** $C(6, 5)$

5. $C(4, 2)$ **6.** $C(8, 4)$ **7.** $C(9, 3)$ **8.** $C(12, 3)$

9. $C(10, 8)$ **10.** $C(9, 5)$ **11.** $C(n, 1)$ **12.** $C(n, 2)$

13. Write the numbers in the $n = 6$ row of Pascal's triangle.

14. Write the numbers in the $n = 7$ row of Pascal's triangle.

15. Evaluate $(x + y)^6$.

16. Evaluate $(x + y)^7$.

17. Evaluate $(3x - y)^4$.

18. Evaluate $(x - 2y)^5$.

19. How many subsets containing four different numbers can be formed from the set $\{1, 2, 3, 4, 5, 6, 7\}$?

20. How many subsets containing eight different letters can be formed from the set $\{a, b, c, d, e, f, g, h, i, j, k, l\}$?

21. How many subsets of $\{b, c, d, f, g, h, j, k, l, m\}$ contain five letters?

22. How many subsets of $\{2, 3, 5, 7, 11, 13, 17, 19, 23\}$ contain four numbers?

23. How many four-element subsets of $\{1, 2, 3, 4, 5, 6, 7, 8, 9, 10, 11, 12\}$ contain no odd numbers?

24. How many three-element subsets of $\{a, b, c, d, e, f, g, h, i, j, k\}$ contain no vowels?

25. Use the binomial theorem to show that $C(n, 0) + C(n, 1) + \ldots + C(n, n) = 2^n$ for all nonnegative integers n.

26. Use (7.1) to verify that $C(n, r) = C(n, n - r)$ for $0 \le r \le n$.

27. Use the binomial theorem to show for $n \ge 0$ that
$C(n, 0) - C(n, 1) + C(n, 2) - C(n, 3) + \ldots + (-1)^n C(n, n) = 0$.

28. Prove that $2^0 C(n, 0) + 2^1 C(n, 1) + \ldots + 2^n C(n, n) = 3^n$ for all positive integers n.

29. Prove that $rC(n, r) = nC(n - 1, r - 1)$ for $1 \le r \le n$.

30. Prove that $2C(n, 2) + n^2 = C(2n, 2)$ for $n \ge 2$.

31. For any positive integer k and any nonnegative integer r, prove that
$C(k, 0) + C(k + 1, 1) + \ldots + C(k + r, r) = C(k + r + 1, r)$.

32. Prove $C(r, r) + C(r + 1, r) + \ldots + C(n, r) = C(n + 1, r + 1)$ for $0 \le r \le n$.
Why do you think the name "hockey stick formula" is used for this result?

33. Prove that $C(2n + 1, 0) + C(2n + 1, 1) + \ldots + C(2n + 1, n) = 2^{2n}$ for all nonnegative integers n.

34. Prove that $1C(n, 0) + 2C(n, 1) + \ldots + (n + 1)C(n, n) = 2^n + n2^{n-1}$ for all positive integers n.

35. Let k and n be nonnegative integers such that $k < \frac{n}{2}$. Prove that $C(n, k) \leq C(n, k + 1)$.

7.2 Three Fundamental Principles

In this section we will introduce three basic principles that will find frequent use throughout this chapter. The first of these is a surprisingly simple existence statement that has many profound consequences.

THEOREM 7.3 *The Pigeonhole Principle* If a set of pigeons is placed into pigeonholes and there are more pigeons than pigeonholes, then some pigeonhole must contain at least two pigeons. More generally, if the number of pigeons is more than k times the number of pigeonholes, then some pigeonhole must contain at least $k + 1$ pigeons.

Proof. The first statement is a special case of the more general result, namely the case in which $k = 1$. So we will prove only the more general result.

Suppose that there are p pigeonholes and q pigeons. If no pigeonhole contains at least $k + 1$ pigeons, then each of the p pigeonholes contains at most k pigeons. So the total number of pigeons cannot exceed kp. Thus, if the number of pigeons is more than k times the number of pigeonholes (that is, if $q > kp$), then some pigeonhole must contain at least $k + 1$ pigeons. ▪

EXAMPLE 7.5 How many people must be selected from a collection of fifteen married couples to ensure that at least two of the persons chosen are married to each other?

It is easy to see intuitively that if we choose any 16 persons from this collection of 15 couples, we must include at least one husband and wife pair. This conclusion is based on the pigeonhole principle. For let us place persons (the pigeons) into sets (the pigeonholes) in such a way that persons cannot belong to the same set unless they are married to each other. Since there are only 15 possible sets, any distribution of 16 persons must place two people in the same set. Thus, there must be at least one married couple included among the 16 persons. Note that if we choose fewer than 16 persons, a married couple may not be included (for instance, if we choose the 15 women).   ■

EXAMPLE 7.6   How many distinct integers must be chosen to assure that there are at least ten having the same congruence class modulo 7?

This question involves placing integers (the pigeons) into congruence classes (the pigeonholes). Recall that there are 7 distinct congruence classes modulo 7, namely [0], [1], [2], [3], [4], [5], and [6]. So if we want to guarantee that there are at least $10 = k + 1$ integers in the same congruence class, the generalized form of the pigeonhole principle states that we must choose more than $7k = 7 \cdot 9 = 63$ distinct integers. Hence, at least 64 integers must be chosen.   ■

In contrast to the pigeonhole principle, which asserts that some pigeonhole contains a certain number of pigeons (an existence statement), the next two results tell us how to count the number of ways that two basic types of selections can be performed. The first result is a restatement of a result from Section 1.2 that enables us to count the number of ways of performing a procedure that consists of a sequence of operations.

THEOREM 7.4   *The Multiplication Principle*   Suppose that a procedure can be divided into a sequence of k steps and that the first step can be performed in n_1 ways, the second step can be performed in n_2 ways, and in general the ith step can be performed in n_i ways. Then the number of different ways in which the entire procedure can be performed is $n_1 \cdot n_2 \cdot \ldots \cdot n_k$.

In using the multiplication principle it is important to note that the number of ways to perform any step must not depend upon the particular choice that is made at a previous step. That is, no matter how the first step is performed there must be n_2 ways of performing the second step, no matter how the first two steps are performed there must be n_3 ways of performing the third step, etc.

To illustrate the multiplication principle, suppose that a couple expecting a child have decided that if it is a girl, they will give it a first name of Jennifer, Karen, or Linda and a middle name of Ann or Marie. Since the process of naming the child can be divided into the two steps of selecting a first name and selecting a middle name, the multiplication principle tells us that there are $3 \cdot 2 = 6$ possible names that can be given. To see that this is the correct answer, we can enumerate the possibilities as in Figure 7.1 on the next page.

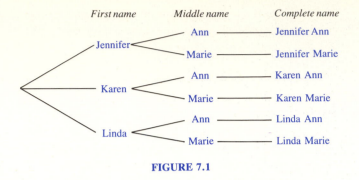

FIGURE 7.1

EXAMPLE 7.7

A particular model of car can be ordered in four different exterior colors and three different interior colors. How many different color combinations are possible for this model of car?

    Since we can regard each color combination as being formed by first choosing an exterior color and then choosing an interior color, the number of different color combinations is $4 \cdot 3 = 12$. ■

EXAMPLE 7.8

A **bit** (or **binary digit**) is a zero or a one. An **n–bit string** is a sequence of n bits. Thus, 01101110 is an 8–bit string. Information is stored and processed in computers in bit strings because a bit string can be regarded as a sequence of on or off settings for switches inside the computer.

    Let us compute the number of 8–bit strings using the multiplication principle. To do so, we will regard the 8–bit string as a sequence of 8 choices (choose the first bit, then choose the second bit, and so forth). Because each bit can be chosen in 2 ways (namely zero or one), the number of possible 8–bit strings is

$$2 \cdot 2 \cdot 2 \cdot 2 \cdot 2 \cdot 2 \cdot 2 \cdot 2 = 2^8 = 256.$$

    Since an 8-bit string can be regarded as the binary representation of a nonnegative integer, this calculation shows that 256 nonnegative integers can be expressed using no more than 8 binary digits. More generally, a similar argument shows that the number of n–bit strings (and, hence, the number of nonnegative integers that can be expressed using no more than n binary digits) is 2^n. ■

    The second basic counting principle is concerned with the number of elements in the union of pairwise disjoint sets.

THEOREM 7.5   *The Addition Principle* Suppose that there are k sets of elements with n_1 elements in the first set, n_2 elements in the second set, etc. If all of the elements are distinct (that is, if all pairs of the k sets are disjoint), then the number of elements in the union of the sets is $n_1 + n_2 + \ldots + n_k$.

To illustrate this result, let $A = \{1, 2, 3\}$ and $B = \{4, 5, 6, 7\}$. Since the elements in A and B are distinct, it follows from the addition principle that the number of elements in the union of sets A and B equals the sum of the number of elements in set A (which is 3) plus the number of elements in set B (which is 4). Therefore, in our example the number of elements in $A \cup B$ is $3 + 4 = 7$. Clearly this answer is correct because $A \cup B = \{1, 2, 3, 4, 5, 6, 7\}$. But note the necessity of distinct elements: If A had been the set $A = \{1, 2, 4\}$, then the answer would no longer have been 7 since in this case $A \cup B = \{1, 2, 4, 5, 6, 7\}$.

EXAMPLE 7.9

Suppose that a couple expecting a child have decided to name it one of six names (Jennifer Ann, Jennifer Marie, Karen Ann, Karen Marie, Linda Ann, or Linda Marie) if it is a girl and one of four names (Michael Alan, Michael Louis, Robert Alan, or Robert Louis) if it is a boy. How many different names can the child receive?

Since the names to be given to a girl are different from those to be given to a boy, the addition principle states that the number of possible names is the sum of the number of girls' names and the number of boys' names. Thus, the answer to the question posed above is $6 + 4 = 10$. ▬

EXAMPLE 7.10

How many integers between 1 and 100 (including 100) are even or end with 5?

Let A denote the set of even integers between 1 and 100, and let B denote the set of integers between 1 and 100 that end with 5. The number of integers between 1 and 100 that are even or end in 5 is then the number of elements in $A \cup B$. Now A contains 50 elements because every other number from 1 to 100 is even. And B contains 10 elements since 5, 15, 25, 35, 45, 55, 65, 75, 85, and 95 are the only integers between 1 and 100 that end in 5. Moreover, the elements in A and B are distinct because a number ending in 5 cannot be even. Thus, it follows from the addition principle that the number of integers between 1 and 100 that are even or end with 5 is $50 + 10 = 60$. ▬

Often the multiplication and addition principles are both needed to solve a problem, as in the following examples.

EXAMPLE 7.11

In the Applesoft BASIC language, the name of a real variable consists of alphanumeric characters beginning with a letter. (An alphanumeric character is a letter A–Z or a digit 0–9.) Although variable names may be as long as 238 characters, they are distinguished by their first two characters only. (Thus, RATE and RATIO are regarded as the same name.) In addition, there are seven reserved words (AT, FN, GR, IF, ON, OR, and TO) that are not legal variable names. We will use the multiplication and addition principles to determine the number of legal variable names that can be distinguished in Applesoft BASIC.

Clearly there are 26 real variable names consisting of a single character, namely A–Z. Any other distinct name will consist of a letter followed by an alphanumeric character. It follows from the multiplication principle that the number of distinguishable names consisting of more than one character is $26 \cdot 36 = 936$, since there are 26 letters and 36 alphanumeric characters. Thus, by the addition principle the number of one-character or two-character names is $26 + 936 = 962$. So there are 962 distinguishable real variable names in Applesoft BASIC, and, hence, $962 - 7 = 955$ legal names that can be distinguished. ■

EXAMPLE 7.12 (a) How many 8–bit strings begin with 1011 or 01?
(b) How many 8–bit strings begin with 1011 or end with 01?

(a) An 8–bit string beginning with 1011 has the form 1011––––, where the dashes denote either zeros or ones. So the number of 8–bit strings that begin with 1011 is equal to the number of ways that bits five through eight can be chosen. By reasoning as in Example 7.8, we find that this number is $2^4 = 16$. Likewise, the number of 8–bit strings that begin with 01 is $2^6 = 64$. Since the set of strings beginning with 1011 is disjoint from the set of strings beginning with 01, the addition principle shows that the number of strings beginning with 1011 or 01 is $16 + 64 = 80$.

(b) Although it is tempting to approach this problem as in part (a), the set of strings beginning with 1011 is not disjoint from the set of strings ending with 01. So the correct answer is not $16 + 64$ as before. Since the addition principle can only be used with disjoint sets, let us define sets of 8–bit strings A, B, and C as follows:

$$A = \{\text{strings beginning with 1011 and not ending with 01}\},$$

$$B = \{\text{strings ending with 01 and not beginning with 1011}\},$$

$$C = \{\text{strings beginning with 1011 and ending with 01}\}.$$

Clearly all pairs of the sets A, B, and C are disjoint and $A \cup B \cup C$ consists of the strings that begin with 1011 or end with 01. Therefore, the addition principle states that the number of elements in $A \cup B \cup C$ is the sum of the sizes of A, B, and C. Now strings in A begin with 1011 and end with 00, 10, or 11. Thus, there is only one way to choose the first four bits of a string in A (namely 1011), there are two ways to choose each of the fifth and sixth bits (0 or 1), and there are three ways to choose the last two bits (00, 10, or 11). Hence, the number of strings in A is $1 \cdot 2 \cdot 2 \cdot 3 = 12$. Similar arguments show that the numbers of strings in B and C are $15 \cdot 2 \cdot 2 \cdot 1 = 60$ and $1 \cdot 2 \cdot 2 \cdot 1 = 4$, respectively. Therefore, the number of strings in $A \cup B \cup C$ is $12 + 60 + 4 = 76$. ■

EXERCISES 7.2

1. How many people must there be in order to assure that at least two of their birthdays fall in the same month?

2. If a committee varies its meeting days, how many meetings must it schedule before we can guarantee that at least two meetings will be held on the same day of the week?

3. A drawer contains unsorted black, brown, blue, and gray socks. How many socks must be chosen in order to be certain of choosing two of the same color?

4. How many words must be chosen in order to assure that at least two begin with the same letter?

5. A conference room contains 8 tables and 105 chairs. What is the smallest possible number of chairs at the table having the most seats?

6. If there are 6 sections of Discrete Math with a total enrollment of 199 students, what is the smallest possible number of students in the section with the largest enrollment?

7. How many books must be chosen from among 24 mathematics books, 25 computer science books, 21 literature books, and 15 economics books in order to assure that there are at least 12 books on the same subject?

8. A sociologist is intending to send a questionnaire to 32 whites, 19 blacks, 27 Hispanics, and 31 Native Americans. How many responses must she receive in order to guarantee that there will be at least 15 responses from the same ethnic group?

9. An automobile can be ordered with any combination of the following options: air conditioning, automatic transmission, bucket seats, rear window defogger, and tape deck. In how many different ways can this car be equipped?

10. How many different pizzas can be ordered if a pizza can be selected with any combination of the following ingredients: anchovies, ham, mushrooms, olives, onion, pepperoni, and sausage?

11. Use the multiplication principle to determine the number of subsets of a set containing n elements.

12. How many different sequences of heads and tails can result if a coin is flipped 20 times?

13. A businessman must fly from Kansas City to Chicago on Monday and from Chicago to Boston on Thursday. If there are 8 daily flights from Kansas City to Chicago and 21 daily flights from Chicago to Boston, how many different routings are possible for this trip?

14. An interior decorator is creating layouts that consist of carpeting and draperies. If there are 4 choices of carpets and 6 choices of draperies, how many layouts must be made to show all of the possibilities?

15. How many different character strings of length three can be formed from the letters A, B, C, D, E, and F if

    (a) letters can be repeated?
    (b) letters cannot be repeated?

16. How many seven-digit numbers end in 5454?

17. In how many different orders can 3 married couples be seated in a row of 6 chairs under the following conditions?

   (a) Anyone may sit in any chair.

   (b) Men must occupy the first and last chairs.

   (c) Men must occupy the first three chairs and women the last three.

   (d) Everyone must be seated beside his or her spouse.

18. On student recognition night a high school will present awards to 4 seniors and 3 juniors. In how many different orders can the awards be presented under the following conditions?

   (a) The awards can be presented in any order.

   (b) Awards are presented to juniors before awards are presented to seniors.

   (c) The first and last awards are presented to juniors.

   (d) The first and last awards are presented to seniors.

19. In the Apple Pascal programming language, identifiers (that is, variable names, file names, and so forth) are subject to the following rules:

   (a) The first character in an identifier must be a letter (capital or lower case).

   (b) Subsequent characters may be letters or digits (0, 1, . . . , 9).

   If the reserved words IF, LN, ON, OR, and TO cannot be used as identifiers, how many different Apple Pascal identifiers contain exactly two characters?

20. In FORTRAN, unless an integer variable is explicitly declared, its name must begin with one of the letters I, J, K, L, M, or N. Subsequent characters can be any letter A, B, . . . , Z or digit 0, 1, . . . , 9. How many such integer variable names contain exactly four characters?

21. A men's clothing store has a sale on selected suits and blazers. If there are 30 suits and 40 blazers on sale, in how many ways may a customer select exactly one item that is on sale?

22. A restaurant offers a choice of 3 green vegetables or a potato prepared in one of 5 ways. How many different choices of vegetable can be made?

23. How many 8–bit strings begin with 1001 or 010?

24. How many 8–bit strings end with 1000 or 01011?

25. In the United States radio station call letters consist of 3 or 4 letters beginning with either K or W. How many different sets of radio station call letters are possible?

26. Suppose that a license plate must contain a sequence of 2 letters followed by 4 digits or 3 letters followed by 3 digits. How many different license plates can be made?

27. From among a group of 4 men and 6 women, 3 persons are to be appointed as a branch manager in different cities. How many different appointments can be made under the following circumstances?

   (a) Any person is eligible for appointment.

   (b) One man and two women are to be appointed.

   (c) At least two men are to be appointed.

   (d) At least one person of each sex is to be appointed.

28. Suppose that 3 freshmen, 5 sophomores, and 4 juniors have been nominated to receive scholarships of $500, $250, and $100. How many different distributions of the scholarships are possible under the following circumstances?

(a) Anyone may receive any scholarship.

(b) The $500 scholarship is to be awarded to a freshman, the $250 scholarship to a sophomore, and the $100 scholarship to a junior.

(c) At least two scholarships are to be awarded to juniors.

(d) One scholarship is to be awarded to someone from each class.

29. How many 8–bit strings begin with 11 or end with 00?

30. How many 8–bit strings begin with 010 or end with 11?

31. Show that if the 26 letters of the English alphabet are written in a circular array in any order whatsoever, there must be 5 consecutive consonants.

32. Prove that in any nonempty list of n integers (not necessarily distinct) there is some nonempty sublist having a sum that is divisible by n.

33. Let $S = \{a_1, a_2, \ldots, a_9\}$ be any set of 9 points in Euclidean space such that all three coordinates of each point are integers. Prove that for some i and j ($i \neq j$) the midpoint of the segment joining a_i and a_j has only integer coordinates.

34. Suppose that there are 15 identical copies of *The Great Gatsby* and 12 distinct biographies on a bookshelf.

(a) How many different selections of 12 books are possible?

(b) How many different selections of 10 books are possible?

7.3   Permutations and Combinations

Two types of counting problems occur so frequently that they deserve special attention. These problems are:

(1) How many different arrangements (ordered lists) of r objects can be formed from a set of n distinct objects?

(2) How many different selections (unordered lists) of r objects can be made from a set of n distinct objects?

In this section we will consider these two questions in the case that repetition of the n distinct objects is not allowed. Section 7.4 will answer these same questions when repetition is permitted.

Recall from Section 1.2 that an arrangement or ordering of n distinct objects is called a **permutation** of the objects. If $r \leq n$, then an arrangement or ordering using r of the n distinct objects is called an **r-permutation.** Thus, 3142 is a permutation of the digits 1, 2, 3, and 4, and 412 is a 3–permutation of these digits.

The number of different r–permutations of a set of n distinct elements is denoted $P(n, r)$. In Theorem 1.2 this number was found to be

$$P(n, r) = \frac{n!}{(n - r)!}. \tag{7.2}$$

Thus, (7.2) gives us the answer to question 1 above.

EXAMPLE 7.13   How many different three-digit numbers can be formed using the digits 5, 6, 7, 8, and 9 without repetition?

This question asks for the number of 3–permutations from a set of 5 digits. This number is $P(5, 3)$. So using (7.2) we see that the answer to the question above is

$$P(5, 3) = \frac{5!}{2!} = \frac{5 \cdot 4 \cdot 3 \cdot 2!}{2!} = 5 \cdot 4 \cdot 3 = 60. \quad \blacksquare$$

EXAMPLE 7.14   In how many different orders can four persons be seated in a row of four chairs?

The answer to this question is the number of permutations of a set of 4 elements. Recalling that $0! = 1$, we see from (7.2) that this number is

$$P(4, 4) = \frac{4!}{0!} = \frac{4!}{1} = 4! = 24. \quad \blacksquare$$

Note that Example 7.14 could also have been answered by appealing to Theorem 1.1, which can be rewritten using our present notation as $P(n, n) = n!$.

Let us now consider the second question above. If $r \leq n$, then an unordered selection of r objects chosen from a set of n distinct objects is called an **r-combination** of the objects. Thus, $\{1, 4\}$ and $\{2, 3\}$ are both 2–combinations of the digits 1, 2, 3, and 4. Note that since combinations are unordered selections, the 2–combinations $\{1, 4\}$ and $\{4, 1\}$ are the same. In fact, an unordered selection of r elements from a set of n distinct elements is just a subset of the set that contains r elements. Thus, the number of different r–combinations of a set of n distinct elements is $C(n, r)$. So using (7.1) we see that the answer to question 2 above is

$$C(n, r) = \frac{n!}{r!(n - r)!}.$$

EXAMPLE 7.15   How many different four-member committees can be formed from a delegation of 7 members?

Since a four-member committee is just a selection of 4 members from the delegation of 7, the answer to this question is $C(7, 4)$. Using (7.1) we find

$$C(7, 4) = \frac{7!}{4! \, 3!} = \frac{7 \cdot 6 \cdot 5 \cdot 4!}{4! \cdot 3 \cdot 2 \cdot 1} = \frac{7 \cdot 6 \cdot 5}{3 \cdot 2 \cdot 1} = 35. \quad \blacksquare$$

EXAMPLE 7.16 How many 8–bit strings contain exactly three 0's?

Note that an 8–bit string containing exactly three 0's is completely determined if we know the positions of the three 0's (since the other five positions must be filled with 1's). Thus, the number of 8–bit strings containing exactly three 0's equals the number of different locations that the three 0's can occupy. But this number is the number of ways to choose three positions from among eight, which is $C(8, 3)$. So the number of 8–bit strings containing exactly three 0's is

$$C(8, 3) = \frac{8!}{3! \ 5!} = \frac{8 \cdot 7 \cdot 6 \cdot 5!}{3 \cdot 2 \cdot 1 \cdot 5!} = \frac{8 \cdot 7 \cdot 6}{3 \cdot 2 \cdot 1} = 56. \quad ■$$

Counting problems often require that permutations or combinations be used together with the multiplication or addition principles. The following examples are of this type.

EXAMPLE 7.17 Three men and three women are going to occupy a row of six seats. In how many different orders can they be seated so that men occupy the two end seats?

We can regard the assigning of seats as a two-step process: First fill the two end seats, and then fill the middle four seats. Since the end seats must be filled by two of the three men, there are $P(3, 2)$ different ways to occupy the end seats. The remaining four persons can fill the middle seats in any order; so there are $P(4, 4)$ different ways to fill the middle seats. Thus, by the multiplication principle the number of ways to fill both the end seats and the middle seats is

$$P(3, 2) \cdot P(4, 4) = 6 \cdot 24 = 144. \quad ■$$

EXAMPLE 7.18 How many different committees consisting of 2 sophomores, 3 juniors, and 4 seniors can be formed from a group of 6 sophomores, 5 juniors, and 8 seniors?

We can regard the construction of a committee containing 2 sophomores, 3 juniors, and 4 seniors as a three-step process by choosing the sophomores first, choosing the juniors second, and choosing the seniors third. Clearly the sophomores can be chosen in $C(6, 2)$ ways, the juniors can be chosen in $C(5, 3)$ ways, and the seniors can be chosen in $C(8, 4)$ ways. So the multiplication principle gives the number of committees consisting of 2 sophomores, 3 juniors, and 4 seniors to be

$$C(6, 2) \cdot C(5, 3) \cdot C(8, 4) = 15 \cdot 10 \cdot 70 = 10,500. \quad ■$$

EXAMPLE 7.19   From among a group of 6 men and 9 women how many three-member committees contain only men or only women?

The number of three-member committees containing only men is $C(6, 3)$, and the number of three-member committees containing only women is $C(9, 3)$. Since the set of committees containing only men is disjoint from the set of committees containing only women, the addition principle shows that the number of three-member committees containing only men or only women is

$$C(6, 3) + C(9, 3) = 20 + 84 = 104. \quad \blacksquare$$

EXAMPLE 7.20   How many 8–bit strings contain six or more 1's?

If an 8–bit string contains six or more 1's, then the number of 1's that it contains must be six, seven, or eight. Reasoning as in Example 7.16, we see that the number of strings containing exactly six 1's is $C(8, 6)$, the number of strings containing exactly seven 1's is $C(8, 7)$, and the number of strings containing exactly eight 1's is $C(8, 8)$. So the addition principle shows that the number of 8–bit strings containing six or more 1's is

$$C(8, 6) + C(8, 7) + C(8, 8) = 28 + 8 + 1 = 37. \quad \blacksquare$$

EXERCISES 7.3

Evaluate the numbers in Exercises 1–12.

1. $C(6, 3)$ **2.** $C(7, 4)$ **3.** $C(5, 2)$ **4.** $C(8, 4)$

5. $P(4, 2)$ **6.** $P(6, 3)$ **7.** $P(9, 5)$ **8.** $P(12, 3)$

9. $P(10, 4)$ **10.** $P(8, 3)$ **11.** $P(n, 1)$ **12.** $P(n, 2)$

13. How many different arrangements are there of the letters a, b, c, and d?

14. How many different arrangements are there of the letters in the word ''number''?

15. How many different four-digit numbers can be formed using the digits 1, 2, 3, 4, 5, and 6 without repetition?

16. How many different ways are there of selecting five persons from a group of seven persons and seating them in a row of five chairs?

17. How many different three-member subcommittees can be formed from a committee with 13 members?

18. How many different 16–bit strings contain exactly four 1's?

19. How many different subsets of $\{1, 2, \ldots, 10\}$ contain exactly six elements?

20. How many different four-person delegations can be formed from a group of twelve people?

21. Five speakers are scheduled to address a convention. In how many different orders can they appear?

22. Six persons are running for four seats on a town council. In how many different ways can these four seats be filled?

23. For marketing purposes a manufacturer wants to test a new product in three areas. If there are nine geographic areas in which to test market the product, in how many different ways can the test areas be selected?

24. An investor intends to buy shares of stock in three companies chosen from a list of twelve companies recommended by her broker. How many different investment options are there under the following circumstances?

   (a) Equal amounts will be invested in each company.

   (b) Amounts of $5000, $3000, and $1000 will be invested in the chosen companies.

25. How many different committees consisting of three representatives of management and two representatives of labor can be formed from among six representatives of management and five representatives of labor?

26. In how many different sequences can we list four novels followed by six biographies if there are eight novels and ten biographies from which to choose?

27. An election will be held to fill three faculty seats and two student seats on a certain college committee. The faculty member receiving the most votes will receive a three-year term, the one receiving the second highest total will receive a two-year term, and the one receiving the third highest total will receive a one-year term. Both of the open student seats are for one-year terms. If there are nine faculty members and seven students on the ballot, how many different election results are possible assuming that ties do not occur?

28. In how many different ways can eight women be paired with eight of twelve men at a dance?

29. Suppose that 3 freshmen, 4 sophomores, 2 juniors, and 3 seniors are candidates for four identical school service awards. In how many ways can the recipients be selected under the following conditions?

   (a) Any candidate may receive any award.

   (b) Only juniors and seniors receive awards.

   (c) One person from each class receives an award.

   (d) One freshman, two sophomores, and one senior receive awards.

30. Suppose that 3 freshmen, 5 sophomores, 4 juniors, and 2 seniors have been nominated to serve on a student advisory committee. How many different committees can be formed under the following circumstances?

   (a) The committee is to consist of any 4 persons.

   (b) The committee is to consist of one freshman, one sophomore, one junior, and one senior.

   (c) The committee is to consist of two persons: one freshman or sophomore and one junior or senior.

   (d) The committee is to consist of three persons from different classes.

31. Prove by a combinatorial argument that $2C(n, 2) + n^2 = C(2n, 2)$ for $n \geq 2$.

32. Prove by a combinatorial argument that $rC(n, r) = nC(n - 1, r - 1)$ for $n \geq r \geq 1$.

33. Prove that $C(1, 1) + C(2, 1) + \ldots + C(n, 1) = C(n + 1, 2)$ for $n \geq 1$ in two ways, algebraically and by a combinatorial argument.

34. For any positive integer n prove that
$C(n, 0)^2 + C(n, 1)^2 + \ldots + C(n, n)^2 = C(2n, n)$.

35. State and prove a generalization of Exercise 33.

7.4   Arrangements and Selections with Repetitions

In this section we will learn how to count the number of arrangements of a collection that includes repeated objects and the number of selections from a set when elements can be chosen more than once. As we will see, both of these counting problems require the use of ideas from the two preceding sections.

Let us consider first the number of arrangements of a collection containing repeated indistinguishable objects. As a simple example of this type of problem, we will count the number of different arrangements of the letters in the word "egg." Since there are only three letters in "egg," it is not difficult to list all of the possible arrangements. These are:

egg   geg   gge.

So there are only 3 arrangements of the letters in "egg" compared to the $P(3, 3) = 6$ arrangements that we would expect if all the letters had been distinct. To see more clearly the effect of the repeated letters, let us capitalize the first "g" in "egg" and regard a capital letter as different from a lower case letter. Then the six possible arrangements of the letters in "eGg" are:

eGg   Geg   Gge

egG   geG   gGe.

Note that because the two g's in the first list are identical, interchanging their positions does not produce different arrangements. But each arrangement in the first list gives rise to two arrangements in the second list, one with "G" preceding "g" and the other with "g" preceding "G." Thus, the number of arrangements in the first list equals the number of arrangements in the second list divided by $P(2, 2) = 2$, the number of permutations of the two g's.

Another way to count the number of arrangements of the letters in "egg" is by thinking of an arrangement as having 3 positions and first choosing positions for the two g's and then choosing a position for the "e". Since positions for the g's can be chosen in $C(3, 2)$ ways and the remaining position for the "e" can be chosen in only $C(1, 1)$ way, the multiplication principle then gives the number of possible arrangements as $C(3, 2) \cdot C(1, 1) = 3 \cdot 1 = 3$. This analysis and the one in the preceding paragraph lead to the same answer (see Exercise 31), which demonstrates the following result.

THEOREM 7.6   Let S be a collection containing n objects of k different types. (Objects of the same type are indistinguishable, and objects of different types are distinguishable.) Suppose that each object is of exactly one type and that there are n_1 objects of type 1, n_2 objects of type 2, and in general n_i objects of type i. Then the number of different arrangements of the objects in S is

$$C(n, n_1) \cdot C(n - n_1, n_2) \cdot C(n - n_1 - n_2, n_3) \cdot \ldots \cdot C(n - n_1 - n_2 - \ldots - n_{k-1}, n_k),$$

which equals $\dfrac{n!}{n_1! n_2! \ldots n_k!}$.

The conclusion of this theorem states that the number of different arrangements of the objects in S equals the number of ways $C(n, n_1)$ to place the n_1 objects of type 1 in n possible locations times the number of ways $C(n - n_1, n_2)$ to place the n_2 objects of type 2 in $n - n_1$ unused locations times the number of ways $C(n - n_1 - n_2, n_3)$ to place the n_3 objects of type 3 in $n - n_1 - n_2$ unused locations, etc. This number can also be written in the form $\dfrac{n!}{n_1! n_2! \ldots n_k!}$. Also note that $n = n_1 + n_2 + \ldots + n_k$ since we are assuming that each of the n elements in S belongs to exactly one of the k types.

EXAMPLE 7.21   How many arrangements are there of the letters in the word "banana"?
Since "banana" is a six-letter word consisting of three types of letters (1 b, 3 a's, and 2 n's), the number of arrangements of its letters is

$$\frac{6!}{1! \, 3! \, 2!} = \frac{6 \cdot 5 \cdot 4 \cdot 3!}{1 \cdot 3! \cdot 2} = \frac{6 \cdot 5 \cdot 4}{2} = 60. \quad \blacksquare$$

EXAMPLE 7.22   Each member of a nine-member committee must be assigned to exactly one of three subcommittees (the executive subcommittee, the finance subcommittee, or the rules subcommittee). If these subcommittees are to contain 3, 4, and 2 members, respectively, how many different subcommittee appointments can be made?
Let us arrange the nine persons in alphabetical order and give each person a slip of paper containing the name of a subcommittee. Then the number of possible subcommittee appointments is the same as the number of arrangements of 9 slips of paper, 3 of which read "executive subcommittee," 4 of which read "finance subcommittee," and 2 of which read "rules subcommittee." By Theorem 7.6 this number is

$$\frac{9!}{3! \, 4! \, 2!} = 1260. \quad \blacksquare$$

Let us now consider the problem of counting the number of selections from a set when elements can be chosen more than once. As an example, suppose that seven persons in a hotel conference room call room service for refreshments. If the choice of refreshments is limited to coffee, tea, or milk, how many different orders for refreshments are possible? Note that since an order is of the form "4 coffees, 1 tea, and 2 milks," we are selecting seven elements from the set {coffee, tea, milk} with repetition allowed.

To answer this question, let us think of each order as a collection of x's on an order form. For instance, an order for 4 coffees, 1 tea, and 2 milks will be recorded as shown.

Coffee	Tea	Milk
xxxx	x	xx

By always listing the beverages in the sequence above, we can simplify the order form by replacing the beverage names with slash marks. So the order for 4 coffees, 1 tea, and 2 milks can be written as xxxx|x|xx, and xxxxx|xx| represents an order for 5 coffees, 2 teas, and 0 milks. Notice that *two* slash marks divide the order form into *three* parts and that any order corresponds uniquely to some pattern of seven x's and two |'s. Hence, the number of different refreshment orders is the same as the number of ways that we can arrange seven x's and two |'s, or equivalently, the number of ways to choose positions for seven x's from nine possible locations. Thus, $C(9, 7) = 36$ different refreshment orders are possible. (Note that since $C(9, 2) = C(9, 7)$, we can also think of the number of different refreshment orders as the number of ways to choose positions for two |'s from nine positions.)

By using the same type of reasoning as in the example above, we obtain the following result. (In the beverage example, $r = 7$ and $n = 3$.)

THEOREM 7.7   If repetition is allowed, the number of selections of r elements that can be made from a set containing n distinct elements is $C(n + r - 1, r)$.

EXAMPLE 7.23   How many different assortments of one dozen donuts can be purchased if there are four different types of donuts from which to choose?

Since we are selecting 12 donuts with repetition from 4 types, we may regard this problem as one of selecting $r = 12$ elements with repetition from a set containing $n = 4$ distinct elements (the types of donuts). Thus, by Theorem 7.7 the number of possible choices is

$$C(4 + 12 - 1, 12) = C(15, 12) = 455. \quad \blacksquare$$

EXAMPLE 7.24 Suppose that we take five coins from a piggy bank containing many pennies, nickels, and quarters. How many different amounts of money might we get?

Note that because we are selecting five coins, each possible choice of coins corresponds to a different amount of money. (This would not be the case if we selected six coins, for 1 quarter and 5 pennies have the same value as 6 nickels.) Thus, by Theorem 7.7 the answer to this question is

$$C(3 + 5 - 1, 5) = C(7, 5) = 21. \quad \blacksquare$$

EXAMPLE 7.25 How many different assortments of one dozen donuts can be purchased if there are four different types of donut from which to choose and we want to include at least one donut of each type?

Since we must include at least one donut of each type, let us first choose one donut of each type. Then the answer to the question above is the number of ways that the remaining 8 donuts can be selected. By Theorem 7.7 this number is

$$C(4 + 8 - 1, 8) = C(11, 8) = 165. \quad \blacksquare$$

Counting problems involving the distribution of objects can be interpreted as problems involving arrangements or selection with repetition. Usually *problems involving the distribution of distinct objects correspond to arrangements with repetition, and problems involving the distribution of identical objects correspond to selections with repetition.* The following examples demonstrate the use of Theorems 7.6 and 7.7 in solving problems involving distributions.

EXAMPLE 7.26 How many distributions of ten different books are possible if Carlos is to receive 5 books, Doris is to receive 3 books, and Earl is to receive 2 books?

Distributing the ten books is equivalent to lining them up in some order and inserting a piece of paper in each book marked with the recipient's name. Then the number of possible distributions is the same as the number of ways of arranging 5 slips of paper marked "Carlos," 3 slips marked "Doris," and 2 slips marked "Earl." Using Theorem 7.6, we see that this number is

$$\frac{10!}{5! \, 3! \, 2!} = 2520.$$

Note the similarity between this solution and that of Example 7.22. $\quad \blacksquare$

EXAMPLE 7.27   If 9 red balloons and 6 blue balloons are to be distributed to four children, how many distributions are possible if every child must receive a balloon of each color?

Let us distribute the red balloons first and the blue balloons second. Since every child must receive a red balloon, we give one to each child. Now we can distribute the remaining 5 balloons in any way whatsoever. To decide who will receive each of these five balloons, we will think of selecting five times with repetition from a set containing the children's names. The number of possible selections is given by Theorem 7.7 to be $C(4 + 5 - 1, 5) = C(8, 5)$. Similar reasoning shows that the number of ways in which the blue balloons can be distributed so that every child receives at least one is $C(4 + 2 - 1, 2) = C(5, 2)$. Thus, by the multiplication principle the number of possible distributions of the balloons in which every child receives a balloon of each color is

$$C(8, 5) \cdot C(5, 2) = 56 \cdot 10 = 560.$$

EXERCISES 7.4

1. How many distinct arrangements of the letters in ''redbird'' are there?

2. How many distinct arrangements of the letters in ''economic'' are there?

3. How many different 7–digit numbers can be formed using the digits in the number 5,363,565?

4. How many different 9–digit numbers can be formed using the digits in the number 277,728,788?

5. How many different fruit baskets containing 8 pieces of fruit can be formed using only apples, oranges, and pears?

6. How many different assortments of six boxes of cereal can be made using packages of corn flakes, shredded wheat, and bran flakes?

7. How many different assortments of one dozen donuts can be purchased from a bakery that makes donuts with chocolate, vanilla, cinnamon, powdered sugar, and glazed icing?

8. How many different boxes containing 10 wedges of cheese can be made using wedges of Cheddar, Edam, Gouda, and Swiss cheese?

9. A box contains 16 crayons, no two having the same color. In how many different ways can they be given to four children so that each child receives 4 crayons?

10. In how many different ways can 15 distinct books be distributed so that Carol receives 6, Don receives 4, and Ellen receives 5?

11. A committee's chairperson and secretary must telephone the other 7 members about a change in the committee's meeting time. In how many different ways can these telephone calls be made if the chairperson calls 3 people and the secretary calls 4?

12. Paula has bought 6 different record albums to give as Christmas gifts. In how many different ways can she distribute the albums so that each of her three boyfriends receives two albums?

13. In how many different ways can 8 identical pieces of construction paper be distributed to 4 children?

$$C(4+8-1, 8) = \binom{11}{8}$$

14. In how many different ways can ten identical quarters be distributed to 5 people?

$$C(5+10-1, 10) = \binom{14}{10}$$

15. In how many different ways can 6 identical sticks of white chalk be distributed to three students so that each student receives at least one stick?

$$C(3+3-1, 3) = \binom{5}{3}$$

16. A father has ten identical life insurance policies. He wants to name one of his three children as the beneficiary of each policy. In how many different ways can the beneficiaries be chosen if each child is to be named a beneficiary on at least two policies?

$$C(3+4-1, 4) = \binom{6}{4}$$

17. A concert pianist is preparing a recital that will consist of 1 Baroque piece, 3 classical pieces, and 3 romantic pieces. Assuming for the sake of programming that pieces of the same period are regarded as indistinguishable, how many different programs containing the seven pieces can the pianist create?

$$\frac{7!}{1!\,3!\,3!}$$

18. In bridge a deal consists of distributing a 52-card deck into four 13-card hands. How many different deals are possible in bridge?

$$\frac{52!}{13!\,13!\,13!\,13!}$$

19. In how many different ways can 8 identical mathematics books and 10 identical computer science books be distributed among 6 students?

$$C(6+8-1, 8) \cdot C(6+10-1, 10) = \binom{13}{8}\binom{15}{10}$$

20. Twelve children are to be divided into groups of three to play different number games. In how many ways can the groups be chosen?

$$\frac{12!}{3!\,3!\,3!\,3!}$$

21. Ten diplomats are awaiting assignments to foreign embassies. If 3 of these diplomats are to be assigned to England, 4 to France, and 3 to Germany, in how many ways can the assignments be made?

$$\frac{10!}{3!\,4!\,3!}$$

22. In order to stagger the terms of service of 12 people elected to a new committee, four members are to be assigned a one-year term, four members are to be assigned a two-year term, and four members are to be assigned a three-year term. In how many different ways can these assignments be made?

23. How many 16-bit strings are there containing six 0's and ten 1's with no consecutive 0's?

24. How many positive integer solutions are there to the equation $x + y + z = 17$?

25. In how many ways can 2 identical teddy bears and 7 distinct Cabbage Patch dolls be distributed to three children so that each child receives 3 gifts?

26. How many numbers greater than 50,000,000 can be formed by rearranging the digits of the number 13,979,397?

27. How many positive integers less than 10,000 are such that the sum of their digits is 8?

28. How many distinct arrangements are there of two a's, one e, one i, one o, and seven x's in which no two vowels are adjacent?

29. In the following segment of a computer program, how many times is the PRINT statement executed?

```
        FOR I : = 1 TO 10
            FOR J : = 1 TO I
                FOR K : = 1 TO J
                    PRINT I, J, K
                NEXT K
            NEXT J
        NEXT I
```

30. A pouch contains $1 in pennies, $1 in nickels, and $1 in dimes. In how many different ways can 12 coins be selected from this pouch? (Assume that coins of the same value are indistinguishable.)

31. Prove that the two expressions in Theorem 7.6 are equal.

32. Use Exercise 31 of Section 7.1 to prove Theorem 7.7 by induction on n.

7.5   Probability

The subject of probability is generally accepted as having begun in 1654 with an exchange of letters between the great French mathematicians Blaise Pascal and Pierre de Fermat. During the next 200 years probability was combined with statistics to form a unified theory of mathematical statistics, and it is in this context that any thorough discussion of probability must occur. Nevertheless, the history of probability is closely related to the history of combinatorics, the branch of mathematics concerned with counting. In this section we will discuss probability as an application of the combinatorial ideas presented in Sections 7.2, 7.3, and 7.4.

Intuitively, probability measures how likely something is to occur. In his important book *Theorie Analytique des Probabilities*, the French mathematician Pierre Simon de Laplace (1749–1827) defined probability as follows: Probability is the ratio of the number of favorable cases to the total number of cases, assuming that all of the various cases are equally possible. Thus, according to Laplace's definition, probability measures the frequency with which a favorable case occurs. In this book we will study probability only in situations where this definition applies. Note that this definition requires that we know the number of favorable cases and the total number of cases and, therefore, requires the use of counting techniques.

By an **experiment** we will mean any procedure that results in an observable outcome. Thus, we may speak of the experiment of flipping a coin (and observing if it falls heads or tails) or the experiment of tossing a die (and noting the number of spots that show). A set consisting of all the possible outcomes of an experiment is called a **sample space** for the experiment. It is important to realize that there may be many possible sample spaces for an experiment. In the experiment of tossing an evenly balanced die, for instance, three possible sample spaces are {1, 2, 3, 4, 5, 6}, {even, odd}, and {perfect square, not a perfect square}. Which of these sample spaces may be most useful depends on the particular type of outcomes that we wish to consider. But in order to use Laplace's definition of probability we must be certain that the outcomes in the sample space are all equally likely to occur. This is the case for the outcomes in the first two sample spaces above, but the outcomes in the third sample space are not equally likely since there are only two perfect squares among the numbers 1 through 6 (namely 1 and 4).

Thus, for our purposes the sample space {perfect square, not a perfect square} will not prove useful.

Any subset of a sample space is called an **event**. Thus, in the die-tossing experiment with sample space {1, 2, 3, 4, 5, 6} the following sets are events:

$$A = \{1, 2, 4, 6\}, B = \{n\colon n \text{ is an integer and } 4 < n \le 6\}, \text{ and}$$

$$C = \{n\colon n \text{ is an even positive integer less than 7}\}.$$

Then for any event E in a sample space S consisting of equally likely outcomes, we will define the **probability** of E, denoted $Pr(E)$, by

$$Pr(E) = \frac{\text{number of elements in } E}{\text{number of elements in } S}. \qquad (7.3)$$

success / *possible outcomes*

So for the events A, B, and C above, we have $Pr(A) = \frac{4}{6} = \frac{2}{3}$, $Pr(B) = \frac{2}{6} = \frac{1}{3}$, and $Pr(C) = \frac{3}{6} = \frac{1}{2}$.

EXAMPLE 7.28   In the experiment of flipping a properly balanced coin three times, what is the probability of obtaining exactly two heads?

Since each flip of the coin has two possible results, heads (H) or tails (T), the multiplication principle shows that there are $2 \cdot 2 \cdot 2 = 8$ possible outcomes for three flips. The set

$$S = \{HHH, HHT, HTH, HTT, THH, THT, TTH, TTT\}$$

is a sample space for this experiment consisting of equally likely outcomes. The event of obtaining exactly two heads is the set $E = \{HHT, HTH, THH\}$. Thus, the probability of obtaining exactly two heads is

$$Pr(E) = \frac{\text{number of elements in } E}{\text{number of elements in } S} = \frac{3}{8}. \qquad \blacksquare \qquad \frac{C(2,3)}{2^3}$$

$Pr(\text{@ least 1 head occurs}) = 1 - Pr(\text{no heads occurring})$

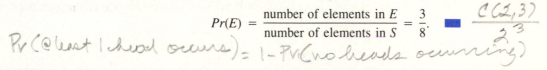

EXAMPLE 7.29   Suppose that there are six applicants for a particular job, four men and two women, who are to be interviewed in a random order. What is the probability that the four men are interviewed before either woman?

To answer this question, we must decide on an appropriate sample space consisting of equally likely outcomes. Since the ordering of the interviews is important, the set S of all possible arrangements of the six interviews is the obvious choice. Let E denote the subset of S in which the men are interviewed before the women. Then the multiplication principle shows that the number of elements in E

equals the number of arrangements of the men times the number of arrangements of the women. So by (7.3) we have

$$Pr(E) = \frac{\text{number of elements in } E}{\text{number of elements in } S}$$

$$= \frac{P(4, 4) \cdot P(2, 2)}{P(6, 6)} = \frac{24 \cdot 2}{720} = \frac{1}{15}.$$ ■

EXAMPLE 7.30

Suppose that there are two defective pens in a box of twelve pens. If we choose three pens at random, what is the probability that we do not select a defective pen?

In this problem the set of all selections of three pens chosen from among the twelve in the box will be our sample space S. The set of all selections of three pens chosen from among the ten nondefective pens is the event E in which we are interested. Thus, by (7.3) we find that

$$Pr(E) = \frac{\text{number of elements in } E}{\text{number of elements in } S}$$

$$= \frac{C(10, 3)}{C(12, 3)} = \frac{120}{220} = \frac{6}{11}.$$ ■

EXAMPLE 7.31

What is the probability that a randomly chosen permutation of the letters in the word "computer" has no adjacent vowels?

Let the sample space be the set S of all permutations of the letters in the word "computer," and let E denote the subset of all such permutations in which no two vowels are adjacent.

To count the permutations in E, we first arrange the five consonants in one of $P(5, 5) = 120$ ways, say

$$\_ \, p \_ t \_ c \_ r \_ m \_.$$

Since no two vowels are adjacent, we must insert at most one vowel in each blank above; the number of ways to choose positions for the three vowels is, therefore, $C(6, 3) = 20$. Finally we arrange the vowels in the chosen positions in $P(3, 3) = 6$ ways. Thus, E contains $120(20)(6) = 14,400$ permutations, and so

$$P(6,3) = 120$$

$$Pr(E) = \frac{\text{number of elements in } E}{\text{number of elements in } S}$$

$$= \frac{14,400}{P(8, 8)} = \frac{14,400}{40,320} = \frac{5}{14}.$$ ■

EXAMPLE 7.32 Suppose that we have ten different novels, five by Hemingway and five by Faulkner, that we want to distribute so that Barbara receives 5, Cathy receives 2, and Darlene receives 3. If the individual novels are distributed at random, what is the probability that Barbara receives all five of the novels by Hemingway?

Here the sample space S is the set of all distributions of 5 novels to Barbara, 2 to Cathy, and 3 to Darlene, and the event of interest is the set E of all such distributions in which Barbara receives all the Hemingway novels. Note that the distributions in which Barbara receives all the Hemingway novels are just those in which the Faulkner novels are distributed so that Cathy receives 2 and Darlene receives 3. So by reasoning as in Example 7.26 we see that

$$Pr(E) = \frac{\text{number of elements in } E}{\text{number of elements in } S}$$

$$= \frac{\dfrac{5!}{2!\,3!}}{\dfrac{10!}{5!\,2!\,3!}} = \frac{5!\,5!}{10!} = \frac{1}{252}.$$

EXERCISES 7.5

1. In the experiment of rolling a die, what is the probability of rolling a number greater than 1?

2. In the experiment of rolling a die, what is the probability of rolling a number divisible by 3?

3. If a coin is tossed 5 times, what is the probability that it will land heads each time?

4. If three dice are rolled, what is the probability that a 1 will appear on each die?

5. If a pair of dice are rolled, what is the probability that the sum of the spots which appear is 11?

6. If four coins are tossed, what is the probability that all of them land with the same side up?

7. If five coins are tossed, what is the probability that exactly three of them land tails?

8. If a coin is tossed 8 times, what is the probability that it will land heads exactly 4 times?

9. If three persons are chosen at random from a set of 5 men and 6 women, what is the probability that 3 women are chosen?

10. Suppose that a 4–digit number is created using the digits 1, 2, 3, 4, and 5 as often as desired. What is the probability that it contains two 1's and two 4's?

11. In a 7–horse race, a bettor bet the trifecta, which requires that the first three horses be identified in order of their finish. What is the probability of winning the trifecta under these conditions by randomly guessing three numbers?

12. If four persons are chosen at random from a class containing 8 freshmen and 12 sophomores, what is the probability that 4 freshmen are chosen?

13. What is the probability that a randomly chosen four-digit number contains no repeated digits?

14. What is the probability that a randomly chosen string of three letters contains no repeated letters?

15. If the letters of "sassafras" are randomly permuted, what is the probability that the four s's are adjacent and the three a's are adjacent?

16. In a consumer preference test 10 people were asked to name their favorite fruit from among apples, bananas, and oranges. If each person named a fruit at random, what would be the probability that no one named bananas?

17. If the personnel files of 5 employees are randomly selected, what is the probability that they are chosen in order of increasing salary? (Assume that no two employees have the same salary.)

18. In a particular group of people, 10 are right-handed and 4 are left-handed. If five of these people are chosen at random, what is the probability that exactly 1 left-handed person is selected?

19. Fifteen sticks of chewing gum are to be given at random to three children. What is the probability that each child receives at least four sticks of gum?

20. Three $10 bills, four $5 bills, and six $1 bills are randomly placed in a stack. What is the probability that the $5 bills are all adjacent?

21. If a five-member committee is selected at random from among 7 faculty and 6 students, what is the probability that it contains exactly 3 faculty and 2 students?

22. Suppose that we randomly distribute five distinct Cabbage Patch dolls and three identical teddy bears to four children. What is the probability that each child receives two gifts?

23. In a small garden there is a row of 8 tomato plants, 3 of which are diseased. Assuming that the disease occurs at random in the plants, what is the probability that the 3 diseased plants are all adjacent?

24. If twenty quarters are distributed at random to four people, what is the probability that everyone receives at least 75 cents?

25. Each of nine different books is to be given at random to Rebecca, Sheila, or Tom. What is the probability that Rebecca receives 2 books, Sheila receives 4, and Tom receives 3?

26. What is the probability that an odd number between 1000 and 9000 contains no repeated digits?

27. What is the probability that a randomly chosen permutation of the letters in the word "determine" has no adjacent e's?

28. A file contains 25 accounts numbered 1–25. If five of these accounts are selected at random for auditing, what is the probability that no two accounts with consecutive numbers are chosen?

7.6   The Principle of Inclusion-Exclusion

The addition principle (Theorem 7.5) tells us how to find the number of elements in the union of pairwise disjoint sets in terms of the number of elements in the individual sets. In this section we will present a similar result that will enable us to count the number of elements in the union of any sets, whether pairwise disjoint or not.

The following simple example will demonstrate the type of counting problem that we will be discussing. Suppose that a certain group of computer science students are all studying logic or mathematics. If 12 are studying logic, 26 are studying mathematics, and 5 are studying both logic and mathematics, how many students are in this group? If we let A denote the set of students studying logic and B denote the set of students studying mathematics, then the answer to this question is the number of elements in the set $A \cup B$. But since A and B are not disjoint, the addition principle cannot be used directly. It is not difficult, however, to see that the set B' of students studying mathematics but not logic contains $26 - 5$ elements. Now A and B' are disjoint and contain all of the students in the group. So the answer to our question is the number of elements in $A \cup B'$, which is $12 + (26 - 5)$ by the addition principle. (See Figure 7.2.)

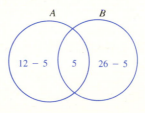

FIGURE 7.2

The example in the preceding paragraph involved finding the number of elements in a set. Recall that the number of elements in a finite set X is denoted by $|X|$. With this notation, our analysis of the example above showed that

$$|A \cup B| = |A| + |B| - |A \cap B|. \qquad (7.4)$$

It is not difficult to see that equation (7.4) holds for any finite sets A and B: The sum $|A| + |B|$ counts the elements of $A \cap B$ twice (once as members of A and once as members of B); so $|A| + |B| - |A \cap B|$ counts each element of $A \cup B$ exactly once.

EXAMPLE 7.33   In Example 7.12 we used the addition principle to count the number of 8–bit strings that begin with 1011 or end with 01. Let us count them again by using (7.4).

Let A and B denote the sets of 8–bit strings that begin with 1011 and end with

01, respectively. Then $A \cap B$ is the set of strings that begin with 1011 and end with 01, that is, strings of the form 1011−−01. Since only the fifth and sixth bits are unspecified, the number of such strings is $2 \cdot 2 = 4$. But since $|A| = 2^4 = 16$ and $|B| = 2^6 = 64$ from Example 7.12, it follows from (7.4) that the number of 8−bit strings that begin with 1011 or end with 01 is

$$|A \cup B| = |A| + |B| - |A \cap B| = 16 + 64 - 4 = 76. \quad \blacksquare$$

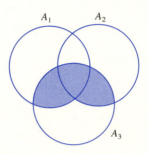

FIGURE 7.3

Our aim in this section is to generalize (7.4) from two sets to r sets, $A_1, A_2, \ldots, A_r$. But let us first consider the case that $r = 3$. It is easy to see in Figure 7.3 that $(A_1 \cup A_2) \cap A_3 = (A_1 \cap A_3) \cup (A_2 \cap A_3)$. By using this fact and (7.4), we can obtain a formula for $|A_1 \cup A_2 \cup A_3|$ as follows.

$$
\begin{aligned}
|A_1 \cup A_2 \cup A_3| &= |(A_1 \cup A_2) \cup A_3| = |A_1 \cup A_2| + |A_3| - |(A_1 \cup A_2) \cap A_3| \\
&= (|A_1| + |A_2| - |A_1 \cap A_2|) + |A_3| - |(A_1 \cap A_3) \cup (A_2 \cap A_3)| \\
&= (|A_1| + |A_2| - |A_1 \cap A_2|) + |A_3| - (|A_1 \cap A_3| + |A_2 \cap A_3| - |A_1 \cap A_3 \cap A_2 \cap A_3|) \\
&= |A_1| + |A_2| + |A_3| - |A_1 \cap A_2| - |A_1 \cap A_3| - |A_2 \cap A_3| + |A_1 \cap A_2 \cap A_3| \\
&= (|A_1| + |A_2| + |A_3|) - (|A_1 \cap A_2| + |A_1 \cap A_3| + |A_2 \cap A_3|) + |A_1 \cap A_2 \cap A_3|
\end{aligned}
$$

In order to generalize (7.4) to r sets, let us define n_s for $1 \le s \le r$ to be the sum of the sizes of all possible intersections of s sets chosen without repetition from among $A_1, A_2, \ldots, A_r$. Thus, if $r = 3$, so that there are only three sets A_1, A_2, and A_3, we have:

$$n_1 = |A_1| + |A_2| + |A_3|,$$

$$n_2 = |A_1 \cap A_2| + |A_1 \cap A_3| + |A_2 \cap A_3|, \text{ and}$$

$$n_3 = |A_1 \cap A_2 \cap A_3|.$$

Notice that with this notation the formula derived in the preceding paragraph can be written $|A_1 \cup A_2 \cup A_3| = n_1 - n_2 + n_3$.

Likewise, if $r = 4$ (there are four sets $A_1, A_2, A_3,$ and A_4), we have:

$$n_1 = |A_1| + |A_2| + |A_3| + |A_4|,$$

$$n_2 = |A_1 \cap A_2| + |A_1 \cap A_3| + |A_1 \cap A_4| + |A_2 \cap A_3| + |A_2 \cap A_4| + |A_3 \cap A_4|,$$

$$n_3 = |A_1 \cap A_2 \cap A_3| + |A_1 \cap A_2 \cap A_4| + |A_1 \cap A_3 \cap A_4| + |A_2 \cap A_3 \cap A_4|, \text{ and}$$

$$n_4 = |A_1 \cap A_2 \cap A_3 \cap A_4|.$$

In this case it can be shown that

$$|A_1 \cup A_2 \cup A_3 \cup A_4| = n_1 - n_2 + n_3 - n_4.$$

Then the desired generalization of (7.4) can be stated as follows.

THEOREM 7.8

The Principle of Inclusion-Exclusion For any finite sets $A_1, A_2, \ldots, A_r$ define n_s for $1 \le s \le r$ to be the sum of the sizes of all possible intersections of s sets chosen without repetition from among $A_1, A_2, \ldots, A_r$. Then

$$|A_1 \cup A_2 \cup \ldots \cup A_r| = n_1 - n_2 + n_3 - n_4 + \ldots + (-1)^{r-1} n_r.$$

Proof. Let $m = |A_1 \cup A_2 \cup \ldots \cup A_r|$. We will show that

$$m - n_1 + n_2 - n_3 + \ldots + (-1)^r n_r = 0.$$

Let $a \in A_1 \cup A_2 \cup \ldots \cup A_r$, and suppose that a belongs to exactly k of the sets A_i. Then a is counted $C(k, 0) = 1$ time in m, $C(k, 1) = k$ times in n_1 (because a belongs to exactly k of the sets A_i), $C(k, 2)$ times in n_2 (because a belongs to exactly $C(k, 2)$ of the intersections $A_i \cap A_j$), $\ldots$, and $C(k, k) = 1$ time in n_k. Furthermore, if $s > k$, then a is not counted at all in n_s because a does not belong to any intersection of more than k of the sets A_i. Hence, the number of times that a is counted in $m - n_1 + n_2 - n_3 + \ldots + (-1)^r n_r$ is

$$C(k, 0) - C(k, 1) + C(k, 2) - C(k, 3) + \ldots + (-1)^k C(k, k),$$

which can be shown to be 0 by taking $x = 1$ and $y = -1$ in the binomial theorem. It follows that

$$m = n_1 - n_2 + n_3 - n_4 + \ldots + (-1)^{r-1} n_r. \quad \blacksquare$$

EXAMPLE 7.34

Among a group of students studying Pascal, COBOL, or FORTRAN, 49 are studying Pascal, 37 are studying COBOL, and 21 are studying FORTRAN. If 9 of these students are studying Pascal and COBOL, 5 are studying Pascal and FORTRAN, 4 are studying COBOL and FORTRAN, and 3 are studying Pascal, COBOL, and FORTRAN, how many students are in this group?

Let us denote the sets of students studying Pascal, COBOL, and FORTRAN by P, C, and F, respectively (instead of A_1, A_2, and A_3). Then the number of students in the group is $|P \cup C \cup F|$. Now

$$n_1 = |P| + |C| + |F| = 49 + 37 + 21 = 107,$$

$$n_2 = |P \cap C| + |P \cap F| + |C \cap F| = 9 + 5 + 4 = 18, \text{ and}$$

$$n_3 = |P \cap C \cap F| = 3.$$

So by the principle of inclusion-exclusion we have

$$|P \cup C \cup F| = n_1 - n_2 + n_3 = 107 - 18 + 3 = 92.$$

Hence, there are 92 students in this group. ■

EXAMPLE 7.35

A bridge hand consists of 13 cards chosen from a standard 52–card deck. How many different bridge hands contain a void suit (that is, contain no cards in a suit)?

Let A_1, A_2, A_3, and A_4 denote the sets of bridge hands that contain no spades, no hearts, no diamonds, and no clubs, respectively. Then the number of bridge hands that contain a void suit is $|A_1 \cup A_2 \cup A_3 \cup A_4|$. We will compute this number using the principle of inclusion-exclusion.

Since a hand that is void in spades must consist of 13 cards chosen from among the 39 hearts, diamonds, and clubs, $|A_1| = C(39, 13)$. Similar reasoning shows that $|A_2| = |A_3| = |A_4| = C(39, 13)$. Hence, $n_1 = 4 \cdot C(39, 13)$.

Likewise, a hand that is void in both spades and hearts must consist of 13 cards chosen from among the 26 diamonds and clubs; so $|A_1 \cap A_2| = C(26, 13)$. Similarly,

$$|A_1 \cap A_3| = |A_1 \cap A_4| = |A_2 \cap A_3| = |A_2 \cap A_4| = |A_3 \cap A_4| = C(26, 13).$$

Thus, $n_2 = 6 \cdot C(26, 13)$.

In addition, a hand that is void in three suits must consist of all of the cards from the remaining suit; so $n_3 = 4$. Finally, no hand can be void in all four suits, so $n_4 = 0$. Therefore, we see that

$$
\begin{aligned}
|A_1 \cup A_2 \cup A_3 \cup A_4| &= n_1 - n_2 + n_3 - n_4 \\
&= 4 \cdot C(39, 13) - 6 \cdot C(26, 13) + 4 - 0 \\
&= 4(8,122,425,444) - 6(10,400,600) + 4 \\
&= 32,427,298,180.
\end{aligned}
$$

So there are 32,427,298,180 different bridge hands containing a void suit. ■

Often the sets $A_1, A_2, \ldots, A_r$ in Theorem 7.8 are subsets of a set U and we are interested in knowing how many elements of U are *not* contained in the set $A_1 \cup A_2 \cup \ldots \cup A_r$. Since this number is clearly

$$|U| - |A_1 \cup A_2 \cup \ldots \cup A_r|,$$

the principle of inclusion-exclusion can also be used to solve this type of problem.

EXAMPLE 7.36

Among a group of 200 college students, 19 study French, 10 study German, and 28 study Spanish. If 3 study both French and German, 8 study both French and Spanish, 4 study both German and Spanish, and 1 studies French, German, and Spanish, how many of these students are not studying French, German, or Spanish?

Let U denote the set of all 200 students and F, G, and S denote the subsets of U consisting of the students who are studying French, German, and Spanish, respectively. Then the number of students in U who are not studying French, German, or Spanish is $|U| - |F \cup G \cup S|$. Now

$$n_1 = |F| + |G| + |S| = 19 + 10 + 28 = 57,$$

$$n_2 = |F \cap G| + |F \cap S| + |G \cap S| = 3 + 8 + 4 = 15, \text{ and}$$

$$n_3 = |F \cap G \cap S| = 1.$$

Thus, by the principle of inclusion-exclusion

$$|F \cup G \cup S| = n_1 - n_2 + n_3 = 57 - 15 + 1 = 43.$$

So $200 - 43 = 157$ students are not studying French, German, or Spanish. ■

A permutation of the integers $1, 2, \ldots, n$ such that no integer occupies its natural position is called a **derangement.** So 41532 is a derangement of the integers 1, 2, 3, 4, 5 because 1 is not the first digit, 2 is not the second digit, and so forth. Counting the number of derangements is a famous problem that requires the use of the principle of inclusion-exclusion.

EXAMPLE 7.37

How many derangements of the integers 1, 2, 3, 4 are there?

Let U denote the set of permutations of 1, 2, 3, 4; and let A_1 denote the set of members of U having a 1 as first digit, A_2 denote the set of members of U having a 2 as second digit, and so forth. Then a derangement of the integers 1, 2, 3, 4 is a member of U that is not in $A_1 \cup A_2 \cup A_3 \cup A_4$.

Note that any permutation in A_1 has the form 1---, where the second, third, and fourth digits can be chosen arbitrarily. So the number of such permutations is $P(3, 3)$. Likewise, $|A_2| = |A_3| = |A_4| = P(3, 3)$.

Permutations in $A_1 \cap A_2$ have the form 12--, and so there are $P(2, 2)$ of them. Likewise, $|A_1 \cap A_3| = |A_1 \cap A_4| = |A_2 \cap A_3| = |A_2 \cap A_4| = |A_3 \cap A_4| = P(2, 2)$.

Similar reasoning shows that $|A_1 \cap A_2 \cap A_3| = |A_1 \cap A_2 \cap A_4| = |A_1 \cap A_3 \cap A_4| = |A_2 \cap A_3 \cap A_4| = P(1, 1)$ and $|A_1 \cap A_2 \cap A_3 \cap A_4| = 1$.

Thus, by the principle of inclusion-exclusion we have

$$|A_1 \cup A_2 \cup A_3 \cup A_4| = 4 \cdot P(3, 3) - 6 \cdot P(2, 2) + 4 \cdot P(1, 1) - P(1, 1)$$

$$= 4 \cdot 6 - 6 \cdot 2 + 4 \cdot 1 - 1 = 15.$$

So the number of derangements of 1, 2, 3, 4 is

$$|U| - |A_1 \cup A_2 \cup A_3 \cup A_4| = P(4, 4) - 15 = 24 - 15 = 9. ■$$

EXERCISES 7.6

1. In a survey of moviegoers it was found that 33 persons liked films by Bergman and 25 liked films by Fellini. If 18 of these persons liked both directors' films, how many liked films by Bergman or Fellini?

2. Among a group of children 88 liked pizza and 27 liked Chinese food. If 13 of these children liked both pizza and Chinese food, how many liked pizza or Chinese food?

3. Among the 318 members of a local union, 127 liked their congressional representative and 84 liked their governor. If 53 of these members liked both their congressional representative and their governor, how many of these union members liked neither their congressional representative nor their governor?

4. In a particular dormitory there are 350 college freshmen. Of these, 312 are taking an English course and 108 are taking a mathematics course. If 95 of these freshmen are taking courses in both English and mathematics, how many are taking a course in neither English nor mathematics?

5. From a group of 350 residents of a city, the following information was obtained:

   > 210 were college-educated.
   > 256 were married.
   > 228 were home-owners.
   > 180 were college-educated and married.
   > 147 were college-educated and home-owners.
   > 166 were married and home-owners.
   > 94 were college-educated, married, and home-owners.

   How many of these residents were not college-educated, not married, and not home-owners?

6. In tabulating the 3681 responses to a questionnaire sent to her constituents, a congresswoman found:

   > 2819 favored tax reform.
   > 2307 favored a balanced budget.
   > 2562 favored a nuclear freeze.
   > 2163 favored tax reform and a balanced budget.
   > 1985 favored tax reform and a nuclear freeze.
   > 1137 favored a balanced budget and a nuclear freeze.
   > 984 favored tax reform, a balanced budget, and a nuclear freeze.

   How many of the respondants opposed tax reform, a balanced budget, and a nuclear freeze?

7. The following data was obtained from the fast-food restaurants in a certain city:

   > 13 served hamburgers.
   > 8 served roast beef sandwiches.
   > 10 served pizza.
   > 5 served hamburgers and roast beef sandwiches.
   > 3 served hamburgers and pizza.

2 served roast beef sandwiches and pizza.
1 served hamburgers, roast beef sandwiches, and pizza.
5 served none of these three foods.

How many fast-food restaurants are there in this city?

8. The following information was found about the residents of a certain retirement community:

> 38 played golf.
> 21 played tennis.
> 56 played bridge.
> 8 played golf and tennis.
> 17 played golf and bridge.
> 13 played tennis and bridge.
> 5 played golf, tennis, and bridge.
> 72 did not play golf, tennis, or bridge.

How many residents are there in this retirement community?

9. Recall that for $1 \le s \le r$ we defined n_s to be the sum of the sizes of s sets chosen without repetition from among $A_1, A_2, \ldots, A_r$. How many terms are there in this sum?

10. List all the derangements of 1, 2, 3, 4.

11. For the graph below, determine the number of ways to assign one of at most k colors to the vertices so that no two adjacent vertices receive the same color.

v_1

v_2 v_3

12. Suppose that three married couples are seated randomly in six chairs around a circular table. What is the probability that no married couple is seated in adjacent seats?

13. How many positive integers less than 101 are square free, that is, divisible by no perfect square greater than 1?

14. How many integers between 1 and 2101 are divisible by 2, 3, 5, or 7?

15. How many five-card poker hands contain at least one card in each suit?

16. How many arrangements of the numbers 1, 1, 2, 2, 3, 3, 4, 4 are there in which no adjacent numbers are equal?

17. How many nonnegative integer solutions of $x_1 + x_2 + x_3 + x_4 = 12$ are there in which no x_i exceeds 4?

18. How many functions are there that map $\{5, 6, 7, 8, 9, 10\}$ onto $\{1, 2, 3, 4\}$?

19. What is the probability that a 13–card bridge hand contains at least one 10, at least one jack, at least one queen, at least one king, and at least one ace?

20. Suppose that five balls numbered 1, 2, 3, 4, and 5 are successively removed from an urn. A rencontre is said to occur if ball number k is the kth ball removed. What is the probability that no rencontres occur?

21. Show that the number of lists of length n having entries chosen from a nonempty set S of m elements and having the property that each element of S appears at least once is

$$C(m, 0)(m - 0)^n - C(m, 1)(m - 1)^n + \ldots + (-1)^{m-1}C(m, m - 1)(1)^n.$$

22. Two integers are called **relatively prime** if 1 is the only positive integer that divides both numbers. Show that if a positive integer n has $p_1, p_2, \ldots, p_k$ as its distinct prime divisors, then the number of positive integers that are less than n and relatively prime to n is

$$n - \frac{n}{p_1} - \frac{n}{p_2} - \ldots + \frac{n}{p_1 p_2} + \frac{n}{p_1 p_3} + \ldots + (-1)^k \frac{n}{p_1 p_2 \ldots p_k}.$$

23. Compute the number D_k of derangements of $1, 2, 3, \ldots, k$.

24. If n is a positive integer, evaluate $D_{n+1} - (n + 1)D_n$, where D_n is as in Exercise 23.

*Define $S(n, m)$ to be the number of ways to distribute n distinguishable balls into m indistinguishable urns with no urns empty. These numbers are named **Stirling numbers of the second kind** after the British mathematician James Stirling (1692–1770).*

25. If n is a positive integer, evaluate $S(n, 0)$, $S(n, 1)$, $S(n, n)$, $S(n, n - 1)$, and $S(n, 2)$.

26. For all positive integers n and m prove that $S(n + 1, m) =$
$C(n, 0)S(0, m - 1) + C(n, 1)S(1, m - 1) + \ldots + C(n, n)S(n, m - 1)$.

27. Prove that $S(n + 1, m) = S(n, m - 1) + m \cdot S(n, m)$ for all positive integers n and m.

28. Use the result of Exercise 27 to describe how the Stirling numbers of the second kind can be computed by a method similar to that involving Pascal's triangle.

29. For all positive integers n and m prove that $S(n, m) =$
$$\frac{1}{m!}[C(m, 0)(m - 0)^n - C(m, 1)(m - 1)^n + \ldots + (-1)^{m-1}C(m, m - 1)(1)^n].$$

30. Let X and Y be sets containing n and m elements, respectively. How many functions are there with domain X and range Y?

7.7   Lexicographic Enumeration of Permutations

Unfortunately there are many practical problems for which no efficient method of solution is known (such as the knapsack problem described in Section 1.3). In such cases the only method of solution may be to perform an exhaustive search, that is, to systematically list and check all of the possibilities. Often, as in Section 1.2, listing all the possibilities amounts to enumerating all of the permutations of a set of elements. In this section we will present a procedure for listing all possible permutations of a set of n objects. For convenience we will assume that the set in question is $\{1, 2, \ldots, n\}$.

The most natural order in which to list permutations is called **lexicographic order** (or **dictionary order**). To describe this order let $p = (p_1, p_2, \ldots, p_n)$ and $q = (q_1, q_2, \ldots, q_n)$ be two different permutations of the integers $1, 2, \ldots, n$. Since p and q are different, they must differ in some entry. Let k denote the smallest index for which $p_k \neq q_k$. Then (reading from left to right) the first $k - 1$ entries of p and q are the same and the kth entries differ. In this case we will say that p is **greater than** q in the lexicographic ordering if $p_k > q_k$. If p is greater than q in the lexicographic ordering, then we will write $p >> q$ or $q << p$. Thus, in the lexicographic order we have $(2, 4, 1, 5, 3) >> (2, 4, 1, 3, 5)$ and $(3, 2, 4, 1, 5, 6) << (3, 2, 6, 5, 1, 4)$.

By using a tree diagram and choosing entries in numerical order, we can list all the permutations of $\{1, 2, \ldots, n\}$ in lexicographic order. For instance, if $n = 3$ we have:

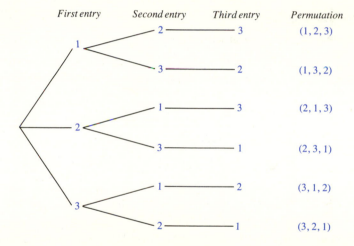

The permutations listed in the last column are in lexicographic order.

In order to have an efficient algorithm for listing permutations in lexicographic order, we must know how to find the successor of a permutation p in the lexicographic order, that is, the first permutation greater than p. Consider, for example, the permutation $p = (3, 6, 2, 5, 4, 1)$ of the integers 1 through 6. Let q denote the successor of p in the lexicographic ordering, and let r denote any permutation greater than q. Since $p << q << r$, q must agree with at least as much of p (from the left) as r does. Thus, q must differ from p as far to the right in its list as possible. Clearly we cannot rearrange the order of the last two entries of p (4 and 1) or the last three entries of p (5, 4, and 1) and obtain a larger number. But we can rearrange the order of the last four entries of p (2, 5, 4, and 1) to get a larger number, and the least such rearrangement is 4, 1, 2, 5. Thus, the successor of p in the lexicographic ordering is $q = (3, 6, 4, 1, 2, 5)$. Notice that the first two entries of q are the same as those of p and that the third entry of q is greater than that of p. Moreover, the third entry of q is the smallest entry of p that exceeds the third entry of p and lies to its right. Finally, note that the entries of q to the right of the third entry are in increasing order.

More generally, consider a permutation $p = (p_1, p_2, \ldots, p_n)$ of the integers 1 through n. The successor of p in the lexicographic ordering will be a permutation $q = (q_1, q_2, \ldots, q_n)$ such that:

(1) The first $k - 1$ entries of q will be the same as in p.

(2) The kth entry of q, q_k, will be the smallest entry of p to the right of p_k that is greater than p_k.

(3) The entries of q that follow q_k will be in increasing numerical order.

Therefore, we can completely determine q from p if we know the value of k, the index of the entry of p to be changed. As we saw in our example, we want k to be chosen as large as possible. So because of condition 2 above, we must choose k to be the largest possible index for which p_k is less than one of the entries that follow it. But then k will be the largest index such that $p_k < p_{k+1}$. Thus, if we examine the entries of p from *right to left,* the entry of p to be changed is the first entry we reach that is less than the number to its right. In addition, since the entries of p to the right of the kth entry are in decreasing order, the value of q_k will be the rightmost entry of p that exceeds p_k. If we now switch p_k with the rightmost entry of p that exceeds it, we obtain a new permutation in which the rightmost entries are the remaining entries of q in reverse order.

EXAMPLE 7.38

Let us determine the permutation q of the integers 1 through 7 that is the successor of $p = (4, 1, 5, 3, 7, 6, 2)$. Scanning p from right to left, we see that the first entry we reach that is less than the number to its right is the fourth entry, which is 3. (So in the notation above, $k = 4$.) Thus, q has the form $(4, 1, 5, ?, ?, ?, ?)$. Moreover, the fourth entry of q will be the rightmost entry of p that exceeds the entry that is being changed (which is 3 in our case). Scanning p again from right to left, we see that the fourth entry of q will be 6. Interchanging the positions of the 3 and 6 in p, we obtain $(4, 1, 5, 6, 7, 3, 2)$. If we now reverse the order of the entries to the right of position k, we will have $(4, 1, 5, 6, 2, 3, 7)$, which is the successor of p. ▪

The algorithm below uses the method described above to list all the permutations of $\{1, 2, \ldots, n\}$.

Algorithm for the Lexicographic Ordering of Permutations

This algorithm accepts as input a positive integer n and prints all the permutations of $\{1, 2, \ldots, n\}$ in lexicographic order. In the algorithm, p denotes the permutation currently being considered.

Step 1 (initialization). For $j = 1$ to n let $p(j) = j$.

Step 2 (output). Print $(p(1), p(2), \ldots, p(n))$.

Step 3 (find the index k of the entry to be changed). Find the largest index k for which $p(k) < p(k + 1)$. If no such k exists, then stop; otherwise go to step 4.

Step 4 (find the index of the entry to interchange with $p(k)$). Find the index j of the smallest entry to the right of $p(k)$ that is larger than $p(k)$.

Step 5 (switch entries). Interchange $p(k)$ and $p(j)$.

Step 6 (determine the successor of p). Reverse the order of $p(k + 1)$, $p(k + 2)$, . . . , $p(n)$, and go to step 2.

Although the lexicographic ordering is the most natural ordering for listing permutations, determining the successor of a given permutation in the lexicographic ordering requires several comparisons. For this reason an algorithm that lists permutations in lexicographic order may be less efficient than one that lists the permutations in a different order. But since there are $n!$ permutations of the integers $1, 2, . . . , n$, the complexity of any algorithm that lists these permutations will be at least $n!$. Readers who are interested in learning more efficient algorithms for listing permutations should consult [3] in the readings at the end of the chapter.

EXERCISES 7.7

For the permutations p and q in Exercises 1–6, determine whether $p \ll q$ or $p \gg q$ in the lexicographic ordering.

1. $p = (3, 2, 4, 1), q = (4, 1, 3, 2)$

2. $p = (2, 1, 3), q = (1, 2, 3)$

3. $p = (1, 2, 3), q = (1, 3, 2)$

4. $p = (2, 1, 3, 4), q = (2, 3, 1, 4)$

5. $p = (4, 2, 5, 3, 1), q = (4, 2, 3, 5, 1)$

6. $p = (2, 5, 3, 4, 1, 6), q = (2, 5, 3, 1, 6, 4)$

In Exercises 7–18 determine the successor of permutation p in the lexicographic ordering of the permutations of $\{1, 2, 3, 4, 5, 6\}$.

7. $p = (2, 1, 4, 3, 5, 6)$

8. $p = (3, 6, 4, 2, 1, 5)$

9. $p = (2, 1, 4, 6, 5, 3)$

10. $p = (3, 6, 5, 4, 2, 1)$

11. $p = (5, 6, 3, 4, 2, 1)$

12. $p = (5, 1, 6, 4, 3, 2)$

13. $p = (6, 5, 4, 3, 2, 1)$

14. $p = (1, 2, 3, 6, 5, 4)$

15. $p = (5, 2, 6, 4, 3, 1)$

16. $p = (4, 5, 6, 3, 2, 1)$

17. $p = (6, 3, 5, 4, 2, 1)$

18. $p = (2, 3, 1, 6, 5, 4)$

19. List the permutations of $\{1, 2, 3, 4\}$ in lexicographic order.

20. Write a computer program implementing the algorithm for listing the permutations of $\{1, 2, . . . , n\}$ in lexicographic order.

Suggested Readings

1. Beckenbach, E. *Applied Combinatorial Mathematics*. New York: John Wiley and Sons, 1964.
2. Even, Shimon. *Algorithmic Combinatorics*. New York: Macmillan, 1973.
3. Nijenhuis, Albert and Herbert S. Wilf. *Combinatorial Algorithms, 2nd ed.* New York: Academic Press, 1978.
4. Ryser, Herbert J. *Combinatorial Mathematics,* Carus Monograph Number 14. New York: Mathematical Association of America, 1963.
5. Whitworth, William Allen. *Choice and Chance*. reprint of the 5th edition. New York: Hafner, 1965.

8

RECURRENCE AND ITS APPLICATIONS

In the preceding chapters we have seen several situations in which we wanted to determine a number associated with a set of n objects, for example, the number of subsets of the set or the number of ways of arranging the objects in the set. Sometimes this number can be related to the corresponding number for a smaller set of objects. For example, in Section 2.6 we saw that the number of subsets of a set with n objects is twice the number of subsets of a set with $n - 1$ objects. Such a relationship can then be exploited to derive a formula for the number we are seeking in terms of n. Techniques for doing this will be explored in this chapter.

8.1   Recurrence Relations

An infinite list of numbers $s_1, s_2, s_3, \ldots, s_n, \ldots$ is called a **sequence.** The number s_n is called the **nth term** of the sequence. In this chapter, we will examine special types of sequences, develop techniques for finding general formulas for their nth terms, and consider a variety of applications.

The sequences we will consider are those where the nth term is a function of earlier terms in the same sequence. In Section 2.5, for instance, we saw that the terms in the sequence of factorials, $0!, 1!, 2!, \ldots, n!, \ldots$, could be defined recursively by specifying that $0! = 1$ and that for $n \geq 1$, $n! = n(n-1)!$. Also, in an arithmetic progression with common difference d, the nth term a_n can be defined in terms of the preceding term a_{n-1} as follows:

$$a_n = a_{n-1} + d.$$

In order to determine the values of the terms in a recursively defined sequence, we must know the values of a specific set of early terms in the sequence. The assignments of values for these earlier terms are called **initial conditions** for the sequence. In the case of the factorials, the initial condition is $0! = 1$. In the case of an arithmetic progression, the initial condition is assigning a value to the first term in the sequence. The equation defining the relationship of the nth term to those terms that precede it is called the **recurrence relation.**

Recurrence relations occur in many different places in mathematics. For example, in Section 2.5 we saw the use of a recurrence relation in the definition of the Fibonacci numbers. The nth Fibonacci number, F_n, is defined as follows:

$$F_n = F_{n-1} + F_{n-2} \text{ for } n > 2.$$

Setting $F_1 = F_2 = 1$ gives the initial conditions for the Fibonacci numbers. This sequence of numbers will be studied extensively in Section 8.3.

In this section we will examine other situations in which recurrence relations occur and illustrate how they can be used to solve problems involving counting.

EXAMPLE 8.1   Let us consider from a recursive point of view the question of how many edges it takes to draw the complete graph K_n on n vertices. Let e_n represent the total number of edges in K_n. We start by considering how many new edges need to be drawn in going from K_{n-1} to K_n. The addition of one more vertex requires the addition of $n - 1$ edges (one to each of the old vertices) as shown in Figure 8.1(a) for $n = 4$ and Figure 8.1(b) for $n = 5$. This results in a recurrence relation for the number of edges in the complete graph K_n on n vertices:

$$e_n = e_{n-1} + (n - 1) \text{ for } n \geq 2. \tag{8.1}$$

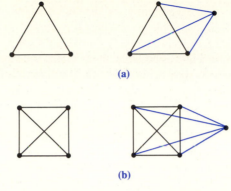

(a)

(b)

FIGURE 8.1

We can use equation (8.1) to evaluate e_n only if we know the value of e_{n-1}. In this case, clearly $e_1 = 0$. Using this we could calculate other values of e_n as follows:

$$e_1 = 0$$

$$e_2 = 0 + (2 - 1) = 1$$

$$e_3 = 1 + (3 - 1) = 3$$

$$e_4 = 3 + (4 - 1) = 6$$

.

.

.

Later, we will consider ways of finding a formula for e_n in terms of n.   ■

EXAMPLE 8.2

Consider the ancient game called the **Towers of Hanoi.** This game is played with a set of disks of graduated size, having holes in their centers, and a playing board having three spokes for holding the disks as shown in Figure 8.2. The object of the game is to move the disks from spoke A to spoke B, moving one disk at a time. A disk cannot be placed on top of a smaller disk. What is the minimum number of moves required if there are n disks?

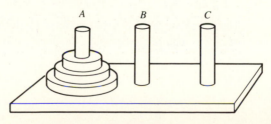

FIGURE 8.2

We represent by m_n the minimum number of moves required to move n disks from one spoke to another. It is easy to see that the minimum number of moves for n disks is obtained by moving the top $n - 1$ disks from spoke A to spoke C as efficiently as possible, then moving the largest disk from spoke A to spoke B, and finally moving the $n - 1$ disks from spoke C to spoke B as efficiently as possible. Now moving the top $n - 1$ disks from the original spoke to spoke C takes m_{n-1} moves. It then takes 1 move to move the remaining disk to spoke B. It will then take m_{n-1} moves to move the pile of $n - 1$ disks on top of the single disk on spoke B, thus completing the task of moving the n disks. This results in a recurrence relation

$$m_n = m_{n-1} + 1 + m_{n-1}$$

or

$$m_n = 2m_{n-1} + 1 \qquad \text{for } n > 1. \tag{8.2}$$

Note that $m_1 = 1$, as only 1 move is required to move one disk from spoke A to spoke B. We can analyze the recurrence relation (8.2) as follows:

$$
\begin{aligned}
m_n &= 2m_{n-1} + 1 \\
&= 2(2m_{n-2} + 1) + 1 \\
&= 2^2 m_{n-2} + 2 + 1 \\
&= 2^2(2m_{n-3} + 1) + 2 + 1 \\
&= 2^3 m_{n-3} + 2^2 + 2 + 1 \\
&\qquad \cdot \\
&\qquad \cdot \\
&\qquad \cdot \\
&= 2^{n-2} m_2 + 2^{n-3} + 2^{n-4} + \ldots + 2 + 1 \\
&= 2^{n-2}(2m_1 + 1) + 2^{n-3} + 2^{n-4} + \ldots + 2 + 1 \\
&= 2^{n-1} m_1 + 2^{n-2} + 2^{n-3} + \ldots + 2 + 1.
\end{aligned}
$$

But, since $m_1 = 1$,

$$m_n = 2^{n-1} + 2^{n-2} + 2^{n-3} + \ldots + 2 + 1.$$

The expression on the right is the sum of a geometric progression with first term 1 and common ratio 2. By Theorem 2.11, this sum is

$$m_n = 2^n - 1.$$

Thus, we have been able to find an explicit formula giving the minimum number of moves required to complete the game with n disks. ∎

EXAMPLE 8.3   Consider a game where Douglas starts with k dollars and Jennifer starts with m dollars. They agree to bet one dollar at a time on the flip of a fair penny and to continue playing until one of them wins all of the other's money. What is the probability that Douglas will win, given that he currently has n dollars?

Let d_n denote the probability that Douglas will win all of Jennifer's dollars if he currently has n dollars. Then $d_0 = 0$, as $n = 0$ indicates that Jennifer has won, and $d_{k+m} = 1$, as $n = k + m$ indicates that Douglas has all of the money. In general, the probability of Douglas winning if he has n dollars is given by

$$d_n = \tfrac{1}{2}d_{n+1} + \tfrac{1}{2}d_{n-1}, \text{ for } 1 \le n \le k + m - 1. \tag{8.3}$$

This equation results from the fact that if Douglas now has n dollars, then on the next flip he will have either $n + 1$ dollars if he wins (with probability $\tfrac{1}{2}$) or $n - 1$ dollars if he loses (also with probability $\tfrac{1}{2}$).

By multiplying both sides of (8.3) by 2 and rearranging the terms, we get

$$d_{n+1} - d_n = d_n - d_{n-1}.$$

Using this equation for $n = k + m - 1, k + m - 2, \ldots, 1$ we obtain $d_{k+m} - d_{k+m-1} = d_{k+m-1} - d_{k+m-2} = \ldots = d_1 - d_0$. Let each of these differences equal the constant c, and then add the resulting $k + m$ equations, noticing that most of the terms on the left side cancel:

$$
\begin{aligned}
d_{k+m} \quad &- d_{k+m-1} = c \\
d_{k+m-1} &- d_{k+m-2} = c \\
d_{k+m-2} &- d_{k+m-3} = c \\
&\quad\cdot \\
&\quad\cdot \\
&\quad\cdot \\
d_2 \quad &- d_1 \quad\;\; = c \\
d_1 \quad &- d_0 \quad\;\; = c \\
\hline
d_{k+m} \quad &- d_0 \quad\;\; = (k + m)c.
\end{aligned}
$$

Since $d_{k+m} = 1$ and $d_0 = 0$, we have $c = \dfrac{1}{k + m}$.

Repeating the process of adding the differences starting with $d_n - d_{n-1}$ and working to $d_1 - d_0$, we get $d_n - d_0 = nc$. Thus, $d_n = \dfrac{n}{k + m}$, and so Douglas's probability of winning all $k + m$ dollars at the point that he has n dollars is $\dfrac{n}{k + m}$. Jennifer's probability of winning all $k + m$ dollars at this point is $1 - \dfrac{n}{k + m}$. ∎

EXAMPLE 8.4

Suppose $1000 is invested in an account at an annual rate of 10% interest compounded annually. Let v_n denote the value of the account at the end of the nth year. Then the value after n years is the value after $n - 1$ years plus 10% interest; that is,

$$v_n = v_{n-1} + 0.1v_{n-1} = 1.1v_{n-1} \quad \text{for } n \geq 1.$$

Knowing that $v_0 = 1000$, we can use this recurrence to develop a formula for calculating v_n.

$$v_0 = 1000$$

$$v_1 = (1.1)(1000)$$

$$v_2 = (1.1)^2(1000)$$

$$v_3 = (1.1)^3(1000)$$

$$\cdot$$
$$\cdot$$
$$\cdot$$

As the above substitutions show, we have a geometric progression with first term $1000 and common ratio 1.1. Thus, the value after n years of interest is

$$v_n = (1.1)^n(1000). \quad \blacksquare\blacksquare$$

EXAMPLE 8.5

Suppose that a certain binary security code for credit cards allows n–digit sequences such that each digit in a given sequence must be either 0 or 1, and no two successive digits can be 0's. How many codewords of length n exist?

Let this number be denoted by w_n. Two one-digit codewords exist: 0 and 1. Thus, $w_1 = 2$. For $n = 2$, the two-digit codewords are 01, 10, and 11. Hence, $w_2 = 3$. For $n = 3$, an analysis shows that a codeword can begin either with 0 or 1. If the word begins with 0, then the second digit must be 1 and the remaining digit could be filled by any one-digit codeword, for a total of 2 codewords. If the three-digit codeword begins with 1, any one of the 3 two-digit codewords can constitute the remaining portion of the word, giving 3 more codewords of length 3. Thus, for $n = 3$ there are $2 + 3 = 5$ codewords. This same pattern of analysis works in general. We can always break the problem into two subproblems by considering the number of n–digit codewords beginning with 0 and the number of n–digit codewords beginning with 1. For $n \geq 3$ the former number is w_{n-2} and the latter is w_{n-1}. Thus, the number of codewords of length n is given by

$$w_n = \begin{cases} 2 & \text{if } n = 1 \\ 3 & \text{if } n = 2 \\ w_{n-2} + w_{n-1} & \text{if } n \geq 3. \end{cases}$$

Notice that this recurrence relation is similar to the formula which generates the Fibonacci numbers. $\quad \blacksquare\blacksquare$

EXAMPLE 8.6   Suppose that a carpenter needs to cover n consecutive 1–foot gaps between the centers of successive roof rafters with 1– and 2–foot long pieces of roofing material, as shown in Figure 8.3. In how many ways can the carpenter complete his task?

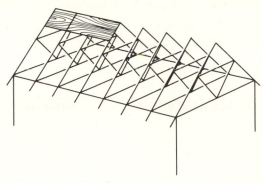

FIGURE 8.3

Let s_n represent the number of ways n gaps can be covered. For a single 1–foot gap, the carpenter can only use a 1–foot board. Hence, $s_1 = 1$. Two of the one-foot gaps can be covered in 2 ways, using two 1–foot boards or one 2–foot board. Hence, $s_2 = 2$. Further experimentation shows that $s_3 = 3$ and $s_4 = 5$.

A solution to this problem requires a look at how one can cover the required n consecutive 1–foot gaps in the roofline. The final board must either be a 1–foot board or a 2–foot board. If the carpenter finishes with a 2–foot board, then the last gap starts at the second-to-last rafter. There are s_{n-2} ways of getting to this rafter. If the last board is a 1–foot board, then the last gap starts on the next-to-last rafter. There are s_{n-1} ways of getting to this rafter. As the carpenter must start the last board at either the second-to-last or last rafter, we have

$$s_n = \begin{cases} 1 & \text{for } n = 1 \\ 2 & \text{for } n = 2 \\ s_{n-1} + s_{n-2} & \text{for } n \geq 3. \end{cases}$$

This recurrence relation is also similar to the Fibonacci relation discussed earlier. ▬

The foregoing examples have shown several different situations in which recurrence relations involving a single variable arise. In three of the cases, the recurrence relations have been converted into expressions showing the "count" or "value" in terms of n. Similar situations involving recursion and the search for associated formulas exist in settings involving two or more variables, such as the combinatorial recurrence relation $C(n, r) = C(n - 1, r) + C(n - 1, r - 1)$ discussed in Section 7.1. Discovering recurrence relations and using them to find formulas for the terms of the corresponding sequences will be the focus for the rest of this chapter.

EXERCISES 8.1

Find the values of s_2 through s_5 in Exercises 1–6.

1. $s_n = 3s_{n-1} + n^2$ for $n \geq 2$, and $s_1 = 2$

2. $s_n = s_{n-1} + s_{n-2} + s_{n-3}$ for $n > 3$, and $s_1 = s_2 = s_3 = 2$

3. $s_n = 7s_{n-1} + 2$ for $n \geq 2$, and $s_1 = 1$

4. $s_n = (-1)^n + s_{n-1}$ for $n \geq 2$, and $s_1 = 1$

5. $s_n = 2s_{n-1} + 1$ for $n \geq 1$, and $s_0 = 1$

6. $s_n = 2ns_{n-1}$ for $n \geq 3$, and $s_1 = 1$, $s_2 = 2$

7. If $s_n = 3(s_{n-1} + s_{n-2})$ for $n \geq 3$, and $s_1 = 1$, $s_2 = 2$, find s_3 and s_4.

8. If $t_{k+1} = 2t_k$ for $k \geq 1$ and $t_1 = 1$, find t_2, t_3, and t_4. Then find an expression for t_k in terms of k.

9. If $s_1 = 1$, $s_2 = 1$, and $s_n = \frac{1}{2}s_{n-1} + \frac{1}{2}s_{n-2}$ for $n \geq 3$, find s_3 and s_4.

10. For s_n as in Exercise 9, is $s_n = \frac{1}{2}F_n$ for $n \geq 3$?

11. Develop a recurrence relation and initial conditions for the number s_n of subsets of a finite set containing n elements.

12. If a data processing position pays a starting salary of \$16,000 and offers yearly raises of \$500 plus a 4 percent cost of living adjustment on the present year's salary, find a recurrence relation and initial conditions for s_n, the salary in the nth year.

13. Suppose that the number of cells in a sample at the beginning of an experiment is 500 and it is known that the number is increasing at a rate of 150 percent per hour. Determine a recurrence relation and initial conditions for a_n, the number of cells in the sample at the beginning of the nth hour. Then find the number of cells at the end of the first 12 hours of the experiment.

14. An ecological group bought an old printing press for \$18,000. If the resale value of the press decreases by 12 percent of its current value per year, write a recurrence relation and initial conditions for the value v_n of the press at the beginning of the nth year.

15. Find a recurrence relation and initial conditions for the value v_n at the start of the nth year of a savings account having an initial principal of \$2000, a yearly deposit of \$100 at the beginning of each successive year, and a yearly interest rate of 6 percent.

16. Find a recurrence relation and initial conditions for the number d_n of possible n–digit sequences employing only the digits $-1, 0$, and 1 that contain no two consecutive -1's and no two consecutive 1's.

17. Suppose that n straight lines are drawn in a plane so that no two lines are parallel and no more than two lines intersect at any point. Let s_n be the number of different regions determined by the lines. Find a recurrence relation and initial conditions for s_n when $n \geq 1$.

18. Find a recurrence relation and initial conditions for p_n, the number of ways to make a sum of n cents using only pennies and nickels.

19. A warehouse manager has a storage bin which contains two types of cartons. One is 8

feet long and the other is 4 feet long. They have the same height and width. If the manager arranges them along the back wall, he has several options on how to proceed. If 10 feet of space is available, he has two choices, one 8–foot carton or two 4–foot cartons. If there are 13 feet of space available, he has two choices, 8–4 or 4–4–4. Develop a recurrence relation and initial conditions for the number c_n of possible combinations of cartons for an opening of n feet.

20. Suppose that the amount of money collected on bad accounts in a given year is projected to be twice the amount collected in the past year minus the amount collected in the year prior to that. Find a recurrence relation for this situation, letting a_n represent the amount collected in year n. Assume that $a_1 = 1000$ and $a_2 = 1000$.

21. A **derangement** of the numbers $1, 2, 3, \ldots , n$ is a rearrangement such that no number appears in its natural position. Find a recurrence relation and initial conditions for the number of derangements, D_n, for $n \geq 2$.

22. Newton's law of cooling states that the change in temperature of a body over time periods of a fixed length is proportional to the difference between the temperature of the body at the beginning of the time period and the surrounding room temperature. If a glass of ice water is placed on a pool deck where the temperature is 90 degrees and it warms from 40 degrees to 42 degrees in 1 minute, find a recurrence relation and initial conditions for t_n, the temperature of the water at the beginning of the nth minute.

23. Radium decays at the rate of approximately one percent every 25 years. If r_0 denotes the original amount of radium and r_n denotes the amount of radium present after $25n$ years, find a recurrence relation and initial conditions for r_n. Use this relation to determine the amount of radium present after 100 years.

24. Let $f(n, k)$ represent the number of ways of selecting k objects, no two of which are consecutive, from a row of n objects. Evaluate $f(n, 1)$ and $f(n, n)$, and show that $f(n, k) = f(n - 1, k) + f(n - 2, k - 1)$ for $0 < k < n$.

25. Let $g(n, k)$ represent the number of ways of selecting k objects, no two of which are consecutive, from n objects arranged in a circle. Show that if $f(n, k)$ is as defined in Exercise 24, then $g(n, k) = f(n - 1, k) + f(n - 3, k - 1)$.

8.2   First-Order Linear Difference Equations

In Section 8.1 we examined several recurrence relations. In this section, we will determine how to find formulas for sequences satisfying one particular type of recurrence relation.

The simplest type of recurrence relation gives s_n as some function of s_{n-1} for $n \geq 2$. If s_1 is assigned some value, then s_2 can be calculated from s_1, s_3 from s_2, and a unique sequence $s_1, s_2, \ldots$ is generated. Often, however, we wish to find a formula for s_n in terms of s_1 and n.

In this section we will focus on finding such a formula for recurrences of the form

$$s_n = as_{n-1} + b,$$

where a and b are constants. Recurrence relations having this form are known as **first-order linear difference equations.** To derive such a formula, we begin using the recurrence relation to express successive values in terms of s_1:

$$s_1 = s_1$$

$$s_2 = as_1 + b$$

$$s_3 = as_2 + b = a^2 s_1 + ab + b$$

$$s_4 = as_3 + b = a^3 s_1 + a^2 b + ab + b$$

$$\cdot$$
$$\cdot$$
$$\cdot$$

$$s_n = as_{n-1} + b = a^{n-1} s_1 + a^{n-2} b + a^{n-3} b + \ldots + ab + b.$$

This last equation can be written

$$s_n = a^{n-1} s_1 + b(a^{n-2} + a^{n-3} + a^{n-4} + \ldots + 1).$$

As the expression in the parentheses is the sum of a geometric progression, we can use Theorem 2.11 to obtain the desired formula, which we state below.

THEOREM 8.1

A first-order linear difference equation of the form $s_n = as_{n-1} + b$ for $n \geq 2$ with first term s_1 has as its nth term

$$s_n = \begin{cases} a^{n-1} \left(s_1 + \dfrac{b}{a-1} \right) - \dfrac{b}{a-1} & \text{if } a \neq 1 \\[2ex] s_1 + (n-1)b & \text{if } a = 1. \end{cases} \tag{8.4}$$

Note that for $a = 1$ the last line of (8.4) is the formula for the nth term of an arithmetic progression and for $b = 0$ the first line of (8.4) gives the nth term of a geometric progression.

EXAMPLE 8.7

Find a formula for s_n if $s_n = 3s_{n-1} + 1$ for $n \geq 2$ and $s_1 = 1$. Here $a = 3$ and $b = 1$. Using Theorem 8.1, we have

$$s_n = 3^{n-1} \left(1 + \frac{1}{3-1} \right) - \frac{1}{3-1}$$

or

$$s_n = \frac{3}{2}(3^{n-1}) - \frac{1}{2}$$

or

$$s_n = \frac{1}{2}(3^n - 1).$$

Substituting $n = 1, 2, 3, \ldots$, we get the following sequence: 1, 4, 13, 40, ... ■

EXAMPLE 8.8 Find a formula for the Tower of Hanoi recurrence relation $m_n = 2m_{n-1} + 1$, where $m_1 = 1$. (See Example 8.2.)

Since the coefficient of the term m_{n-1} does not equal 1, we apply the first formula in Theorem 8.1. This gives us

$$m_n = 2^{n-1}\left(1 + \frac{1}{2 - 1}\right) - \frac{1}{2 - 1}$$

or

$$m_n = 2^n - 1.$$

This is the same formula we derived in Example 8.2. ■

EXAMPLE 8.9 Suppose a woman invests $1000 at 12% interest compounded annually, and adds $3000 to the account each year on the anniversary of its opening. Find a recurrence relation and initial conditions that describe this situation, and then give a formula for the value of this account after n years.

If we represent the amount after n years by s_n, we get, as in Example 8.4, the recurrence relation

$$s_n = \begin{cases} 1.12s_{n-1} + 3000 & \text{for } n \geq 1 \\ 1000 & \text{for } n = 0. \end{cases}$$

As the recurrence relation is a first-order linear difference equation, we can apply Theorem 8.1. In the present case,

$$a = 1.12, b = 3000, \text{ and } s_1 = 1.12s_0 + 3000 = 4120.$$

Thus,

$$s_n = (1.12)^{n-1}\left(4120 + \frac{3000}{1.12 - 1}\right) - \frac{3000}{1.12 - 1}$$

$$= (1.12)^{n-1}\left(4120 + \frac{3000}{0.12}\right) - \frac{3000}{0.12}$$

$$= (1.12)^{n-1}(4120 + 25{,}000) - 25{,}000$$

$$= 29{,}120(1.12)^{n-1} - 25{,}000. \ ■$$

EXERCISES 8.2

In Exercises 1–10 compute the first 5 terms of the sequences using the information given.

1. $s_n = 4s_{n-1}, s_1 = 1$

2. $s_n = 8s_{n-1}, s_1 = 2$

3. $s_n = -s_{n-1} + 3, s_1 = -1$

4. $s_n = -s_{n-1} - 3, s_1 = -1$

5. $s_n = s_{n-1} + 4, s_1 = 5$

6. $s_n = 4s_{n-1}, s_1 = 3$

7. $s_n = -s_{n-1}, s_1 = -1$

8. $s_n = s_{n-1} + 4, s_1 = 2$

9. $s_n = 2s_{n-1} - 1, s_1 = 2$

10. $s_n = -s_{n-1} - 1, s_1 = 3$

In Exercises 11–15 use Theorem 8.1 to find formulas for the first-order linear difference equations given.

11. $s_n = 4s_{n-1} + 5$ for $n \geq 2$ and $s_1 = 1$

12. $s_n = s_{n-1} + 4$ for $n \geq 2$ and $s_1 = 5$

13. $s_n = -3s_{n-1}$ for $n \geq 2$ and $s_1 = -1$

14. $s_n = 5s_{n-1} + 5$ for $n \geq 2$ and $s_1 = 10$

15. $s_n - 2s_{n-1} + 2 = 0$ for $n \geq 2$ and $s_1 = 3$

In Exercises 16–21 find formulas for each of the recurrences in the specified exercises from Section 8.1.

16. Exercise 8

17. Exercise 3

18. Exercise 12

19. Exercise 13

20. Exercise 14

21. Exercise 15

22. Suppose that the number of insects in a large container is 1000, and this number increases by 250 percent every 24 hours. Write a recurrence equation and initial condition to describe the number i_n of insects in the container after n days. Then find a formula for i_n.

23. Suppose that a home owner has a $48,000 home mortgage with 12 percent annual interest. Interest is charged monthly on the unpaid balance of the mortgage. If the buyer makes $800 monthly payments on the loan, find a recurrence relation, initial condition, and formula for u_n, the unpaid balance at the beginning of the nth month. Also find u_{12}, u_{60}, and u_{90}.

24. Suppose that the number of bad data points in a given data set increases seven-fold with each new calculation. After two calculations, the number of bad data points was 98. Find a recurrence relation describing this situation and use it to find a formula for e_n, the number of bad data points after n calculations.

8.3   Second-Order Homogeneous Linear Difference Equations

Recurrence relations of the form $s_n = as_{n-1} + bs_{n-2}$, where a and b are constants, are known as **second-order homogeneous linear difference equations.** They are called second-order because two successive terms are needed to compute the next. The word ''homogeneous'' indicates that there is no constant term in the difference equation.

The Fibonacci recurrence is perhaps the most famous of all second-order homogeneous linear difference equations. It appears in a variety of applications, often in places where one least expects to find it. We will discuss several of its properties, related sequences, and a formula for the terms of the Fibonacci sequence.

The Fibonacci sequence, first presented in Section 2.5, is defined by

$$F_n = \begin{cases} 1 & \text{for } n = 1 \\ 1 & \text{for } n = 2 \\ F_{n-1} + F_{n-2} & \text{for } n \geq 3. \end{cases}$$

This sequence of numbers is named after Leonardo Fibonacci of Pisa (c. 1170–1250), the most famous European mathematician of the Middle Ages. The sequence first appeared in his text *Liber Abaci,* in the following problem:

> A man has one male-female pair of rabbits in a hutch entirely surrounded by a wall. We wish to know how many pairs can be bred from it in one year, if the nature of these rabbits is such that every month they breed one other male-female pair which begin to breed in the second month after their birth. Assume that no rabbits die during the year.

The diagram in Figure 8.4, with the letter "*M*" for a mature pair and the letter "*I*" for an immature pair, shows the pattern of reproduction described in the problem.

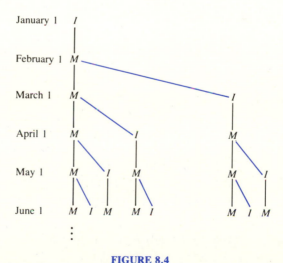

FIGURE 8.4

The number of pairs of rabbits at the beginning of a month is the number of pairs at the beginning of the previous month plus the number of new pairs, which equals the number of pairs at the beginning of the month previous to that. Thus, the number of rabbits satisfies the Fibonacci recurrence. Consequently, after one year, there would be $F_{13} = 233$ pairs of rabbits in the hutch!

If one changes the values of the two initial terms but retains the method of generating the subsequent terms in the sequence, the resulting sequence is known as a **Lucas sequence,** in honor of the 19th century French mathematician E. Lucas. The general Lucas sequence can be defined as

$$L_n = \begin{cases} p & \text{for } n = 1 \\ q & \text{for } n = 2 \\ L_{n-1} + L_{n-2} & \text{for } n \geq 3. \end{cases}$$

For example, if the first value is 3 and the second value is 4, then the resulting sequence is

$$3, 4, 7, 11, 18, 29, 47, \ldots .$$

Such Lucas sequences bear an interesting relationship to the Fibonacci sequence. The only thing different about their formulation is the change of initial conditions.

Adding successive terms in the general Lucas sequence, we see that the sequence has the terms

$$p, q, p + q, p + 2q, 2p + 3q, 3p + 5q, 5p + 8q, \ldots .$$

These values are combinations of p with q with Fibonacci numbers for coefficients. The relationship between the terms of a Lucas sequence and the Fibonacci sequence is as follows.

THEOREM 8.2   If L_n is a Lucas sequence with initial conditions $L_1 = p$ and $L_2 = q$, then

$$L_n = qF_{n-1} + pF_{n-2} \qquad \text{for all } n \geq 3.$$

The Lucas recurrence relation is also an example of a second-order homogeneous linear difference equation. Both the Fibonacci and Lucas recurrences have many interesting properties. Some of these are found in courses on number theory.

Obtaining a formula for second-order homogeneous linear difference equations is more complicated than for first-order equations, and so the following theorem is given without proof.

THEOREM 8.3   Consider the second-order homogeneous linear difference equation

$$s_n = as_{n-1} + bs_{n-2} \text{ for } n \geq 3$$

with initial values s_1 and s_2. Let r_1 and r_2 be roots of $x^2 - ax - b = 0$. Then:

(a) If $r_1 \neq r_2$, there exist constants k_1 and k_2 such that $s_n = k_1 r_1^n + k_2 r_2^n$.

(b) If $r_1 = r_2 = r$, there exist constants k_1 and k_2 such that $s_n = k_1 r^n + k_2 n r^n$.

EXAMPLE 8.10   Find a formula for the recurrence relation $s_n = 3s_{n-1} - 2s_{n-2}$ for $n \geq 3$, where $s_1 = s_2 = 1$.

Using Theorem 8.3 with $a = 3$ and $b = -2$, we set up the quadratic equation $x^2 - 3x + 2 = 0$ and find its roots $r_1 = 1$ and $r_2 = 2$. As these roots are distinct, we employ part (a) of Theorem 8.3 to get a formula for s_n. We now must find the values for the constants k_1 and k_2 by using the initial values of the recurrence relation. This gives us

$$s_1 = 1 = k_1(1)^1 + k_2(2)^1 \quad \text{and}$$

$$s_2 = 1 = k_1(1)^2 + k_2(2)^2.$$

These have the solution $k_1 = 1$ and $k_2 = 0$. Thus, $s_n = 1(1)^n + 0(2)^n = 1$. ◼

EXAMPLE 8.11   Find a formula for the sequence satisfying the initial conditions $s_1 = 1$ and $s_2 = 2$ and the relation $s_n = 2s_{n-1} - s_{n-2}$ for $n \geq 3$.

Here $a = 2$ and $b = -1$, and so we solve $x^2 - 2x + 1 = 0$, or $(x - 1)^2 = 0$. As we have a double root of 1, we must use part (b) of Theorem 8.4. Substituting $n = 1$ and $n = 2$ gives

$$s_1 = 1 = k_1 + k_2 \quad \text{and}$$

$$s_2 = 2 = k_1 + 2k_2.$$

Thus, $k_1 = 0$ and $k_2 = 1$, and hence $s_n = n$. It is easy to check this formula by substituting $s_k = k$ into the original difference equation. ◼

EXAMPLE 8.12   Use Theorem 8.3 to find a formula for the Fibonacci recurrence.

Given that $F_n = F_{n-1} + F_{n-2}$ for $n \geq 3$ and $F_1 = F_2 = 1$, we begin by solving the quadratic equation involving the coefficients of the terms in the Fibonacci recurrence relation. This equation is $x^2 - x - 1 = 0$ and has two distinct roots,

$$r_1 = \frac{1 + \sqrt{5}}{2} \quad \text{and} \quad r_2 = \frac{1 - \sqrt{5}}{2}.$$

Theorem 8.3 states that there exist constants k_1 and k_2 such that we can represent the nth term of the Fibonacci sequence as

$$F_n = k_1 \left(\frac{1 + \sqrt{5}}{2} \right)^n + k_2 \left(\frac{1 - \sqrt{5}}{2} \right)^n.$$

To determine the values of k_1 and k_2, we use the initial values F_1 and F_2 to obtain

$$1 = k_1 \left(\frac{1 + \sqrt{5}}{2} \right)^1 + k_2 \left(\frac{1 - \sqrt{5}}{2} \right)^1 \quad \text{and}$$

$$1 = k_1 \left(\frac{1 + \sqrt{5}}{2} \right)^2 + k_2 \left(\frac{1 - \sqrt{5}}{2} \right)^2.$$

Solving this system of two equations for k_1 and k_2, we get $k_1 = \frac{1}{\sqrt{5}}$ and $k_2 = \frac{-1}{\sqrt{5}}$. Substituting these values into the formula for F_n above gives

$$F_n = \frac{1}{\sqrt{5}} \left(\frac{1 + \sqrt{5}}{2} \right)^n - \frac{1}{\sqrt{5}} \left(\frac{1 - \sqrt{5}}{2} \right)^n. \tag{8.5}$$

The ability to take a recurrence relation, such as the Fibonacci recurrence, and use it to obtain a formula for the sequence is clearly a powerful tool, since it allows us to evaluate a particular term without iterating the recurrence forward from the initial conditions.

EXAMPLE 8.13   A bank pays 11 percent yearly interest, but only on money it has held for two years. That is, no interest is paid on money during the first year it is held. Assume an initial deposit of $300 is made. What is the value of this account after n years?

If s_n is the amount of money in the account after n years, then $s_0 = 300$, $s_1 = 300$, and for $n \geq 2$

$$s_n = s_{n-1} + 0.11 s_{n-2}.$$

The equation $x^2 - x - 0.11 = 0$ has the roots 1.1 and -0.1. Making the appropriate substitutions in part (a) of Theorem 8.3, we get the equations

$$s_1 = 300 = k_1 (1.1)^1 + k_2 (-0.1)^1 \quad \text{and}$$

$$s_2 = 333 = k_1 (1.1)^2 + k_2 (-0.1)^2.$$

These equations imply that $k_1 = 275$ and $k_2 = 25$. Thus, the value of the account at the end of the nth year is

$$s_n = 275 (1.1)^n + 25 (-0.1)^n.$$

Examining the behavior of this formula, we note that the contribution of the term $25(-0.1)^n$ becomes very small as n grows large. Hence, for a long term investment, the value of the account is approximately $275(1.1)^n$. Thus, in the long run, the value of $300 invested under the bank's plan is similar to the value of $275 invested under a plan of 10% interest compounded yearly.   ■

Example 8.3 in Section 8.1 (in which d_n represented the probability that a person with n dollars wins in a coin flipping game) also leads to a second-order homogeneous linear difference equation. The equation

$$d_n = \tfrac{1}{2}d_{n+1} + \tfrac{1}{2}d_{n-1}$$

derived there can be written as

$$d_{n+1} = 2d_n - d_{n-1}.$$

So in the notation of Theorem 8.3 we have $a = 2$ and $b = -1$. Exactly as in Example 8.11 the corresponding equation $x^2 - 2x + 1 = 0$ has the double root $r = 1$, and so the theorem tells us that $d_n = k_1(1)^n + k_2 n(1)^n = k_1 + nk_2$ for some constants k_1 and k_2.

In Example 8.3 we determined that $d_0 = 0$ and $d_{k+m} = 1$, where $k + m$ was the total number of dollars the players started with. Thus,

$$d_0 = 0 = k_1 + 0k_2 \quad \text{and}$$

$$d_{k+m} = 1 = k_1 + (k + m)k_2.$$

From these equations we see that $k_1 = 0$ and $k_2 = \dfrac{1}{k + m}$. We conclude that

$$d_n = 0 + \frac{n}{k + m} = \frac{n}{k + m}, \text{ as before.}$$

EXERCISES 8.3

In Exercises 1–6 use the information given to compute the first 5 terms of the sequences defined by the difference equations.

1. $s_n = s_{n-1} + s_{n-2}$ for $n \geq 3$, $s_1 = 1$, and $s_2 = 2$

2. $s_n = 3s_{n-1} - s_{n-2}$ for $n \geq 3$, $s_1 = 0$, and $s_2 = 2$

3. $s_n = 4s_{n-1} - 4s_{n-2}$ for $n \geq 3$, $s_1 = 4$, and $s_2 = 8$

4. $s_n = 8s_{n-1} - 16s_{n-2}$ for $n \geq 3$, $s_1 = -1$, and $s_2 = 0$

5. $s_n = -2s_{n-1} + s_{n-2}$ for $n \geq 3$, $s_1 = 1$, and $s_2 = 2$

6. $s_n = -s_{n-1} - s_{n-2}$ for $n \geq 3$, $s_1 = 1$, and $s_2 = 2$

In Exercises 7–12 use Theorem 8.3 to find a formula for s_n.

7. $s_n = 4s_{n-1} - 4s_{n-2}$ for $n \geq 3$, $s_1 = 4$, and $s_2 = 8$

8. $s_n = 3s_{n-1} - 2s_{n-2}$ for $n \geq 3$, $s_1 = 0$, and $s_2 = 2$

9. $s_n = 4s_{n-1} + 5s_{n-2}$ for $n \geq 3$, $s_1 = 0$, and $s_2 = 3$

10. $s_n = 2s_{n-1} - s_{n-2}$ for $n \geq 3$, $s_1 = 1$, and $s_2 = 1$

11. $s_n = 6s_{n-1} + 27s_{n-2}$ for $n \geq 3$, $s_1 = 1$, and $s_2 = 1$

12. $s_n + 2s_{n-1} + s_{n-2} = 0$ for $n \geq 3$, $s_1 = 0$, and $s_2 = 1$

13. Calculate the first 20 Fibonacci numbers.

14. Compute the first 10 terms of the Lucas sequence with first two terms 1 and 3. Verify the relationship in Theorem 8.2.

15. Compute the quotient, to three decimal places, of $\dfrac{F_{i+1}}{F_i}$ for $1 \leq i \leq 6$. Compare the resulting quotients to the value $\dfrac{1 + \sqrt{5}}{2}$, the golden ratio.

16. Prove that the sum of the first n Fibonacci numbers is equal to $F_{n+2} - 1$, i.e.,
$$1 + 1 + 2 + 3 + \ldots + F_n = F_{n+2} - 1.$$

17. Consider the 8×8 square puzzle given. Make a copy of it, cut out the pieces indicated by the lines, and then reassemble the pieces to form the rectangle shown. Compare the area of the rectangular region with the area of the square region given at the start. Can you explain the results?

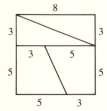
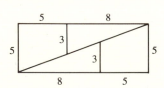

18. Verify that $F_{n-1} \times F_{n+1} - F_n^2 = (-1)^n$ for $n \geq 2$. What does this result have to do with the puzzle presented in Exercise 17? Can you create another similar puzzle?

19. Suppose that Douglas and Lisa play the game in Example 8.3 with an unfair coin so that Douglas wins on $\dfrac{2}{3}$ of the tosses. Let d_n be Douglas's probability of winning when he has n dollars. Write a recurrence relation and initial conditions for d_n.

20. Use Theorem 8.3 to find a formula for d_n in Exercise 19.

21. Verify (8.5) for $n = 1, 2,$ and 3.

22. Prove (8.5) by induction.

23. Prove Theorem 8.2.

24. Solve the recurrence relation $\sqrt{a_n} = \sqrt{a_{n-1}} + 2\sqrt{a_{n-2}}$, where $a_1 = a_2 = 1$.

8.4   Sorting and Searching Algorithms

One application of recursion is in sorting or searching a list of items. Algorithms for doing these tasks efficiently occupy a great deal of the literature in computer science.

The Bubble Sort

The most widely known sorting algorithm is the bubble sort. This sorting procedure is so named because of the similarity between its actions and the movement of a bubble to the surface in a glass of water. In this algorithm, the smaller items "bubble" to the beginning of the list. In order to keep the discussion simple, we will assume that the items $a_1, a_2, \ldots, a_n$ in the list to be sorted are real numbers, although our algorithm can be easily adapted to any objects subject to a suitable order relation (for example, names and alphabetic order).

Starting with a list of n numbers, we first consider the last two items in the list, a_{n-1} and a_n. Comparing them, we exchange their positions in the list if a_n is less than a_{n-1}. We next consider the items in the $n-2$ and $n-1$ positions. We compare them, exchanging them if the $n-1$ item is less than the $n-2$ item. This process of comparing two adjacent items continues until the comparison, and possible exchange, of the items in the first two positions.

At this point, the smallest element in the list has been brought to the first position in the list. We now start over again, this time operating on the smaller list consisting of the elements in the second through nth positions. We begin with the comparison of the last two items, a possible exchange, the comparison of the items in the third from last and next-to-last positions, a possible exchange, and so forth, until the items in the second and third positions are compared and exchanged, if necessary. This will bring the smallest of the items in the second through nth positions into the second position in the list. This process continues until all of the elements in the original list have been arranged in nondecreasing order.

Bubble Sort Algorithm

This algorithm places the numbers in the list $a_1, a_2, \ldots, a_n$ in nondecreasing order.

Step 1   (check for 1–item list). If $n = 1$, then stop.

Step 2   (mark the beginning of the unsorted portion of the list). Set $I = 1$.

Step 3   (initialize comparison loop). Set $J = n - 1$.

Step 4   (compare and exchange). If a_{J+1} is less than a_J, exchange a_{J+1} and a_J.

Step 5   (end comparison loop). If $J > I$, then decrease J by 1 and return to step 4.

Step 6   (reset the beginning of the unsorted list). If $I = n - 1$, then stop; otherwise increase I by 1 and return to step 3.

EXAMPLE 8.14   We will use the bubble sort to order the list 7, 6, 14, 2. The chart below shows the positions of the numbers in the list as step 4 of the algorithm is performed. The circled numbers are those being compared.

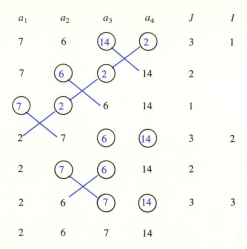

Thus, 6 steps are required to sort the given list into nondecreasing order.   ▬

To measure how efficient the bubble sort is as a sorting algorithm, we count the number of comparisons required to get the n items in order. The first pass through the list requires $n - 1$ comparisons; this moves the smallest element to the front of the list. The second pass, using the items in the second through last positions, requires $n - 2$ comparisons. This pattern continues until the final pass, which compares only the items in the last two positions of the altered list. In all, there are

$$(n - 1) + (n - 2) + (n - 3) + \ldots + 3 + 2 + 1$$

comparisons. Using the formula for summing an arithmetic progression, we see that the bubble sort requires

$$\frac{n^2 - n}{2}$$

comparisons in order to place n items in nondecreasing order. Thus, the bubble sort requires approximately $\frac{n^2}{2}$ comparisons to sort a list with n items.

Binary Search

In order to find a more efficient method of sorting, we must consider a special class of algorithms involving recursion. These algorithms, called **divide-and-conquer** algorithms, are characterized by the process of splitting a problem into several smaller but more readily solved problems. The splitting is carried out as often as necessary until the smaller problems can be easily solved, and their solutions are then reassembled to provide a solution to the original problem.

EXAMPLE 8.15   Consider the problem of determining the number of rounds necessary to determine a winner in a tennis tournament involving 32 players.

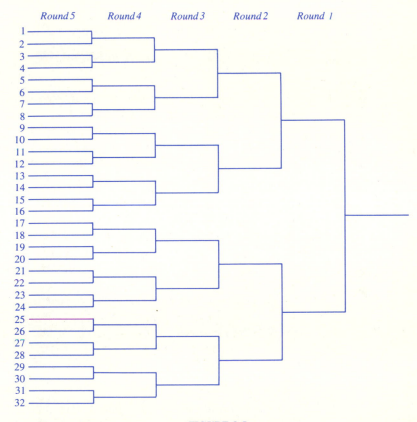

FIGURE 8.5

Figure 8.5 shows the 5 rounds needed. Notice that if there are only 16 players (half as many), then one fewer round is necessary.

If we represent the number of rounds needed to complete a tournament containing n players by s_n, where n is a power of 2, then in general s_n is one more than the number of rounds required for half as many players, that is,

$$s_n = s_{n/2} + 1.$$

Note that when there are only two players in the tournament, we need only one round. Thus, $s_2 = 1$. Hence, we can calculate the first few values for this recurrence relation:

$$s_2 = 1$$
$$s_4 = 2$$
$$s_8 = 3$$
$$s_{16} = 4$$
$$s_{32} = 5.$$

Thus if $n = 2^k$, we have $s_n = k$, or, since $n = 2^k$ is equivalent to $k = \log_2 n$,

$$s_n = \log_2 n. \quad \blacksquare$$

In Example 2.45 we used a divide-and-conquer approach to search for an unknown integer among the numbers 1, 2, . . . , 64. We will now present an algorithm that formalizes this searching process. The algorithm uses the concept of the *floor* of a real number, which is defined to be the greatest integer less than or equal to that number. We will denote the floor of x by $\lfloor x \rfloor$.

EXAMPLE 8.16   Find the floors of the following numbers:

$$312.5, \frac{10}{3}, 7, -3.6, \text{ and } \sqrt{1000}.$$

Since the floor of a number is the greatest integer less than or equal to the number, we have $\lfloor 312.5 \rfloor = 312, \left\lfloor \dfrac{10}{3} \right\rfloor = 3, \lfloor 7 \rfloor = 7, \lfloor -3.6 \rfloor = -4$, and, since $31 < \sqrt{1000} < 32, \lfloor \sqrt{1000} \rfloor = 31. \quad \blacksquare$

The following search algorithm is more general than that used in Example 2.45 because it allows the search of any nondecreasing list of real numbers. For example, it enables us to determine whether a particular credit card number is present in a list of such numbers.

Binary Search Algorithm

This algorithm searches a list of n numbers $s_1 \le s_2 \le \ldots \le s_n$ for a given target number t. In the algorithm L and R mark the beginning and ending of the sublist of $s_1, s_2, \ldots, s_n$ currently being searched.

Step 1 (initialization). Set $L = 1$ and $R = n$.

Step 2 (determine the middle of the sublist). Set $I = \lfloor \frac{1}{2}(L + R) \rfloor$.

Step 3 (t is found). If $t = s_I$, then stop; t is in the original list.

Step 4 (determine the sublist boundaries). If $t < s_I$, then set $R = I - 1$; otherwise set $L = I + 1$.

Step 5 (t is not found). If $L > R$, then stop; t is not in the list. Otherwise go to Step 2.

EXAMPLE 8.17

Consider the application of the binary search algorithm to determine whether the target number 253 is in the list of even integers from 2 to 1000. In the notation of the algorithm, $t = 253$, $n = 500$, and $s_k = 2k$ for $k = 1, 2, \ldots, 500$.

Our first value of I is $\lfloor \frac{1}{2}(1 + 500) \rfloor = \lfloor 250.5 \rfloor = 250$. In steps 3 and 4 we find that $t \neq s_{250} = 500$, and in fact $t < 500$. Now we change R to 249; L remains 1.

The working of the algorithm can be exhibited in a table as a sequence of questions and answers as shown.

L	R	I	s_I	Is $t = s_I$?
1	500	$\lfloor \frac{1}{2}(1 + 500) \rfloor = 250$	500	no; less
1	249	$\lfloor \frac{1}{2}(1 + 249) \rfloor = 125$	250	no; greater
126	249	$\lfloor \frac{1}{2}(126 + 249) \rfloor = 187$	374	no; less
126	186	$\lfloor \frac{1}{2}(126 + 186) \rfloor = 156$	312	no; less
126	155	$\lfloor \frac{1}{2}(126 + 155) \rfloor = 140$	280	no; less
126	139	$\lfloor \frac{1}{2}(126 + 139) \rfloor = 132$	264	no; less
126	131	$\lfloor \frac{1}{2}(126 + 131) \rfloor = 128$	256	no; less
126	127	$\lfloor \frac{1}{2}(126 + 127) \rfloor = 126$	252	no; greater
127	127	$\lfloor \frac{1}{2}(127 + 127) \rfloor = 127$	254	no; less
127	126			

Since in the last line of the table $L > R$, we find that the target number is not in the list. ■

As we saw in Theorem 2.14, at most $k + 1$ comparisons are required by the binary search algorithm to determine if a particular number is present in a list of 2^k numbers that are in nondecreasing order. Thus, for an input list containing n items the order of the binary search algorithm is at most $1 + \log_2 n$.

The Merge Sort

We saw above that the bubble sort algorithm requires approximately $\dfrac{n^2}{2}$ comparisons to sort a list of n numbers. We will now discuss a more efficient algorithm called the **merge sort** that makes use of the divide-and-conquer approach. This sorting process involves two major parts. The first portion of the algorithm consists of breaking the original list of n items into smaller lists using a divide-and-conquer approach. These smaller lists are then in turn broken into yet smaller lists, and so forth, until we reach the stage of one-item lists. These lists, by definition, are in nondecreasing order. They are then merged together to form the desired sorted list of the n items. This merging of two smaller ordered lists into one larger ordered list is the second part of the process. This merging procedure was described in Example 2.46. We now present a formal description of this procedure.

Merging Algorithm

This algorithm merges two nondecreasing lists

$$A: a_1, a_2, a_3, \ldots, a_m$$

$$B: b_1, b_2, b_3, \ldots, b_n$$

into a nondecreasing list

$$C: c_1, c_2, c_3, \ldots, c_{m+n}.$$

Step 1 (initialize variables). Set $I = 1, J = 1$, and $K = 1$.

Step 2 (check to see if either A or B is copied). If $I > m$ or $J > n$, then go to step 5.

Step 3 (find next member of C). If $a_I < b_J$, then set $c_K = a_I$ and add 1 to I; otherwise set $c_K = b_J$ and add 1 to J.

Step 4 (increment K). Add 1 to K and go to step 2.

Step 5 (is all of A copied into C?). If $I > m$, then go to step 7.

Step 6 (add end of A to C). Set $c_K = a_I$, add 1 to both I and K, and go to step 5.

Step 7 (is all of B copied into C?). If $J > n$, stop.

Step 8 (add end of B to C). Set $c_K = b_J$, add 1 to both J and K, and go to step 7.

This algorithm takes at most $m + n - 1$ comparisons for two lists of lengths m and n, as was shown in Theorem 2.15.

We now build a procedure that uses the merging algorithm in conjunction with a divide-and-conquer procedure to sort a list of n elements. This procedure divides the n elements into two sublists. It then checks to see if the two sublists contain single elements. If not, the divide-and-conquer procedure is used again and again until the resulting sublists are all one-item lists. The sublists are then merged to get the desired nondecreasing list. The algorithm is given for the special case where $n = 2^k$ for some positive integer k.

Merge Sort Algorithm

This algorithm places the numbers in the list $s_1, s_2, \ldots, s_n$ in nondecreasing order. It requires an input list of n elements, where $n = 2^k$.

Step 1 Given a list $s_1, s_2, \ldots, s_m$, divide it into two lists $s_1, s_2, \ldots, s_{m/2}$ and $s_{m/2+1}, s_{m/2+2}, \ldots, s_m$.

Step 2 Given a sequence of lists $L_1, L_2, \ldots, L_r$, if each list has 1 element, then go to step 3. Otherwise, perform step 1 on each of the lists.

Step 3 Given lists $L_1, L_2, \ldots, L_r$, use the merging algorithm to merge L_1 and L_2, L_3 and $L_4, \ldots, L_{r-1}$ and L_r.

Step 4 If there is a single list, stop. Otherwise, go to step 3.

EXAMPLE 8.18

We will use the merge sort algorithm to sort the items

$$14, 19, 11, 18, 7, 30, 6, 2$$

into a nondecreasing list. Steps 1 and 2 divide the list into eight one-item sublists

$$(14), (19), (11), (18), (7), (30), (6), (2).$$

Step 3 of the algorithm then combines pairs of lists using the merging algorithm. This results in the two-item sublists

$$(14, 19), (11, 18), (7, 30), (2, 6).$$

The merging algorithm is again used to merge these two-item lists, giving

$$(11, 14, 18, 19) \text{ and } (2, 6, 7, 30).$$

One more use of the merging algorithm gives the original list in nondecreasing order:

$$2, 6, 7, 11, 14, 18, 19, 30. \quad \blacksquare$$

EXAMPLE 8.19   This process can be viewed through the use of a tree diagram to illustrate the dividing and an inverted tree diagram to show the resulting list merging. The diagrams in Figure 8.6 show the application of the merge sort algorithm to the set of items $23, 6, -7, 5, 13, 98, 3, 35$.

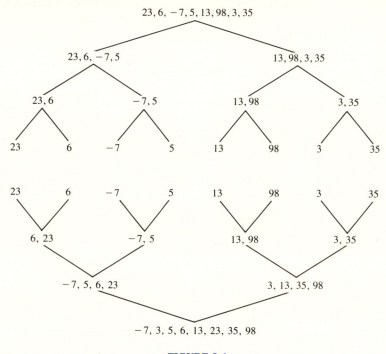

FIGURE 8.6

The final row in Figure 8.6 provides the output of the merge sort algorithm.

The analysis of this algorithm requires techniques beyond those available in this text. But it can be shown that for a list of n items, its order is at most $n \log_2 n$, the smallest order possible for a sorting algorithm.

EXERCISES 8.4

In Exercises 1–8 find the floor of the given number.

1. 243

2. -34.5

3. $\dfrac{34}{7}$

4. 28.963

5. 0.871

6. 2483

7. $\dfrac{-(-34 + 2)}{2}$

8. $\dfrac{-343}{26}$

In Exercises 9–16 illustrate as in Example 8.14 the use of the bubble sort algorithm to sort each given list of numbers.

9. 13, 56, 87, 42 **10.** 42, 87, 56, 13 **11.** 13, 42, 56, 87 **12.** 87, 56, 42, 13

13. 3, 4, 3, 5, 6 **14.** 6, 4, 3, 5, 3 **15.** 4, 3, 2, 5, 2 **16.** 6, 5, 4, 3, 2

In Exercises 17–20 make a table as in Example 8.17 showing the working of the binary search algorithm.

17. $t = 83, n = 100, s_i = i$ for $i = 1, 2, \ldots, 100$

18. $t = 17, n = 125, s_i = i$ for $i = 1, 2, \ldots, 125$

19. $t = 400, n = 300, s_i = 3i$ for $i = 1, 2, \ldots, 300$

20. $t = 305, n = 100, s_i = 2i + 100$ for $i = 1, 2, \ldots, 100$

In Exercises 21–26 draw two diagrams as in Example 8.19 illustrating the application of the merge sort algorithm to the given list of numbers.

21. 13, 56, 87, 42

22. 42, 87, 56, 13

23. 13, 42, 56, 87

24. 95, 87, 56, 54, 42, 23, 16, 15

25. 34, 81, 46, 2, 53, 5, 4, 8, 26, 1, 0, 45, 26, 5, 48, 35

26. 34, 67, 23, 54, 92, 18, 34, 54

In Exercises 27–29 prove the given statement true for all real numbers x and y.

27. $\lfloor x \rfloor \leq x < \lfloor x \rfloor + 1$.

28. $\lfloor x + y \rfloor \geq \lfloor x \rfloor + \lfloor y \rfloor$. Show by example that equality need not hold.

29. $\lfloor x \rfloor + \lfloor y \rfloor \geq \lfloor x + y \rfloor - 1$. Show by example that equality need not hold.

30. Show that if x is any real number and if n is an integer, then $\lfloor x + n \rfloor = \lfloor x \rfloor + n$.

31. If in using the bubble sort algorithm on a list of n items no exchange takes place in the comparison at the stage $J = I$ for a fixed I, what can we conclude about the number of items that move to their final positions in the list during the comparisons for that value of I? How many comparisons (values of J) could be deleted for the next value of I in the algorithm?

32. Suppose that one sorting algorithm requires $\dfrac{n^2}{4}$ comparisons to complete and a second sorting algorithm requires $n \log_2 n$ comparisons to complete. How large must n be for the second algorithm to be more efficient?

33. Suppose that one sorting algorithm requires n^2 comparisons to complete and a second algorithm requires $n \log_2 n$ comparisons to complete. For which n is each algorithm more efficient?

34. Explain how the merge sort algorithm treats equal items appearing in the two lists being merged.

35. Another way to sort a list of 2^k items is to merge the first and second, the third and fourth, etc., then repeat this process with the 2–element lists generated, etc. Find a recurrence relation for the number, c_n, of comparisons required to complete this process for a list of $n = 2^k$ items.

36. Solve the recurrence relation obtained in Exercise 35.

Suggested Readings

1. Goldberg, Samuel. *Introduction to Difference Equations*. New York: John Wiley, 1958.
2. Horowitz, Ellis and Sartaj Sahni. *Fundamentals of Computer Algorithms*. Rockville, MD: Computer Science Press, 1984.
3. Levy, H. and F. Lessman. *Finite Difference Equations*. New York: Macmillan, 1961.
4. Ryser, Herbert John. *Combinatorial Mathematics*. Washington, DC: Mathematical Association of America, 1963.
5. Stanat, Donald F. and David F. McAllister. *Discrete Mathematics in Computer Science*. Englewood Cliffs, NJ: Prentice-Hall, 1977.

9

COMBINATORIAL CIRCUITS AND FINITE STATE MACHINES

Today tiny electronic devices called **microprocessors** are found in such diverse places as automobiles, digital watches, missiles, electronic games, compact disc players, and toasters. These devices control the larger machines in which they are embedded by responding to a variety of inputs according to a preset pattern. How they do this is determined by the circuits they contain. This chapter will provide an introduction to the logic of such circuits.

9.1   Logical Gates

The sensitive electronic equipment in the control room of a recording studio needs to be protected from both high temperatures and excess humidity. An air conditioner is provided which must go on whenever either the temperature exceeds 80° or the humidity exceeds 50%. What is required is a control device that has two inputs, coming from a thermostat and a humidistat, and one output, going to the air conditioner. It must perform the function of turning on the air conditioner if it gets a yes signal from either of the input devices, as summarized in the following table.

Temperature > 80°?	Humidity > 50%?	Air conditioner on?
no	no	no
no	yes	yes
yes	no	yes
yes	yes	yes

We will follow the usual custom of using x and y to label our two inputs and 1 and 0 to stand for the input or output signals yes and no, respectively, Thus, x and y can assume only the values 0 and 1; such variables are called **Boolean variables.** These conventions give our table a somewhat simpler form.

x	y	Output
0	0	0
0	1	1
1	0	1
1	1	1

This device is an example of a **logical gate,** and the particular one whose working we have just described is called an **OR-gate,** since its output is 1 whenever either x *or* y is 1. We will denote the output of an OR-gate with inputs x and y by $x \vee y$, so that

$$x \vee y = \begin{cases} 1 & \text{if } x = 1 \text{ or } y = 1 \\ 0 & \text{otherwise.} \end{cases}$$

We will not delve into the actual internal workings of the devices we call logical gates, but merely describe how they function. A logical gate is an electronic device that has either 1 or 2 inputs and a single output. These inputs and output are in one of two states, which we denote by 0 and 1. For example, the two states might be a low and high voltage.

Logical gates are represented graphically by standard symbols established by the Institute of Electrical and Electronics Engineers. The symbol for an OR-gate is shown in Figure 9.1.

FIGURE 9.1

We will study only two other logical gates, the **AND-gate** and the **NOT-gate.** Their symbols are shown in Figure 9.2. Notice that the symbols for the OR-gate and AND-gate are quite similar, so care must be taken to distinguish between them.

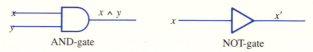

FIGURE 9.2

The output of an AND-gate with inputs x and y is 1 only when both x *and* y are 1. This output is denoted by $x \wedge y$, so that the values of $x \wedge y$ are given by the following table, which, as in logic, is called a **truth table.**

x	y	$x \wedge y$
0	0	0
0	1	0
1	0	0
1	1	1

EXAMPLE 9.1   A dot-matrix printer attached to a personal computer will print only when the "on-line" button on its case has been pressed and a paper sensor tells it that there is paper in the printer. We can represent this as an AND-gate as in Figure 9.3. ■

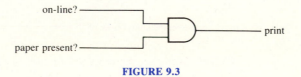

FIGURE 9.3

The other logical gate we will consider is the NOT-gate, which has only a single input. Its output is always exactly the opposite from its input. If the input is x, then the output of a NOT-gate, which we will denote by x', is as follows.

x	x'
0	1
1	0

EXAMPLE 9.2   A rental truck is equipped with a governor. If the speedometer exceeds 70 miles per hour the ignition of the truck is cut off. We can describe this with a NOT-gate as in Figure 9.4. ▬

speed greater than 70? — ignition

FIGURE 9.4

The reader familiar with logic will notice the similarity between the three gates we have described and the logical operators "or," "and," and "not." Although other logical gates may be defined, by appropriately combining the three gates we have introduced we can simulate any logical gate that has no more than two inputs.

EXAMPLE 9.3   A home gas furnace is attached to two thermostats, one in the living area of the house, and the other in the chamber where the furnace heats air to be circulated. If the first thermostat senses that the temperature in the house is below 68°, it sends a signal to the furnace to turn on. On the other hand, if the thermostat in the heating chamber becomes hotter than 150°, it sends a message to the furnace to turn off. This signal is for reasons of safety and should be obeyed no matter what message the house thermostat is sending.

One arrangement of gates giving the desired output is shown in Figure 9.5. It

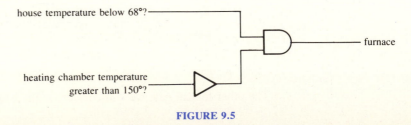

house temperature below 68°? —

heating chamber temperature greater than 150°? —

— furnace

FIGURE 9.5

is easier to check the effect of this arrangement if we denote the signals from the two thermostats by x and y as in Figure 9.6. We can compute the value of $x \wedge y'$

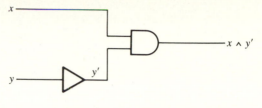

FIGURE 9.6

for the possible values of x and y by means of a truth table.

x	y	y'	$x \wedge y'$
0 (room ok)	0 (chamber ok)	1	0 (furnace off)
0 (room ok)	1 (chamber hot)	0	0 (furnace off)
1 (room cold)	0 (chamber ok)	1	1 (furnace on)
1 (room cold)	1 (chamber hot)	0	0 (furnace off)

Notice that the furnace will run only when the house is cold and the heating chamber is not too hot. ▬

Figure 9.6 shows an example of combining logical gates to produce what is called a **combinatorial circuit,** which we will usually refer to simply as a "circuit." More than two independent inputs are allowed, and an input may feed into more than one gate. A more complicated example is shown in Figure 9.7, in which the inputs are denoted by x, y, and z.

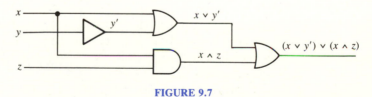

FIGURE 9.7

We will only consider circuits which have a single output, and we will not allow circuits such as shown in Figure 9.8, in which the output of the NOT-gate doubles back to be an input for the previous AND-gate. (We leave for the exercises the precise formulation of this condition.)

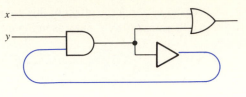

FIGURE 9.8

In Figure 9.7 the input x splits at the heavy dot. In order to simplify our diagrams we may instead label more than one original input with the same variable. Thus, Figure 9.9 is simply another way to draw Figure 9.7.

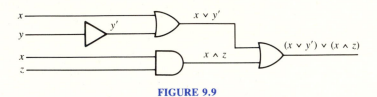

FIGURE 9.9

The effect of complicated circuits can be computed by successively evaluating the output of each gate for all possible values of the input variables, as in the following truth table.

x	y	z	y'	$x \lor y'$	$x \land z$	$(x \lor y') \lor (x \land z)$
0	0	0	1	1	0	1
0	0	1	1	1	0	1
0	1	0	0	0	0	0
0	1	1	0	0	0	0
1	0	0	1	1	0	1
1	0	1	1	1	1	1
1	1	0	0	1	0	1
1	1	1	0	1	1	1

TABLE 9.1

The strings of symbols heading the columns of our table are examples of Boolean expressions. In general, given a finite set of Boolean variables x_1, x_2, ..., x_n, by a **Boolean expression** we mean any of these Boolean variables, either of the constants 0 and 1 (which represent variables with the constant value 0 or 1, respectively) and any subsequently formed expressions

$$B \lor C, \ B \land C, \text{ or } B',$$

where B and C are Boolean expressions.

EXAMPLE 9.4   Which of the following are Boolean expressions for the set of Boolean variables x, y, z?

$$x \vee (y \wedge (x \wedge z')') \qquad 1 \wedge y \qquad z$$
$$(x \wedge 'z) \vee y \qquad \vee y' \wedge 0$$

The first three are Boolean expressions, but not the last two, since neither $\wedge\,'$ nor $\vee\, y$ make sense. ▪

Just as a combinatorial circuit leads to a Boolean expression, likewise each Boolean expression corresponds to a circuit, which can be found by taking the expression apart from the outside. Consider the first expression of Example 9.4, $x \vee (y \wedge (x \wedge z')')$. This corresponds to a circuit with an OR-gate having inputs

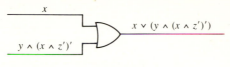

FIGURE 9.10

x and $y \wedge (x \wedge z')'$, as in Figure 9.10. By continuing to work backwards in this way we find the circuit shown in Figure 9.11.

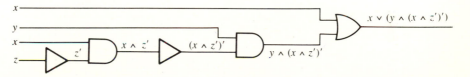

FIGURE 9.11

It may be that different circuits produce the same output for each combination of values of the input variables. For example, if we examine Table 9.1 in which we analyzed the effect of the circuit in Figure 9.9, we may notice that the output is 1 exactly when x is 1 or y is 0. Thus, the circuit has exactly the same effect as

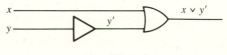

FIGURE 9.12

that corresponding to $x \vee y'$, shown in Figure 9.12. Since this circuit is much simpler, manufacturing it rather than the circuit of Figure 9.9 will be cheaper. A simpler circuit will also usually run faster. Some integrated circuits contain more than 100,000 logical gates in an area of one square centimeter, and so their efficient use is very important.

Circuits that give the same output for all possible values of their input variables are said to be **equivalent,** as are their corresponding Boolean expressions. Thus, $(x \vee y') \vee (x \wedge z)$ is equivalent to $x \vee y'$, as can be confirmed by comparing the following table to the truth table for $(x \vee y') \vee (x \wedge z)$ (Table 9.1).

x	y	z	y'	$x \vee y'$
0	0	0	1	1
0	0	1	1	1
0	1	0	0	0
0	1	1	0	0
1	0	0	1	1
1	0	1	1	1
1	1	0	0	1
1	1	1	0	1

Since the circuits corresponding to equivalent Boolean expressions have exactly the same effect, we will write an equal sign between such expressions. For example, we will write

$$(x \vee y') \vee (x \wedge z) = x \vee y'$$

since the truth tables of the two expressions are the same. In subsequent sections we will study how we can reduce Boolean expressions to simpler equivalent expressions to improve circuit design.

EXERCISES 9.1

In Exercises 1–8 write the Boolean expression associated with each circuit.

1.

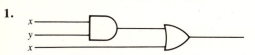

2.

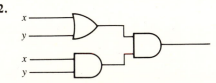

3.

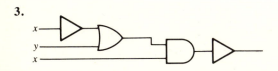

4.

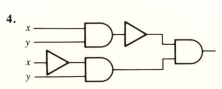

5.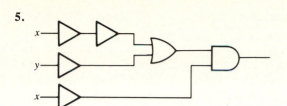

6.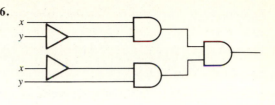

7.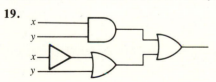

8.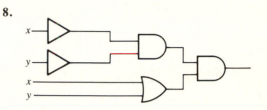

In Exercises 9–14 draw a circuit representing the given Boolean expression.

9. $(x \wedge y) \vee (x' \vee y)$

10. $(x' \wedge y) \vee [x \wedge (y \wedge z)]$

11. $[(x \wedge y') \vee (x' \wedge y')] \vee [x' \wedge (y \vee z)]$

12. $(w \wedge x) \vee [(x \vee y') \wedge (w' \vee x')]$

13. $(y' \wedge z') \vee [(w \wedge x') \wedge y']'$

14. $[x \wedge (y \wedge z)] \wedge [(x' \wedge y') \vee (z \wedge w')]$

In Exercises 15–18 give the output value for the Boolean expression with the given input values.

15. $(x \vee y) \wedge (x' \vee z)$ for $x = 1, y = 1, z = 0$

16. $[(x \wedge y) \vee z] \wedge [x \vee (y' \wedge z)]$ for $x = 0, y = 1, z = 1$

17. $(x \wedge (y \wedge z))'$ for $x = 0, y = 1, z = 0$

18. $[(x \wedge (y \wedge z')) \vee ((x \wedge y) \wedge z)] \wedge (x \vee z')$ for $x = 0, y = 1, z = 0$

In Exercises 19–22 construct a truth table for the circuit shown.

19.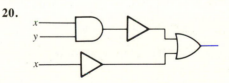

20.

21.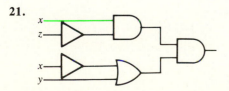

22.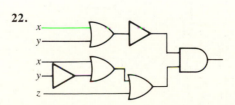

In Exercises 23–28 construct a truth table for the given Boolean expression.

23. $x \wedge (y \vee x')$

24. $(x \vee y')' \vee x$

25. $(x \wedge y) \vee (x' \wedge y')'$

26. $x \vee (x' \wedge y)$

27. $(x \vee y') \vee (x \wedge z')$

28. $[(x \wedge y) \wedge z] \vee [x \wedge (y \wedge z')]$

In Exercises 29–36 use truth tables to determine which pairs of circuits are equivalent.

29.

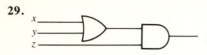   and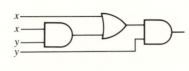

30.

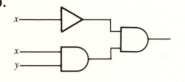   and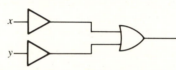

31.

   and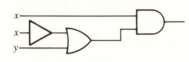

32.

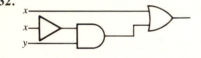   and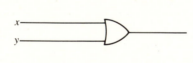

33.

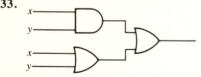   and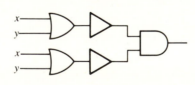

34.

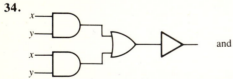   and

35.

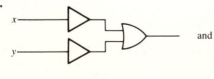   and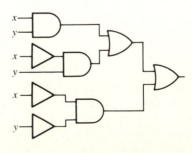

36.

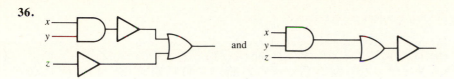

and

In Exercises 37–42 use truth tables to determine whether or not the Boolean expressions given are equivalent.

37. $x \vee (x \wedge y)$ and x

38. $x \wedge (x' \wedge y)$ and $x \wedge y$

39. $[(x \vee y) \wedge (x' \vee y)] \wedge (y \vee z)$ and $(x \vee y) \wedge (x' \vee z)$

40. $(x \wedge (y \wedge z)) \vee [x' \vee ((x \wedge y) \wedge z')]$ and $x' \vee y$

41. $y' \wedge (y \vee z')$ and $y' \wedge x'$

42. $x \wedge [w \wedge (y \vee z)]$ and $(x \wedge w) \wedge (y \vee z)$

43. A home security alarm is designed to alert the police department if a window signal is heard or if a door is opened without someone first throwing a safety switch. Draw a circuit for this situation, describing the meaning of your input variables.

44. The seat belt buzzer for the driver's side of an automobile will sound if the belt is not buckled, the weight sensor indicates someone is in the seat, and the key is in the ignition. Draw a circuit for this situation, describing the meaning of your input variables.

45. Prove that equivalence of Boolean expressions using a fixed finite set of Boolean variables is an equivalence relation as defined in Chapter 2.

46. Define the directed graph associated with a combinatorial circuit. State a condition on this directed graph that excludes circuits similar to that shown in Figure 9.8.

47. What is the output of the illegal circuit shown, for $x = 0$ and 1?

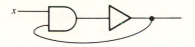

9.2   Creating Combinatorial Circuits

In Section 9.1 we saw how each combinatorial circuit corresponds to a Boolean expression, and found that sometimes we could simplify a circuit by finding a simpler equivalent Boolean expression. One way to simplify Boolean expressions is by using standard identities, much in the way that the algebraic expression $(a + b)^2 - b(b - 3a)$ can be reduced to $a(a + 5b)$ by using the rules of algebra. Some of these identities for Boolean expressions are listed on the following page.

THEOREM 9.1   For any Boolean expressions X, Y, and Z,

(a) $X \wedge Y = Y \wedge X$ and $X \vee Y = Y \vee X$
(b) $(X \wedge Y) \wedge Z = X \wedge (Y \wedge Z)$ and $(X \vee Y) \vee Z = X \vee (Y \vee Z)$
(c) $X \wedge (Y \vee Z) = (X \wedge Y) \vee (X \wedge Z)$ and $X \vee (Y \wedge Z) = (X \vee Y) \wedge (X \vee Z)$
(d) $X \vee (X \wedge Y) = X \wedge (X \vee Y) = X$
(e) $X \vee X = X \wedge X = X$
(f) $X \vee X' = 1$ and $X \wedge X' = 0$
(g) $X \vee 0 = X \wedge 1 = X$
(h) $X \wedge 0 = 0$ and $X \vee 1 = 1$
(i) $(X')' = X$, $0' = 1$, and $1' = 0$
(j) $(X \vee Y)' = X' \wedge Y'$ and $(X \wedge Y)' = X' \vee Y'$

Many of these identities have the same form as familiar algebraic rules. For example, rule (a) says that the operations $\vee$ and $\wedge$ are commutative, and rule (b) is an associative law for these operations. In spite of rule (b), $x \vee (y \wedge z)$ is not equivalent to $(x \vee y) \wedge z$.

Rule (c) gives two distributive laws. For example, if in the first equation of rule (c) we substitute multiplication for $\wedge$ and addition for $\vee$, we get

$$X(Y + Z) = (XY) + (XZ),$$

which is the distributive law of ordinary algebra. Making the same substitution in the second equation, however, produces

$$X + (YZ) = (X + Y)(X + Z),$$

which is not true in ordinary algebra. Thus, these rules must be used with care; one should not jump to conclusions about how Boolean expressions may be manipulated based on rules for other algebraic systems.

The equations of rule (j) are known as the **De Morgan laws;** compare them to the rules for the complements of set unions and intersections in Theorem 2.2. The validity of all these identities can be proved by computing the truth tables for the expressions that are claimed to be equivalent.

EXAMPLE 9.5   Prove rule (d).

We compute truth tables for the expressions $X \vee (X \wedge Y)$ and $X \wedge (X \vee Y)$ as follows.

X	Y	$X \wedge Y$	$X \vee (X \wedge Y)$	$X \vee Y$	$X \wedge (X \vee Y)$
0	0	0	0	0	0
0	1	0	0	1	0
1	0	0	1	1	1
1	1	1	1	1	1

Since the first, fourth, and sixth columns of this table are identical, rule (d) is proved. ■

As an example of the use of our rules we will prove that the expressions $(x \lor y') \lor (x \land z)$ and $x \lor y'$ are equivalent without computing, as we did in Section 9.1, the truth table of each expression. We will start with the more complicated expression and use our rules to simplify it.

$$
\begin{aligned}
(x \lor y') \lor (x \land z) &= (y' \lor x) \lor (x \land z) \quad \text{(rule (a))} \\
&= y' \lor (x \lor (x \land z)) \quad \text{(rule (b))} \\
&= y' \lor x \quad \text{(rule (d))} \\
&= x \lor y' \quad \text{(rule (a))}
\end{aligned}
$$

EXAMPLE 9.6 Simplify the expression $x \lor (y \land (x \land z')')$, which corresponds to the circuit shown in Figure 9.11.

$$
\begin{aligned}
x \lor (y \land (x \land z')') &= x \lor (y \land (x' \lor z'')) \quad \text{(rule (j))} \\
&= x \lor (y \land (x' \lor z)) \quad \text{(rule (i))} \\
&= (x \lor y) \land (x \lor (x' \lor z)) \quad \text{(rule (c))} \\
&= (x \lor y) \land ((x \lor x') \lor z) \quad \text{(rule (b))} \\
&= (x \lor y) \land (1 \lor z) \quad \text{(rule (f))} \\
&= (x \lor y) \land 1 \quad \text{(rule (h))} \\
&= x \lor y \quad \text{(rule (g))}
\end{aligned}
$$

We see that the complex circuit of Figure 9.11 can be replaced by a circuit having only one gate. ■

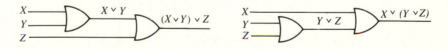

FIGURE 9.13

Because of rule (b) we can use expressions such as $X \lor Y \lor Z$ without ambiguity, since the result is the same no matter whether we evaluate $X \lor Y$ or $Y \lor Z$ first. In terms of circuits, this means that the two circuits in Figure 9.13 are equivalent. Thus we will use the diagram of Figure 9.14 to represent either of the circuits in Figure 9.13; its output is 1 when any of X, Y, or Z is 1. We use the same convention for more than 3 inputs. For example, the circuit shown in

FIGURE 9.14

Figure 9.15 represents any of the equivalent circuits corresponding to a Boolean expression formed by putting parentheses in $W \wedge X \wedge Y \wedge Z$; one such expression is $(W \wedge X) \wedge (Y \wedge Z)$, another is $((W \wedge X) \wedge Y) \wedge Z$.

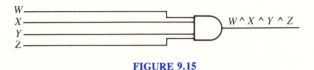

FIGURE 9.15

Of course before we can simplify a circuit we must *have* a circuit. Thus, we must consider the problem of finding a circuit that will accomplish the particular job we have in mind. Whether the circuit we find is simple or complicated is of secondary importance. There is always the possibility of simplifying a complicated circuit by reducing its corresponding Boolean expression.

As an example we will consider the three-person finance committee of a state senate. The committee must vote on all revenue bills, and of course 2 or 3 yes votes are necessary for a bill to clear the committee. We will design a circuit that will take the three senators' votes as inputs and yield whether the bill passes or not as output. (Ours will be a scaled-down version of the electronic voting devices used in some legislatures.) If we denote yes votes and the passage of a bill by 1, we desire a circuit with the following truth table.

	x	y	z	Pass?
	0	0	0	0
	0	0	1	0
	0	1	0	0
*	0	1	1	1
	1	0	0	0
*	1	0	1	1
*	1	1	0	1
*	1	1	1	1

We have marked the rows which have 1's in the output column because these rows will be used to construct a Boolean expression with this truth table. Consider, for example, the fourth row of the table. Since there is a 1 in the output column in this row, when x is 0 and y and z are 1, our Boolean expression should have a value 1. But x is 0 if and only if x' is 1, so this row corresponds to the condition that x', y, and z all have value 1. This happens exactly when $x' \wedge y \wedge z$ has value 1. The other marked rows indicate that the output is 1 also when $x \wedge y' \wedge z$,

$x \wedge y \wedge z'$, or $x \wedge y \wedge z$ have value 1. Thus, we want an output of 1 exactly when $(x' \wedge y \wedge z) \vee (x \wedge y' \wedge z) \vee (x \wedge y \wedge z') \vee (x \wedge y \wedge z)$ has value 1, and this is the Boolean expression we seek. The circuit corresponding to this expression is shown below.

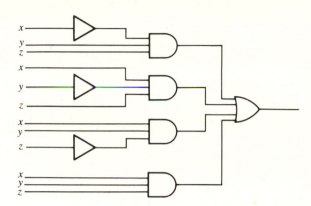

Notice that we have designed a crude arithmetic computer, since our circuit counts the number of yes votes and tells us whether there are 2 or more.

Now we summarize our method of finding a Boolean expression corresponding to a given truth table. Let us suppose the input variables are $x_1, x_2, \ldots, x_n$. If all outputs are 0, then the desired Boolean expression is 0. Otherwise we proceed as follows.

Step 1 Identify the rows of the truth table having output 1. For each row form the Boolean expression

$$y_1 \wedge y_2 \wedge \ldots \wedge y_n,$$

where y_i is taken to be x_i if there is a 1 in the x_i column, and y_i is taken to be x_i' if there is a 0 in the x_i column. The expressions thus formed are called **minterms.**

Step 2 If $B_1, B_2, \ldots, B_k$ are the minterms formed in step 1, form the expression

$$B_1 \vee B_2 \vee \ldots \vee B_k.$$

This Boolean expression has a truth table identical to the one with which we started.

EXAMPLE 9.7   A garage light is to be controlled by three switches, one inside the kitchen to which the garage is attached, one at the garage door, and one at a back door to the garage. It should be possible to turn the light on or off with any of these switches, no matter what the positions of the other switches are. Design a circuit to make this possible.

The inputs are the 3 switches, which we will label 1 or 0 according to whether they are in an up or down position. We will design a circuit that turns the light on whenever the number of inputs equal to 1 is odd, since flipping any switch will change whether this number is odd or even. We want a circuit with the following truth table.

	x	y	z	Number of 1's	Output
	0	0	0	0	0
*	0	0	1	1	1
*	0	1	0	1	1
	0	1	1	2	0
*	1	0	0	1	1
	1	0	1	2	0
	1	1	0	2	0
*	1	1	1	3	1

The rows having output 1 are marked, and the required Boolean expression is $(x' \wedge y' \wedge z) \vee (x' \wedge y \wedge z') \vee (x \wedge y' \wedge z') \vee (x \wedge y \wedge z)$. The corresponding circuit is shown in Figure 9.16. ▪

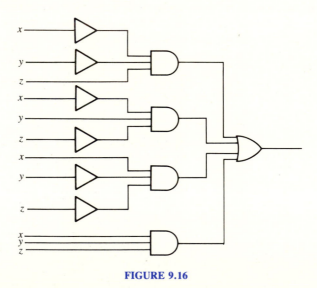

FIGURE 9.16

The Boolean expressions our method produces tend to be complicated, and so correspond to complicated circuits. The circuit shown in Figure 9.16 is actually more complex than it appears, since if it were expressed using only our original three logical gates, each of the gates on the left of the diagram with three inputs would have to be replaced by two standard 2–input AND-gates, and the gate on the right with four inputs would have to be replaced by three standard 2–input OR-gates. Thus, the circuit of Figure 9.16 requires 6 NOT-gates, 8 AND-gates, and 3 OR-gates, for a total of 17 elementary gates. Although we might simplify the corresponding Boolean expression using the rules given at the beginning of this section, it is not clear how to do this. In the next section we will consider a method for simplifying Boolean expressions in an organized way.

EXERCISES 9.2

In Exercises 1–8 prove the equivalence using truth tables.

1. $x \wedge y = y \wedge x$

2. $x \wedge (y \vee z) = (x \wedge y) \vee (x \wedge z)$

3. $x \vee x = x$

4. $(x')' = x$

5. $(x \wedge y)' = x' \vee y'$

6. $x' \wedge y' = (x \vee y)'$

7. $x \wedge x' = 0$

8. $x \wedge (y \wedge z) = (x \wedge y) \wedge z$

In Exercises 9–18 establish the validity of the equivalence using Theorem 9.1. List by letter the rules you use, in order. Start with the expression on the left side.

9. $(x \wedge y) \vee (x \wedge y') = x$

10. $x \vee (x' \wedge y) = x \vee y$

11. $x \wedge (x' \vee y) = x \wedge y$

12. $[(x \wedge y) \vee (x \wedge y')] \vee [(x' \wedge y) \vee (x' \wedge y')] = 1$

13. $(x' \vee y)' \vee (x \wedge y') = x \wedge y'$

14. $[(x \vee y) \wedge (x' \vee y)] \wedge [(x \vee y') \wedge (x' \vee y')] = 0$

15. $(x \wedge y)' \vee z = x' \vee (y' \vee z)$

16. $((x \vee y) \wedge z)' = z' \vee (x' \wedge y')$

17. $(x \wedge y) \wedge [(x \wedge w) \vee (y \wedge z)] = (x \wedge y) \wedge (w \vee z)$

18. $(x \vee y)' \vee (x \wedge y)' = (x \wedge y)'$

In Exercises 19–22 show that the Boolean expressions given are not equivalent.

19. $x \wedge (y \vee z)$ and $(x \wedge y) \vee z$

20. $(x \wedge y)'$ and $x' \wedge y'$

21. $(x \wedge y) \vee (x' \wedge z)$ and $(x \vee x') \wedge (y \vee z)$

22. $(1 \vee x) \vee x$ and x

In Exercises 23–28 find a Boolean expression of minterms which has the given truth table. Then draw the corresponding circuit.

23.

x	y	*Output*
0	0	0
0	1	1
1	0	1
1	1	0

24.

x	y	*Output*
0	0	0
0	1	1
1	0	0
1	1	1

25.

x	y	z	*Output*
0	0	0	0
0	0	1	0
0	1	0	0
0	1	1	1
1	0	0	0
1	0	1	1
1	1	0	1
1	1	1	0

26.	x	y	z	Output
	0	0	0	1
	0	0	1	0
	0	1	0	0
	0	1	1	1
	1	0	0	1
	1	0	1	0
	1	1	0	0
	1	1	1	1

27.	x	y	z	Output
	0	0	0	0
	0	0	1	0
	0	1	0	0
	0	1	1	1
	1	0	0	0
	1	0	1	1
	1	1	0	0
	1	1	1	1

28.	x	y	z	Output
	0	0	0	1
	0	0	1	0
	0	1	0	0
	0	1	1	0
	1	0	0	0
	1	0	1	1
	1	1	0	0
	1	1	1	1

In Exercises 29–34 describe the number of AND-, OR-, and NOT-gates with one or two inputs it would take to represent the given circuits.

29.

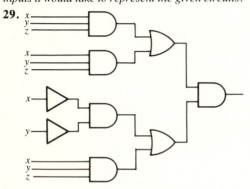

30.

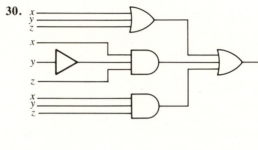

31.

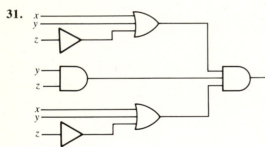

32.

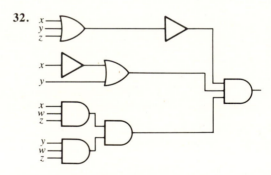

33.

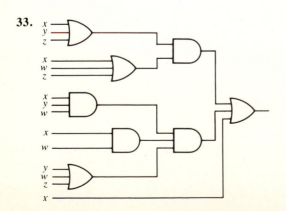

34.

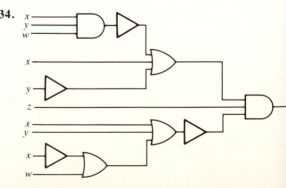

35. Suppose a company wishes to manufacture logical devices having inputs x and y and with output equivalent to the value of the logical statement $\sim (x \rightarrow y)$, where 0 corresponds to T and 1 to F. Draw a circuit using AND-, OR-, and NOT-gates that will do this.

36. A security network for a three-guard patrol at a missile base is set up so that an alarm is sounded if guard one loses contact and at least one of the other two guards is not in contact, or if guard one and guard two are in contact but guard three loses contact. Find a Boolean expression that has value 1 exactly when the alarm sounds. Let the input 1 correspond to losing contact.

37. An inventory control system for a factory recognizes an error in an order if it contains part A and part B but not part C; if it contains parts B or C, but not part D; or if it contains parts A and D. Find a Boolean expression in the variables a, b, c, and d that is 1 exactly when an error is recognized. Let a be 1 if part A is present, etc.

38. Which of the rules in Theorem 9.1 hold if X, Y, and Z stand for real numbers and we make the substitutions of multiplication for $\wedge$, addition for $\vee$, and $-X$ for X'?

39. Which of the rules in Theorem 9.1 hold if X, Y, and Z stand for subsets of a set U and we make the substitutions $\cap$ for $\wedge$, $\cup$ for $\vee$, $\overline{A}$ (the complement of A) for A', U for 1, and $\varnothing$ (the empty set) for 0?

We define a **Boolean algebra** to be a set B satisfying the following conditions:

(i)   For each pair of elements a and b in B there are defined unique elements $a \vee b$ and $a \wedge b$ in B.

(ii)   If a and b are in B, then $a \vee b = b \vee a$ and $a \wedge b = b \wedge a$.

(iii) If a, b, and c are in B, then $a \vee (b \vee c) = (a \vee b) \vee c$ and $a \wedge (b \wedge c) = (a \wedge b) \wedge c$.

(iv)   If a, b, and c are in B, then $a \vee (b \wedge c) = (a \vee b) \wedge (a \vee c)$ and $a \wedge (b \vee c) = (a \wedge b) \vee (a \wedge c)$.

(v)   There exist distinct elements 0 and 1 in B such that if $a \in B$, then $a \vee 0 = a$ and $a \wedge 1 = a$.

(vi)   If $a \in B$, there is defined a unique element $a' \in B$.

(vii) If $a \in B$, then $a \vee a' = 1$ and $a \wedge a' = 0$.

In Exercises 40–45 assume that B is a Boolean algebra. Exercises 41–45 show that the rules of Theorem 9.1 hold in any Boolean algebra.

40. Show that if a and b are in B and $a \vee b = 1$ and $a \wedge b = 0$, then $b = a'$. (*Hint:* Show that $b = b \wedge (a \vee a') = b \wedge a' = a' \wedge (a \vee b)$.)

41. Show that if $a \in B$, then $a \wedge 0 = 0$ and $a \vee 1 = 1$. (*Hint:* Compute $a \wedge (0 \vee a')$ and $a \vee (1 \wedge a')$ two ways.)

42. Show that if a and b are in B, then $a \vee (a \wedge b) = a \wedge (a \vee b) = a$. (*Hint:* Compute $a \wedge (1 \vee b)$ and $a \vee (0 \wedge b)$ two ways.)

43. Show that if $a \in B$, then $a \vee a = a \wedge a = a$. (*Hint:* Compute $(a \vee a) \wedge (a \vee a')$ and $(a \wedge a) \vee (a \wedge a')$ two ways.)

44. Show that if $a \in B$, then $a'' = a$, $0' = 1$, and $1' = 0$. (*Hint:* Use Exercise 40.)

45. Show that if a and b are in B, then $(a \vee b)' = a' \wedge b'$ and $(a \wedge b)' = a' \vee b'$. (*Hint:* Use Exercise 40.)

9.3 Karnaugh Maps

In the previous section we saw how to create a Boolean expression, and so a logical circuit that corresponds to any given truth table. The circuits we created, however, were usually quite complicated. We will show how to create simpler circuits by, in effect, making a picture of the truth table. Of course, "simpler" has not been defined precisely, and, in fact, various definitions might be appropriate. For compactness and economy of manufacture we might want to consider one circuit better than another if it contains fewer gates. For speed of operation, however, we might prefer a circuit such that the maximal number of gates between any original input and the output is as small as possible. The method we will describe will lead to circuits that are in general much simpler than those we learned to create at the end of the previous section, although they will not necessarily be simplest by either of the criteria just suggested. We will only treat the cases of 2, 3, or 4 Boolean variables as inputs, although there are methods for dealing with more than 4 input variables. (See suggested reading [7].)

We will show how to produce a simple Boolean expression that has a prescribed truth table. A circuit can then be constructed from this expression. The truth table we start with may represent the desired output of a circuit we are designing, or it may have been computed from an existing circuit or Boolean expression that we wish to simplify.

We will start with the following truth table.

x	y	Output
0	0	1
0	1	0
1	0	1
1	1	1

For this truth table, our previous method gives the Boolean expression $(x' \wedge y') \vee (x \wedge y') \vee (x \wedge y)$ and the circuit of Figure 9.17. To find a simpler

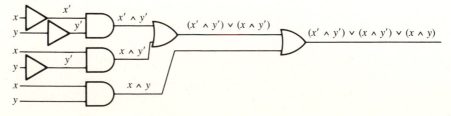

FIGURE 9.17

circuit we will represent our truth table graphically as in Figure 9.18(a). Each cell in the grid shown corresponds to a row of the truth table, with the rows of the grid

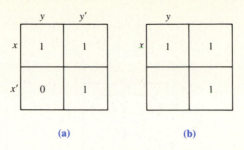

(a) **(b)**

FIGURE 9.18

corresponding to x and x' and the columns to y and y'. For example, the top left cell corresponds to the row of the truth table with $x = 1$ and $y = 1$, and the 1 in that cell tells us that there is a 1 in the output column of this row. Since in the grid each row is labeled either x or x', each column either y or y', and there is either a 0 or 1 in each cell, from now on we will save time by omitting the labels x' and y' and 0's, as in Figure 9.18(b). This is called the **Karnaugh map** of the truth table.

Each cell in the Karnaugh map corresponds to a minterm, as shown in Figure

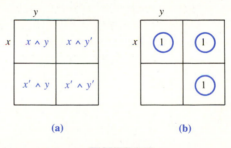

(a) **(b)**

FIGURE 9.19

9.19(a). Thus, we can create a Boolean expression having the truth table we started with by joining with the symbol $\vee$, the minterms in cells containing a 1, as circled in Figure 9.19(b). This amounts to the method of the previous section, and produces the Boolean expression

$$(x \wedge y) \vee (x \wedge y') \vee (x' \wedge y').$$

The key to our method is to notice that groups of adjacent cells may have even simpler Boolean expressions. For example, the two cells in the top row of the grid can be expressed simply as x. This can be confirmed using Theorem 9.1 as follows.

$$(x \wedge y) \vee (x \wedge y') = x \wedge (y \vee y') \quad \text{(rule (c))}$$
$$= x \wedge 1 \quad \text{(rule (f))}$$
$$= x \quad \text{(rule (g))}$$

Other such groups of two cells and the corresponding Boolean expressions are shown in Figures 9.20(a) and (b), where the ovals outline the cell groups named.

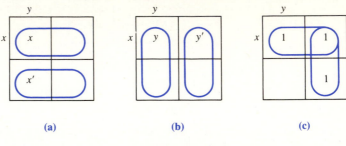

(a)   (b)   (c)

FIGURE 9.20

In Figure 9.20(c) we see that the three cells with 1's can be characterized as those cells in either the x oval or the y' oval, and so correspond to the Boolean expression $x \vee y'$. This is the simpler expression we have been seeking.

It is easily checked that $x \vee y'$ has the desired truth table. The corresponding circuit is shown in Figure 9.21. Comparison with the circuit of Figure 9.17 shows that it is simpler by any reasonable criterion.

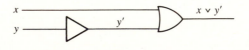

FIGURE 9.21

Since the case of two input variables is fairly straightforward, we shall proceed to three input variables, say x, y, and z. The grid we will use is shown in Figure 9.22(a). Recall the convention that the unmarked second row corresponds to x'. Likewise, columns 3 and 4 correspond to y' and columns 1 and 4 to z'. The minterms for each cell are shown in Figure 9.22(b).

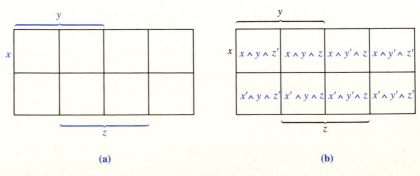

(a)   (b)

FIGURE 9.22

We will make a somewhat technical definition. We define two cells to be **adjacent** in case the minterms to which they correspond differ in only a single variable. A pair of adjacent cells can be described by a Boolean expression with one variable fewer than a minterm. For example, the two cells in the second row and third and fourth columns correspond to

$$(x' \wedge y' \wedge z) \vee (x' \wedge y' \wedge z') = (x' \wedge y') \wedge (z \vee z')$$
$$= (x' \wedge y') \wedge 1$$
$$= x' \wedge y'.$$

Any two cells next to each other in a row or column are adjacent and have a 2–variable Boolean expression, as shown in Figure 9.23.

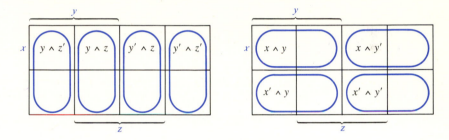

FIGURE 9.23

There are also two pairs of adjacent cells that wrap around the sides of our grid; these are shown in Figure 9.24, along with their simplified Boolean expressions.

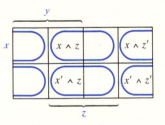

FIGURE 9.24

There are also groups of four cells with single-variable Boolean expressions. These are shown in Figure 9.25 on the following page. The student should not try

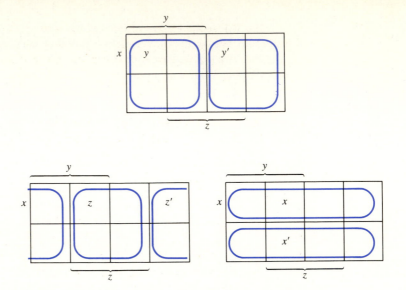

FIGURE 9.25

to memorize the Boolean expression for the groups of cells outlined in Figures 9.23, 9.24, and 9.25, but rather should study them to understand the principles behind them.

The method for constructing a simple Boolean expression corresponding to a truth table will be similar to the two-variable case. We draw the Karnaugh map for the truth table, then enclose the cells containing 1's (and only those cells) in ovals corresponding to Boolean expressions. Since larger groups of cells have simpler Boolean expressions, we use them whenever possible; and we try not to use more groups than necessary. We then join these expressions by $\vee$ to form a Boolean expression with the required truth table.

Consider, for example, the two Karnaugh maps shown in Figure 9.26. The

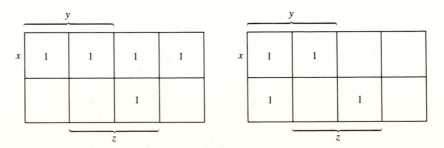

FIGURE 9.26

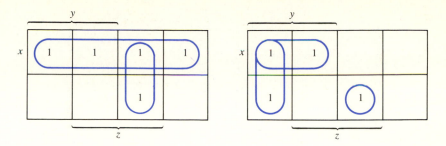

FIGURE 9.27

appropriate groups of cells are shown in Figure 9.27. The corresponding Boolean expressions are

$$x \vee (y' \wedge z) \quad \text{and} \quad (y \wedge z') \vee (x \wedge y) \vee (x' \wedge y' \wedge z),$$

respectively. Notice that the cell in the second row and third column of the second Karnaugh map is adjacent to no other cell with a 1 and so its 3–term minterm must be used.

EXAMPLE 9.8   Simplify the voting-machine circuit shown in Figure 9.16.

Since the machine is to give output 1 when at least two of x, y, and z are 1, the corresponding Karnaugh map is shown in Figure 9.28. Using the ovals indi-

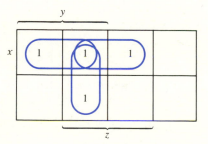

FIGURE 9.28

cated, we write the Boolean expression $(x \wedge y) \vee (x \wedge z) \vee (y \wedge z)$. The corresponding circuit is shown in Figure 9.29 on the following page. This circuit is considerably simpler than the one of Figure 9.16. In fact, if only gates with no more than two inputs are used, the previous circuit contains 14 while our new version has only 5. ∎

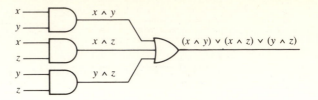

FIGURE 9.29

EXAMPLE 9.9

Simplify the expression $x \vee (y \wedge (x \wedge z')')$ of Example 9.6.
We compute the following truth table.

x	y	z	z'	$x \wedge z'$	$(x \wedge z')'$	$y \wedge (x \wedge z')'$	$x \vee (y \wedge (x \wedge z')')$
0	0	0	1	0	1	0	0
0	0	1	0	0	1	0	0
0	1	0	1	0	1	1	1
0	1	1	0	0	1	1	1
1	0	0	1	1	0	0	1
1	0	1	0	0	1	0	1
1	1	0	1	1	0	0	1
1	1	1	0	0	1	1	1

This leads to the Karnaugh map of Figure 9.30. Using the indicated groups of

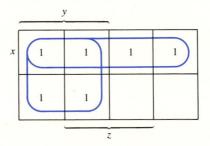

FIGURE 9.30

cells produces the same Boolean expression $x \vee y$ that was derived using the rules of Theorem 9.1 in Example 9.6. ■

Finally, we consider Karnaugh maps for circuits with four inputs w, x, y, and z. We will use a 4–by–4 grid, labeled as in Figure 9.31(a). For example, the cell marked (1) corresponds to the minterm $w \wedge x' \wedge y \wedge z'$, and the cells marked (2) and (3) to the minterms $w \wedge x \wedge y' \wedge z$ and $w' \wedge x' \wedge y \wedge z$, respectively.

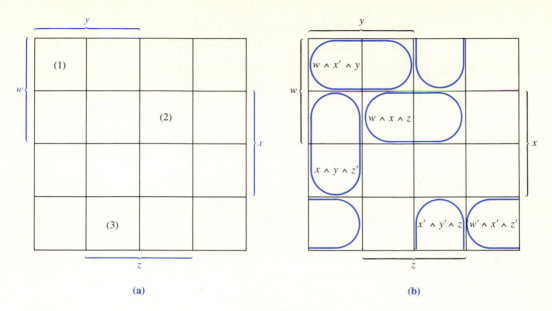

(a)

(b)

FIGURE 9.31

Figure 9.31(b) shows various groups of two adjacent cells and their Boolean expressions. Of course, there are many more such groups. Examples of groups of four cells and their 2–variable Boolean expressions are shown in Figure 9.32. Notice that they can wrap around either horizontally or vertically. There are also 8–cell groups whose Boolean expressions have a single variable; some of these are shown in Figure 9.33 on the following page.

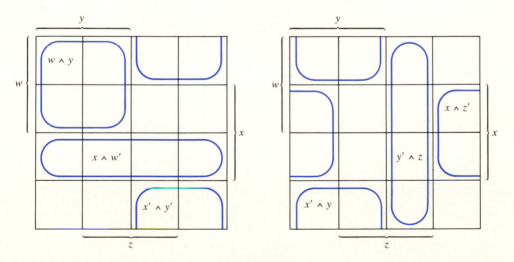

FIGURE 9.32

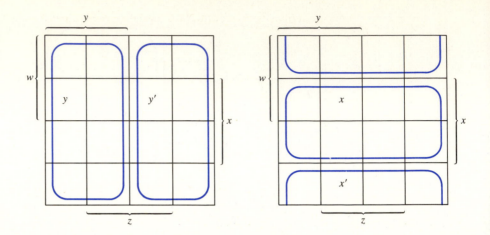

FIGURE 9.33

As before, given a truth table we form its Karnaugh map and then enclose its 1's (and only its 1's) in rectangles of 1, 2, 4, or 8 cells that are as large as possible. The required Boolean expression is formed by joining the expressions for these rectangles with $\vee$.

EXAMPLE 9.10   Find a circuit having the following truth table.

w	x	y	z	Output
0	0	0	0	1
0	0	0	1	0
0	0	1	0	1
0	0	1	1	1
0	1	0	0	1
0	1	0	1	1
0	1	1	0	1
0	1	1	1	0
1	0	0	0	1
1	0	0	1	0
1	0	1	0	1
1	0	1	1	0
1	1	0	0	1
1	1	0	1	1
1	1	1	0	1
1	1	1	1	1

The Karnaugh map for this table is shown in Figure 9.34. Using the rectangles of

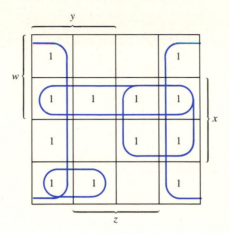

FIGURE 9.34

cells shown yields the expression $z' \vee (w \wedge x) \vee (x \wedge y') \vee (w' \wedge x' \wedge y)$. Figure 9.35 shows the corresponding circuit. ▪

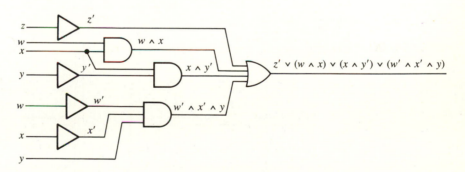

FIGURE 9.35

EXAMPLE 9.11   Use Karnaugh maps to simplify the circuit of Figure 9.36(a).

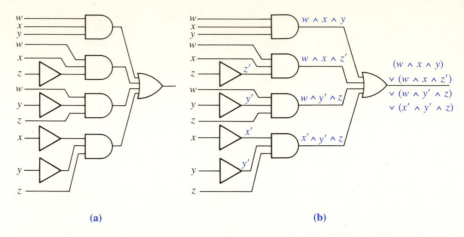

FIGURE 9.36

We compute the Boolean expression

$$(w \wedge x \wedge y) \vee (w \wedge x \wedge z') \vee (w \wedge y' \wedge z) \vee (x' \wedge y' \wedge z)$$

for the circuit as shown in Figure 9.36(b). The terms separated by $\vee$'s in this expression correspond to the four rectangles marked in Figure 9.37(a). The same

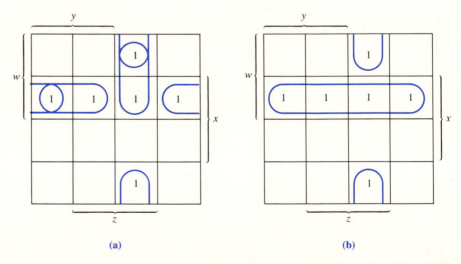

FIGURE 9.37

cells can be enclosed by two rectangles, as shown in Figure 9.37(b). These yield the Boolean expression $(w \wedge x) \vee (x' \wedge y' \wedge z)$ and the circuit of Figure 9.38. ■

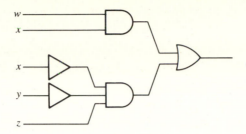

EXERCISES 9.3

In Exercises 1–6 find a Boolean expression of minterms which has the given truth table.

1.

x	y	Output
0	0	1
0	1	1
1	0	0
1	1	1

2.

x	y	z	Output
0	0	0	1
0	0	1	0
0	1	0	0
0	1	1	1
1	0	0	1
1	0	1	0
1	1	0	1
1	1	1	1

3.

x	y	z	Output
0	0	0	1
0	0	1	1
0	1	0	1
0	1	1	1
1	0	0	0
1	0	1	0
1	1	0	0
1	1	1	0

4.

x	y	z	Output
0	0	0	0
0	0	1	1
0	1	0	0
0	1	1	1
1	0	0	1
1	0	1	1
1	1	0	1
1	1	1	0

5.

w	x	y	z	Output
0	0	0	0	1
0	0	0	1	0
0	0	1	0	1
0	0	1	1	0
0	1	0	0	1
0	1	0	1	0
0	1	1	0	1
0	1	1	1	0
1	0	0	0	0
1	0	0	1	0
1	0	1	0	0
1	0	1	1	0
1	1	0	0	0
1	1	0	1	1
1	1	1	0	0
1	1	1	1	0

6.

w	x	y	z	Output
0	0	0	0	1
0	0	0	1	0
0	0	1	0	0
0	0	1	1	0
0	1	0	0	0
0	1	0	1	1
0	1	1	0	0
0	1	1	1	1
1	0	0	0	1
1	0	0	1	0
1	0	1	0	0
1	0	1	1	0
1	1	0	0	0
1	1	0	1	1
1	1	1	0	0
1	1	1	1	1

In Exercises 7–12 write the Boolean expression corresponding to the ovals in the Kar-naugh map.

7.

8.

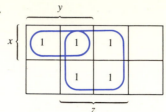

9.

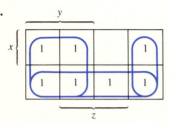

10.

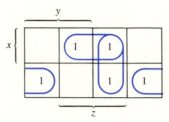

11.

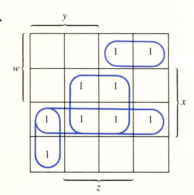

12.

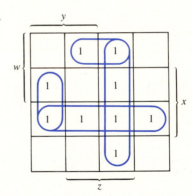

In Exercises 13–18 draw a Karnaugh map for the Boolean expression of the indicated exercise.

13. Exercise 1   **14.** Exercise 2   **15.** Exercise 3

16. Exercise 4   **17.** Exercise 5   **18.** Exercise 6

In Exercises 19–24 use the Karnaugh map method to simplify the Boolean expression in the indicated exercise. Then draw a circuit representing the simplified Boolean expression.

19. Exercise 1   **20.** Exercise 2   **21.** Exercise 3

22. Exercise 4   **23.** Exercise 5   **24.** Exercise 6

In Exercises 25–32 use the Karnaugh map method to simplify the expression.

25. $(x' \wedge y' \wedge z) \vee (x' \wedge y \wedge z) \vee (x \wedge y' \wedge z)$

26. $(x' \wedge y \wedge z) \vee (x' \wedge y' \wedge z') \vee (x \wedge y \wedge z) \vee (x \wedge y' \wedge z')$

27. $(x' \wedge y' \wedge z) \vee (x' \wedge y \wedge z) \vee (x \wedge y' \wedge z')$

28. $[(x \vee y') \wedge (x' \wedge z')] \vee y$

29. $[x \wedge (y \vee z)] \vee (y' \wedge z')$

30. $(x \wedge y \wedge z) \vee (x \wedge y' \wedge z) \vee (x' \wedge y' \wedge z)$

31. $(w \wedge x \wedge y) \vee (w \wedge x \wedge z) \vee (w \wedge y' \wedge z') \vee (y' \wedge z')$

32. $(w' \wedge x' \wedge y') \vee (w' \wedge y' \wedge z) \vee (w \wedge y \wedge z) \vee (w \wedge x \wedge z') \vee (w \wedge y' \wedge z') \vee$
$(w \wedge x' \wedge y \wedge z) \vee (w' \wedge x \wedge y \wedge z')$

In Exercises 33 and 34 use Karnaugh maps to simplify the given circuit.

33.   **34.**

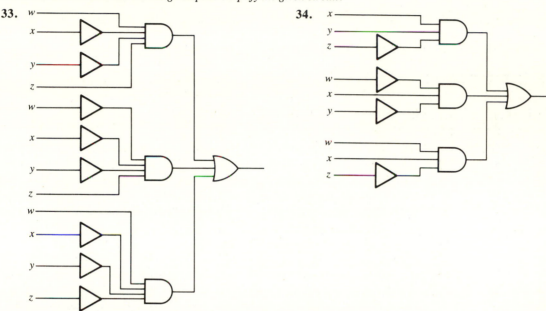

35. How many groups of two adjacent cells are there in a Karnaugh map grid for 4 Boolean variables?

36. How many 4–element square groups of adjacent cells are there in a Karnaugh map grid for 4 Boolean variables?

Although part (b) of Theorem 9.1 suggests that we get equal expressions for any two ways we insert parentheses in $x_1 \vee x_2 \vee \ldots \vee x_n$, we have not given a formal proof of this fact. (We will only treat $\vee$; $\wedge$ could be handled in the same way.) Define $x_1 \vee x_2 \vee \ldots \vee x_n$ recursively as follows:

$$x_1 \vee x_2 \vee \ldots \vee x_n = \begin{cases} x_1 \text{ if } n = 1 \\ (x_1 \vee x_2 \vee \ldots x_{n-1}) \vee x_n \text{ for } n > 1. \end{cases}$$

37. Prove by induction on n that $(x_1 \vee x_2 \vee \ldots \vee x_m) \vee (y_1 \vee y_2 \vee \ldots \vee y_n) = x_1 \vee \ldots \vee x_m \vee y_1 \vee \ldots \vee y_n$ for any positive integers m and n.

38. Prove that any two expressions formed by inserting parentheses in $x_1 \vee x_2 \vee \ldots \vee x_n$ are equal.

39. Prove by induction that $(x_1 \vee x_2 \vee \ldots \vee x_n)' = x_1' \wedge x_2' \wedge \ldots \wedge x_n'$ for all positive integers n, where the definition of the latter expression is similar to that for $\vee$.

40. Denote by q_n the number of ways of inserting $n - 2$ sets of parentheses in $x_1 \vee x_2 \vee \ldots \vee x_n$ so that the order in which the $\vee$'s are applied is unambiguous. For example, $q_3 = 2$ counts the expressions $(x_1 \vee x_2) \vee x_3$ and $x_1 \vee (x_2 \vee x_3)$. Likewise, $q_4 = 5$. Show that $q_n = q_1 q_{n-1} + q_2 q_{n-2} + \ldots + q_{n-1} q_1$ for $n > 1$.

41. Let r_n be the number of ways of listing $x_1, x_2, \ldots, x_n$ joined by $\vee$'s in any order and with parentheses. For example, $r_1 = 1$, $r_2 = 2$ counts the two expressions $x_1 \vee x_2$ and $x_2 \vee x_1$, and $r_3 = 12$. Show that $r_{n+1} = (4n - 2)r_n$ for all positive integers n.

42. Show that $r_n = \dfrac{(2n - 2)!}{(n - 1)!}$ and $q_n = \dfrac{(2n - 2)!}{n!(n - 1)!}$ for all positive integers n, where r_n and q_n are defined as in Exercises 40 and 41.

9.4   Finite State Machines

In this section we will study devices, like computers, that have not only inputs and outputs, but also a finite number of internal states. What the device does when presented with a given input will depend not only upon that input, but also upon the internal state that the device is in at the time. For example, if a person types "RUN" into a personal computer, what happens will depend on the internal state of the computer—in particular the language the computer recognizes, and the program, if any, in the computer's memory.

Such devices are called finite state machines. Various formal definitions of a finite state machine may be given. We will study two types, one simple and the other somewhat more complicated. Our main concern will be with understanding what such machines are and how they operate, rather than with how to construct finite state machines to do specific tasks.

A simple example of a finite state machine is a newspaper vending machine. Such a vending machine has two states, locked and unlocked, which we will denote by L and U. We will consider a machine that only accepts quarters, the price of a paper. Two inputs are possible, to put a quarter into the machine (q), and to try to open and shut the door to get a paper (d). Putting in a quarter unlocks the machine, after which opening and shutting the door locks it again. Of course, putting a quarter into a machine that is already unlocked does not change the state of the machine, nor does trying to open the door of a locked machine.

There are various ways we can represent this machine. One way is to make a table showing how each input affects the state the machine is in.

		Present state	
		L	U
Input	q	U	U
	d	L	L

Here the entries in the body of the table show the next state the machine enters, depending on the present state (column) and input (row). For example, the colored entry means that if the machine is in state L and the input is q, it changes to state U. Since this table gives a state for each ordered pair (i, s), where i is an input and s is a state, it describes a function with the Cartesian product $\{q, d\} \times \{U, L\}$ as its domain and the set of states $\{U, L\}$ as its codomain. (The reader may want to review the concepts of Cartesian product and function in Sections 2.1 and 2.4.) Such a table is called the **state table** of the machine.

We can also represent our machine graphically, as in Figure 9.39. Here the

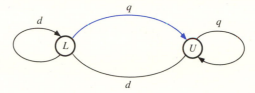

FIGURE 9.39

states L and U are shown as circles, and labeled arrows indicate the effect of each input when the machine is in each state. For example, the colored arrow indicates that a machine in state L with input q moves to state U. This diagram is called the **transition diagram** of the machine. (In the language of Section 3.4 the transition diagram is a directed multigraph.)

We will generally use the pictorial representation for finite state machines since our examples will be fairly simple. For a machine with many inputs and states the picture may be so complicated that a state table is preferable.

A Parity Checking Machine

Before we give a formal definition of a finite state machine, we will give one more example. Data sent between electronic devices is generally represented as a sequence of 0's and 1's. Some way of detecting errors in transmission is desirable. We will describe one simple means of doing this. Before a message is sent, the number of 1's in the message is counted. If this number is odd, a single 1 is added to the end of the message, and if it is even, a 0 is added. Thus, all transmissions will contain an even number of 1's.

After a transmission is received the 1's are counted again to determine whether there are an even or odd number of them. This is called a **parity check.** If there are an odd number of 1's, then some error must have occurred in transmission. In this case a repeat of the message can be requested. Of course, if there are two or more errors in transmission, a parity check may not tell the receiver so. But if the transmission of each digit is reliable and the message not too long, this may be far less likely than a single error. If the received transmission passes the parity check, its last digit is discarded to regain the original message.

Actually it is not necessary to count the number of 1's in a message to tell if this number is odd or even. Figure 9.40 represents a device that can be used to do

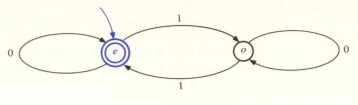

FIGURE 9.40

this job. Here the states are e (even) and o (odd), and the inputs are 0 and 1. The corresponding state table is as follows.

	State	
	e	o
Input 0	e	o
1	o	e

We can use this device to determine whether the number of 1's in a string of 0's and 1's is even or odd by starting in state e and using each successive digit as a new input. For example, if the message 11010001 is used as input (reading from left to right), the machine starts in state e and moves to state o because the first input is 1. The second input is also 1, putting the machine back in state e, where it stays after the third input, 0. The way the machine moves from state to state is summarized in the following table.

Input:	Start	1	1	0	1	0	0	0	1
State:	e	o	e	e	o	o	o	o	e

If 11010001 is received, we would presume that no error occurred in transmission and that the original message was 1101000.

Two new symbolisms appear in Figure 9.40. One is the arrow pointing into state e. This indicates that we must start in state e for our device to work properly. The other is the double circle corresponding to state e. This indicates that this state is a desirable final state; otherwise in our example some error has occurred.

Now we formally define a **finite state machine** to consist of a finite set of states S, a finite set of inputs I, and a function f with $I \times S$ as its domain and S as its codomain such that if $i \in I$ and $s \in S$, then $f(i, s)$ is the state the machine moves to when it is in state s and is given input i. We may also, depending upon the application, specify an **initial state** s_0, as well as a subset S' of S. The elements of S', called **accepting states,** are the states we would like to end in.

Thus, our parity checking machine is a finite state machine with $S = \{e, o\}$, $I = \{0, 1\}$, $s_0 = e$, and $S' = \{e\}$. The function f is specified by

$$f(0, e) = e, \quad f(0, o) = o,$$
$$f(1, e) = o, \quad f(1, o) = e,$$

which corresponds to our previous state table.

A **string** is a finite sequence of inputs, such as 11010001 in our last example. Suppose, given the string $i_1 i_2 \ldots i_n$ and the initial state s_0, we successively compute $f(i_1, s_0) = s_1$, then $f(i_2, s_1) = s_2$, etc., finally ending up with state s_n. This amounts to starting in the initial state, applying the inputs of the string from left to right, and ending up in state s_n. If s_n is in S', the set of accepting states, then we say that the string is **accepted;** otherwise it is **rejected.** In the parity check example rejected transmissions contain some error, while accepted transmissions are presumed to be correct.

EXAMPLE 9.12   Figure 9.41 shows a finite state machine with input set $I = \{0, 1\}$ that accepts a string precisely when it ends with the triple 100. Here $S = \{A, B, C, D\}$, $s_0 = A$,

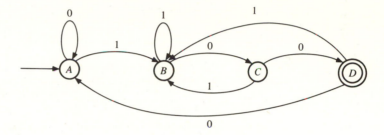

FIGURE 9.41

$S' = \{D\}$, and the function f is as indicated by the labeled arrows in the diagram. For example, if the string 101010 is input, we move through the states $ABCBCBC$, and since C is not in S' the string is rejected. On the other hand, if 001100 is input, we move through the states $AAABBCD$, and the string is accepted because D is an accepting state.

To see that the machine of Figure 9.41 does what we claim, the reader should first check that no matter what state we are in, if the string 100 is input we are taken to state D. This shows that all strings ending in 100 will be accepted by the machine. It remains to show that an accepted string must end in 100. Since we start in state A, an accepted string clearly must contain at least three digits. Since when 1 is input we move to state B no matter what the present state is, the accepted string must end in 0. Likewise, the reader should check that any string ending in 10 leaves the machine in state C. Thus, our accepted string must end in two 0's. Finally the reader should check that any string ending in 000 puts the machine in state A. Thus, any accepted string must end in 100.   ■

One important application of machines that accept certain strings and reject others is in compilers for computer languages. Before a program is run each statement must be checked to see whether it conforms to the syntax of the language being used.

Finite State Machines with Output

Now we consider a slightly more complicated type of device. We start with an example more sophisticated than a newspaper vending machine, namely, a gum machine. Our gum machine accepts only quarters, which is the price of a pack of gum. Three varieties are available: Doublemint (denoted by D), Juicy Fruit (J), and Spearmint (S), which can be chosen by pressing buttons d, j, or s, respectively. The internal states of the machine are locked (denoted by L) and unlocked (U); and if the machine is unlocked it will return any extra quarters put into it. The inputs are q (quarter), d, j, and s. A diagram showing some of the action of the machine is given in Figure 9.42(a). Figure 9.42(b) shows a more compact way of

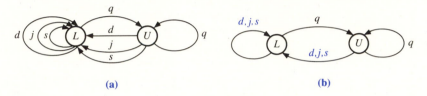

(a)   (b)

FIGURE 9.42

indicating multiple arrows going between the same two states; here, for example, the three arrows from U to L in Figure 9.42(a) have been replaced by a single arrow and the corresponding inputs separated by commas.

This diagram does not tell the whole story, however. Nowhere does it show that if we press the d button on a machine in state U we get a pack of Doublemint. Neither does it show that excess quarters are returned. We need to introduce the additional concept of **outputs** of the machine. In this example the possible outputs are D, J, S and also Q (an excess quarter returned) and $\emptyset$, which we will use to stand for no output, as for example when a button is pressed while the machine is in state L.

Notice that the output may depend upon both the input and the state of the machine. The inputs d and j produce the distinct outputs D and J when the machine is in state U. Likewise, the input d produces the outputs $\emptyset$ or D depending on whether the machine is in state L or U. Another function is involved here, having the Cartesian product of the set of inputs and the set of states as its domain and the set of outputs as its codomain. Since each arrow in our diagram stands for the result of an input being applied to a particular state, we can also label these arrows to show the corresponding outputs. This is done in Figure 9.43.

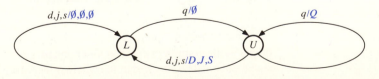

FIGURE 9.43

We will use slashes to separate the input and output labels on each arrow. Thus, in Figure 9.43 the $q/\varnothing$ on the arrow from L to U indicates that there is no output when we put a quarter in a locked machine; and the $d, j, s/D, J, S$ on the arrow from U to L indicates the outputs $D, J,$ and S, respectively, when we push buttons $d, j,$ and s on an unlocked machine.

In general we define a **finite state machine with output** to consist of finite sets S of states, I of inputs, and O of outputs, along with a function $f: I \times S \rightarrow S$ such that $f(i, s)$ is the state the machine goes to from state s when the input is i, and another function $g: I \times S \rightarrow O$ such that $g(i, s)$ is the output corresponding to input i when the machine is in state s. Depending on the application, we may again designate a particular state s_0 as the **initial state.**

In the gum machine example we have $S = \{L, U\}$, $I = \{q, d, j, s\}$, and $O = \{D, J, S, Q, \varnothing\}$. The functions f and g are indicated in Figure 9.43, but they can also be described, as before, using tables.

<table>
<tr><td></td><td></td><td colspan="2" align="center">State</td></tr>
<tr><td></td><td></td><td>L</td><td>U</td></tr>
<tr><td rowspan="4">Input</td><td>q</td><td>U</td><td>U</td></tr>
<tr><td>d</td><td>L</td><td>L</td></tr>
<tr><td>j</td><td>L</td><td>L</td></tr>
<tr><td>s</td><td>L</td><td>L</td></tr>
</table>

<table>
<tr><td></td><td></td><td colspan="2" align="center">State</td></tr>
<tr><td></td><td></td><td>L</td><td>U</td></tr>
<tr><td rowspan="4">Input</td><td>q</td><td>$\varnothing$</td><td>Q</td></tr>
<tr><td>d</td><td>$\varnothing$</td><td>D</td></tr>
<tr><td>j</td><td>$\varnothing$</td><td>J</td></tr>
<tr><td>s</td><td>$\varnothing$</td><td>S</td></tr>
</table>

The first table, which gives the values of f, is still called the state table of the machine, while the second, which gives the values of g, is called the **output table.**

If a string of inputs is fed into a finite state machine with output, a corresponding sequence of outputs is produced, called the **output string.** This is illustrated in the next example.

EXAMPLE 9.13 Figure 9.44 shows the transition diagram of a **unit delay** machine. This is a finite state machine with output in which $I = \{0, 1\}$, $S = \{A, B, C\}$, $O = \{0, 1\}$, and the initial state is A. Note that the first output is always 0, while any input of 0 puts

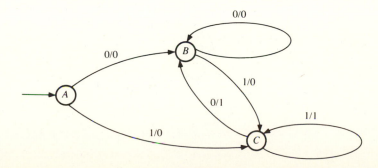

FIGURE 9.44

the machine in state B, from which the next output will be 0. Likewise any input of 1 puts it in state C, from which the next output will be 1. Thus, each output after the first is always the same as the input one step previously. An input string $i_1 i_2 \ldots i_n$ produces the output string $0 i_1 i_2 \ldots i_{n-1}$. For example, the input string 1100111 produces the output string 0110011. If it is desired to copy an entire input string, then a 0 must be appended to it before the string is input. ▪

EXAMPLE 9.14   Draw the transition diagram for the finite state machine with output that has the following state and output tables, and describe what the machine does to an input string of x's and y's. The initial state is A.

		State						State				
		A	B	C	D	E	F	A	B	C	D	E	F
Input	x	A	C	C	E	E	F	0	1	1	2	2	3
	y	B	B	D	D	F	F	1	1	2	2	3	3

The transition diagram is shown in Figure 9.45. Notice that once an input x

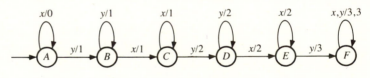

FIGURE 9.45

or y puts the machine into one of the states A, B, C, D, or E, the machine stays in that state until the input changes. The output is 0 or 1 according to whether the first input is x or y, and increases by one whenever the input changes from x to y. Thus, the output at any time counts the number of groups of consecutive y's in the input string, up to three such groups. For example, the input string $xxyxxxyyyxx$ produces the output string 00111122222; and the last 2 counts the two groups of y's (y and yyy) in the input string. ▪

EXERCISES 9.4

In Exercises 1–6 draw the transition diagram for the finite state machine with the given state table.

1.

	A	B
0	A	A
1	A	B

2.

	A	B	C
0	B	C	A
1	A	C	B

3.

	x	y	z
0	y	z	z
1	x	x	y

Initial state x, accepting state z

4.

	A	B
x	B	A
y	A	A
z	B	B

Initial state A,
accepting state A

5.

	A	B	C	D
a	B	A	D	C
b	C	C	A	A

Initial state B,
accepting states C, D

6.

	u	v	w
0	u	w	v
1	u	w	w
2	w	v	u

Accepting states u, v

In Exercises 7–10 give the state table for the finite state machine with the given transition diagram. List the initial and accepting states, if any.

7.
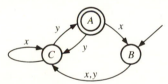

8.
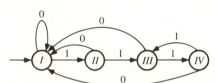

9.
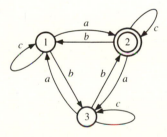

10.
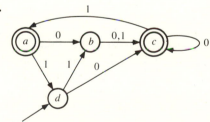

For the finite state machine and input string in Exercises 11–14, determine the state that the machine ends in if it starts at the initial state.

11. Input string 1011001, machine of Exercise 3

12. Input string xyyzzx, machine of Exercise 4

13. Input string yxxxy, machine of Exercise 7

14. Input string 0100011, machine of Exercise 8

In Exercises 15–18 tell whether the given input string would be accepted by the indicated finite state machine.

15. Input string xyzxyzx, machine of Exercise 4

16. Input string aabbaba, machine of Exercise 5

17. Input string xyxxyy, machine of Exercise 7

18. Input string 0011010, machine of Exercise 10

In Exercises 19–22 draw the transition diagram for the finite state machine with output whose state and output tables are given.

19.

	A	B	A	B
0	B	A	x	y
1	A	B	z	x

20.

	1	2	3	1	2	3
red	2	3	1	A	B	A
blue	1	1	3	A	A	B

21.

	A	B	A	B
0	A	A	x	y
1	B	B	w	x
2	A	B	y	w

Initial state A

22.

	00	01	10	11	00	01	10	11
A	11	10	01	00	1	−1	0	1
B	01	10	11	11	−1	0	1	−1

Initial state 10

In Exercises 23–26 give the state and output tables of the pictured finite state machine with output. Name the initial state, if any.

23.

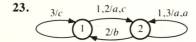

24.

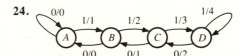

25.

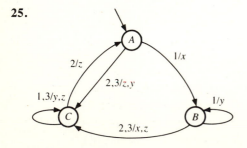

26.

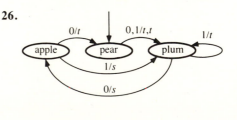

In Exercises 27–30 give the output string for the given input string and finite state machine with output.

27. Input string 2101211, machine of Exercise 21

28. Input string *BAABBB*, machine of Exercise 22

29. Input string 322113, machine of Exercise 25

30. Input string 10100110, machine of Exercise 26

In Exercises 31–34 describe which input strings of 0's and 1's are accepted by the pictured finite state machine.

31.

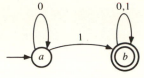

32.

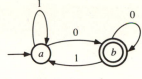

33.

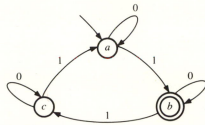

34.

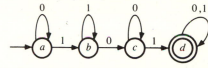

In Exercises 35–38 assume the set of inputs is {0, 1}.

35. Design a finite state machine that accepts a string if and only if it ends with two 1's.

36. Design a finite state machine that accepts a string if and only if it does not contain two consecutive 0's.

37. Design a finite state machine with output such that, given an input string, its last output is the remainder when the number of 1's in the input string is divided by 3.

38. Design a finite state machine with output such that its output string contains as many 1's as there are pairs of consecutive 0's or 1's in the input string.

39. Let F and G be finite state machines. We say that F and G are **equivalent** if they have the same set of inputs and if, whenever a string is accepted by either of the machines, it is also accepted by the other. Let I and S be sets. Show that equivalence of finite state machines is an equivalence relation on the set of finite state machines having input sets and state sets that are subsets of I and S, respectively.

Suggested Readings

1. Dornhoff, Larry L. and Franz E. Hohn. *Applied Modern Algebra*. New York: Macmillan, 1978.

2. Fisher, James L. *Application-Oriented Algebra*. New York: Crowell, 1977.

3. Friedman, Arthur D. and Premachandran R. Menon. *Theory & Design of Switching Circuits*. Rockville, MD: Computer Science Press, 1975.

4. Liu, C. L. *Elements of Discrete Mathematics*. 2nd ed. New York: McGraw-Hill, 1985.

5. Stanat, Donald F. and David F. McAllister. *Discrete Mathematics in Computer Science*. Englewood Cliffs, NJ: Prentice-Hall, 1977.

6. Stone, Harold S. *Discrete Mathematical Structures and Their Applications*. Chicago: Science Research Associates, 1973.

7. Tremblay, J. P. and R. Manohar. *Discrete Mathematical Structures with Applications to Computer Science*. New York: McGraw-Hill, 1975.

APPENDIX:
AN INTRODUCTION
TO LOGIC AND PROOF

It is essential that persons in such fields as mathematics, physics, and computer science understand basic principles of logic so that they are able to recognize valid and invalid arguments. In this appendix we present an informal introduction to logic and proof that provides a sufficient working knowledge of these subjects for students of computer science, mathematics, and the sciences. In Chapter 9 we explore an application of logic to the design of circuits such as those found in computers.

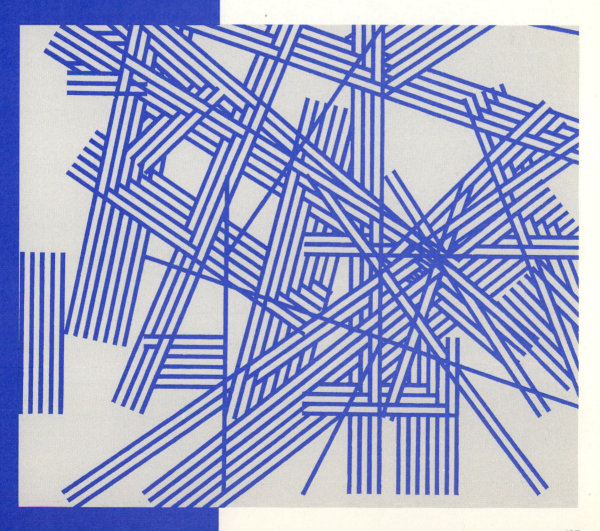

A.1   Statements and Connectives

One aspect of logic involves determining the truth or falsity of meaningful assertions. By a **statement** we will mean any sentence that is either true or false, but not both. For example, each of the following is a statement.

(1) Millard Fillmore was the thirteenth president of the United States.
(2) Baltimore is the capital of Maryland.
(3) $6 + 3 = 9$.
(4) Texas has the largest area of any state in the United States.
(5) All dogs are animals.
(6) Some species of birds migrate.
(7) Every even integer greater than 2 is the sum of two primes.

In the sixth statement above the word "some" appears. In logic we interpret the word "some" to mean "at least one." Thus, the sixth statement means that at least one species of birds migrates or that there is a species of birds that migrates.

The first, third, fifth, and sixth of the statements above are true, and the second and fourth are false. At this time, however, it is not known whether the seventh statement is true or false. (This statement is a famous unsolved mathematical problem called Goldbach's conjecture.) Nevertheless, it is a statement because it must be true or false but not both.

On the other hand, the following sentences are *not* statements.

(1) Why should we study logic?
(2) Eat at the cafeteria.
(3) Enjoy your birthday!

The reason that these fail to be statements is that none of them can be judged to be true or false.

It is possible that a sentence is a statement and yet we are unable to ascertain its truth or falsity because of an ambiguity. The following statements are of this type.

(1) Yesterday it was cold.
(2) He thinks New York is a wonderful city.
(3) There is a number x such that $x^2 = 5$.
(4) Lucille is a brunette.

In order to decide the truth of the first statement we need to specify what we mean by the word "cold." Similarly, in the second statement we need to know whose opinion is being considered in order to decide whether this statement is true or false. Whether the third statement is true depends upon what types of numbers we allow, and the truth of the fourth statement depends on which Lucille we have in mind. Hereafter we will not consider such ambiguous statements.

EXAMPLE A.1 The following sentences are statements.

(a) On December 4, 1985, the temperature dropped below freezing in Miami, Florida.
(b) In the opinion of Mayor Koch, New York is a wonderful city.
(c) There is an integer x such that $x^2 = 5$.
(d) Comedienne Lucille Ball is a brunette. ■

We will be interested in studying the truth or falsity of statements formed from other statements using the expressions below. These expressions are called **connectives.**

Connective	Symbol	Name
not	$\sim$	negation
and	$\wedge$	conjunction
or	$\vee$	disjunction
if . . . then . . .	$\rightarrow$	conditional
if and only if	$\leftrightarrow$	biconditional

The use of the connective "not" in logic is the same as in standard English; that is, its use denies the statement to which it applies. It is easy to form the negation of most simple statements, as we see in the following example.

EXAMPLE A.2 Consider the following statements.

(a) Today is Friday.
(b) Los Angeles is not the capital of California.
(c) $3^2 = 9$.
(d) It is not true that I went to the movies today.
(e) The temperature is above 60° Fahrenheit.

The negations of the statements above are

(a) Today is not Friday.
(b) Los Angeles is the capital of California.
(c) $3^2 \neq 9$.
(d) It is true that I went to the movies today.
(e) The temperature is less than or equal to 60° Fahrenheit. ■

However, the negation of statements containing words such as "some," "all," and "every" requires more care. Consider, for instance, the statement s below.

s: Some bananas are blue.

Since "some" means "at least one," the negation of s is the statement

$\sim s$: No bananas are blue.

Likewise, the negation of the statement

t: Every banana is yellow.

is the statement

$\sim t$: Some bananas are not yellow.

As these examples suggest, we can negate a statement involving the word "some" by changing "some" to "no," and we can negate a sentence involving the words "all," "each," or "every" by changing these words to "some . . . not . . .".

EXAMPLE A.3

Negate each of the following statements.

(a) Some cowboys live in Wyoming.
(b) There are movie stars who are not famous.
(c) No integers are divisible by 5.
(d) All doctors are rich.
(e) Every college football player weighs at least 200 pounds.

The negations of these statements are given below.

(a) No cowboys live in Wyoming.
(b) No movie stars are not famous. (Or, all movie stars are famous.)
(c) Some integers are divisible by 5.
(d) Some doctors are not rich.
(e) Some college football players do not weigh at least 200 pounds. ■

It is obvious that the negation of a true statement is false, and the negation of a false statement is true. We can record this information in the following table, called a **truth table.**

p	$\sim p$
T	F
F	T

Here p denotes a statement and $\sim p$ denotes its negation. The letters T and F signify that the indicated statement is true or false, respectively.

The **conjunction** of two statements is formed by joining the statements with the word "and." For example, the conjunction of the statements

p: Today is Monday   and q: I went to school

is the statement

$p \wedge q$: Today is Monday, and I went to school.

This statement is true only when both of the original statements p and q are true. Thus, the truth table for the connective "and" is as shown below.

p	q	$p \wedge q$
T	T	T
T	F	F
F	T	F
F	F	F

The **disjunction** of two statements is formed by joining the statements with the word "or." For example, the disjunction of the statements p and q above is

$p \vee q$: Today is Monday, or I went to school.

This statement is true when at least one of the original statements is true. For example, the statement $p \vee q$ is true in each of the following cases:

(1) Today is not Monday and I went to school.
(2) Today is Monday and I did not go to school.
(3) Today is Monday and I went to school.

Thus, the truth table for the connective "or" is as shown below.

p	q	$p \vee q$
T	T	T
T	F	T
F	T	T
F	F	F

The connectives "if . . . then . . . ." and "if and only if" occur rather infrequently in ordinary discourse, but they are used very often in mathematics. A statement containing the connective "if . . . then . . . ." is called a **conditional statement** or, more simply, a **conditional.** For example, suppose x is a real number and p and q are the statements

$p: x > 3$ and $q: x > 0.$

Then the conditional $p \rightarrow q$ is the statement

$p \rightarrow q$: If $x > 3$, then $x > 0$.

Another way of reading the statement "if p, then q" is "p implies q." In the conditional statement "if p, then q," statement p is called the **premise** and statement q is called the **conclusion.**

It is important to note that conditional statements should not be interpreted in terms of cause and effect. Thus, when we say ''if p, then q,'' we do not mean that the premise p causes the conclusion q, but only that when p is true, q must be true also.

In Normal, Illinois there is a city ordinance designed to aid city street crews in removing snow. This ordinance states: If there is a snowfall of two or more inches, then cars cannot be parked overnight on city streets. This regulation is phrased as a conditional statement with premise

p: There is a snowfall of two or more inches

and conclusion

q: Cars cannot be parked overnight on city streets.

Let us consider under what circumstances the conditional statement $p \rightarrow q$ is false, that is, under what circumstances the ordinance has been violated. The ordinance is clearly violated if there is a snowfall of two or more inches and cars are parked overnight on city streets, that is, if p is true and q is false. Moreover, if there is a snowfall of two or more inches and cars are not parked overnight on city streets (that is, if both p and q are true), then the ordinance has been followed. If there is no snowfall of two or more inches (that is, if p is false), then the ordinance does not apply. Hence, in this case the ordinance is not violated whether there are cars parked overnight on city streets or not. Thus, the ordinance is violated only when the premise is true and the conclusion is false.

It may seem unnatural to regard a conditional statement $p \rightarrow q$ as being true whenever p is false. Indeed it seems reasonable to regard a conditional statement as being not applicable when the premise is false. But then the conditional $p \rightarrow q$ would be neither true nor false when p is false, and so $p \rightarrow q$ would no longer be a statement by our definition. For this reason logicians regard a conditional statement to be true if its premise is false. Therefore the truth table for a conditional statement is as shown below.

p	q	$p \rightarrow q$
T	T	T
T	F	F
F	T	T
F	F	T

The **biconditional** statement $p \leftrightarrow q$ means that $p \rightarrow q$ and $q \rightarrow p$. Thus, a biconditional statement is the conjunction of two conditional statements. We read the biconditional statement $p \leftrightarrow q$ as ''p if and only if q'' or ''p is necessary and sufficient for q.'' For instance, the following statements are biconditional statements:

An integer is even if and only if it is divisible by two.

For real numbers, $x = 0$ is necessary and sufficient for $x^2 = 0$.

We can obtain the truth table for $p \leftrightarrow q$ from the tables for $p \to q$ and $q \to p$.

p	q	$p \to q$	$q \to p$	$(p \to q) \wedge (q \to p)$
T	T	T	T	T
T	F	F	T	F
F	T	T	F	F
F	F	T	T	T

Thus, we see that the conditional statements $p \to q$ and $q \to p$ are both true only when p and q are both true or both false. Hence, the truth table for a biconditional statement is as shown below.

p	q	$p \leftrightarrow q$
T	T	T
T	F	F
F	T	F
F	F	T

In the first table of the preceding paragraph we can see that the conditional statements $p \to q$ and $q \to p$ do not always have the same truth values. Unfortunately it is a common mistake to confuse these two conditionals and to assume that one is true if the other is. Although these two statements are different, they are obviously related because both involve the same p and q. We call the statement $q \to p$ the **converse** of $p \to q$. There are two other conditional statements that are related to the conditional $p \to q$. The statement $\sim p \to \sim q$ is called the **inverse** of $p \to q$, and the statement $\sim q \to \sim p$ is called the **contrapositive** of $p \to q$.

EXAMPLE A.4

Form the converse, inverse, and contrapositive of the following statement about the real number x: If $x > 3$, then $x > 0$.

The given conditional statement is of the form $p \to q$, where p and q are the statements below.

$$p: x > 3 \quad \text{and} \quad q: x > 0$$

The converse, inverse, and contrapositive of the given statement are

If $x > 0$, then $x > 3$. (converse)

If $x \leq 3$, then $x \leq 0$. (inverse)

If $x \leq 0$, then $x \leq 3$. (contrapositive)

In this case the given statement and its contrapositive are true, but the converse and inverse are false. ■

EXAMPLE A.5   Form the converse, inverse, and contrapositive of the statement: If it isn't raining today, then I am going to the beach.
The desired statements are

> If I am going to the beach today, then it isn't raining. (converse)

> If it is raining today, then I am not going to the beach. (inverse)

> If I am not going to the beach today, then it is raining. (contrapositive)

In this case we must be careful not to read more into the given statement than it says. It is tempting to regard the given statement as a biconditional statement meaning that I am going to the beach today if it isn't raining and not going if it is. However, the given statement does *not* say that I am not going to the beach if it is raining. This is the inverse of the given statement. ■

EXERCISES A.1

In Exercises 1–12 determine if each sentence is a statement. If so, determine whether the statement is true or false.

1. Georgia is the southernmost state in the United States.

2. E.T., phone home.

3. If $x = 3$, then $x^2 = 9$.

4. Cats can fly.

5. What's the answer?

6. Ottawa is the capital of Canada.

7. Five is an odd integer, and seven is an even integer.

8. Six is an even number or seven is an even number.

9. Please be quiet until I am finished or leave the room.

10. California is the largest state in the United States and Vermont is the smallest.

11. Five is a positive number or zero is a positive number.

12. Go home and leave me alone.

Write the negations of the statements in Exercises 13–24.

13. $4 + 5 = 9$.

14. Christmas is celebrated on December 25.

15. Miss Piggy is not married to Kermit the Frog.

16. Elizabeth Taylor has not been married.

17. All birds can fly.

18. Some people are rich.

19. There is a man who weighs 400 pounds.

20. Every millionaire pays taxes.

21. Some students do not pass calculus.

22. All residents of Chicago love the Cubs.

23. Everyone enjoys cherry pie.

24. There are no farmers in South Dakota.

For each of the given pairs of statements p and q in Exercises 25–32, write: (a) the conjunction and (b) the disjunction. Then indicate which, if either, of these statements are true.

25. *p*: One is an even integer. *q*: Nine is a positive integer.

26. *p*: Oregon is in the midwest. *q*: Egypt is in the Orient.

27. *p*: The Atlantic is an ocean. *q*: The Nile is a river.

28. *p*: Cardinals are red. *q*: Robins are blue.

29. *p*: Birds have four legs. *q*: Rabbits have wings.

30. *p*: Oranges are fruit. *q*: Potatoes are vegetables.

31. *p*: Flutes are wind instruments. *q*: Timpani are string instruments.

32. *p*: Algebra is an English course. *q*: Accounting is a business course.

For each statement in Exercises 33–36 write: (a) the converse, (b) the inverse, and (c) the contrapositive.

33. If this is Friday, then I will go to the movies.

34. If I complete this assignment, then I will take a break.

35. If Hart doesn't run for the Senate, then he will run for President.

36. If I get an A on the final exam, then I'll get a B for the course.

A.2   Logical Equivalence

When analyzing a complicated sentence involving connectives, it is often useful to consider the simpler sentences that form it. The truth or falsity of the complicated sentence can then be determined by considering the truth or falsity of the simpler sentences. Consider, for instance, the sentence

$$x > 0 \text{ implies } y < 5 \text{ if and only if } y < 5 \text{ or } x \leq 0.$$

This sentence is formed from the two simpler sentences

$$p: x > 0 \quad \text{and} \quad q: y < 5.$$

We can write the given sentence symbolically as $(p \rightarrow q) \leftrightarrow (q \vee \sim p)$. Let us analyze the truth of this sentence in terms of the truth of p and q. This analysis can be conveniently carried out in the truth table shown below, where each row corresponds to a different pair of truth values for p and q.

p	q	$p \to q$	$\sim p$	$q \lor \sim p$	$(p \to q) \leftrightarrow (q \lor \sim p)$
T	T	T	F	T	T
T	F	F	F	F	T
F	T	T	T	T	T
F	F	T	T	T	T

Thus, we see that the original sentence

$$x > 0 \text{ implies } y < 5 \text{ if and only if } y < 5 \text{ or } x \le 0$$

is always true, regardless of the truth or falsity of the sentences $x > 0$ and $y < 5$.

EXAMPLE A.6

Use a truth table to analyze the symbolic sentence $p \lor [(p \land \sim q) \to r]$.

The truth table below shows that the sentence $p \lor [(p \land \sim q) \to r]$ is always true.

p	q	r	$\sim q$	$p \land \sim q$	$[(p \land \sim q) \to r]$	$p \lor [(p \land \sim q) \to r]$
T	T	T	F	F	T	T
T	T	F	F	F	T	T
T	F	T	T	T	T	T
T	F	F	T	T	F	T
F	T	T	F	F	T	T
F	T	F	F	F	T	T
F	F	T	T	F	T	T
F	F	F	T	F	T	T

In Example A.6 the symbolic sentence $p \lor [(p \land \sim q) \to r]$ is formed by joining the symbols p, q, and r by one or more connectives. The symbols p, q, and r that compose this sentence are called the **sentence variables.** Thus, for instance, the symbolic sentence

$$(r \land \sim t) \to (\sim s \lor q)$$

has q, r, s, and t as its sentence variables.

Sentences such as the one in Example A.6 that are true no matter what the truth values of the sentence variables are of special interest because of their use in constructing valid arguments. Such a sentence is said to be a **tautology.** Likewise, it is possible for a sentence to be false no matter what the truth values of the sentence variables; such a sentence is called a **contradiction.** Obviously the negation of a tautology is a contradiction and vice versa.

EXAMPLE A.7

As we can see in the truth table below, the symbolic sentence $(p \wedge \sim q) \wedge (\sim p \vee q)$ is a contradiction.

p	q	$\sim p$	$\sim q$	$(p \wedge \sim q)$	$(\sim p \vee q)$	$(p \wedge \sim q) \wedge (\sim p \vee q)$
T	T	F	F	F	T	F
T	F	F	T	T	F	F
F	T	T	F	F	T	F
F	F	T	T	F	T	F

Thus, $\sim[(p \wedge \sim q) \wedge (\sim p \vee q)]$, the negation of the given expression, is a tautology. ■

Two sentences with the same sentence variables are called **logically equivalent** if they have the same truth values for all possible truth values of the sentence variables. Thus, two sentences S and T with the same sentence variables are logically equivalent if and only if the biconditional $S \leftrightarrow T$ is a tautology. For example, we saw in the first truth table in this section that the biconditional $(p \to q) \leftrightarrow (q \vee \sim p)$ is a tautology. Therefore, the sentences $p \to q$ and $q \vee \sim p$ are logically equivalent.

EXAMPLE A.8

Show that the symbolic sentences $\sim(p \vee q)$ and $(\sim p \wedge \sim q)$ are logically equivalent. (This fact is called De Morgan's law.)

In order to prove that the two sentences are logically equivalent, it is sufficient to show that the columns in a truth table corresponding to these sentences are identical. Since this is the case in the truth table below, we conclude that $\sim(p \vee q)$ and $(\sim p \wedge \sim q)$ are logically equivalent.

p	q	$p \vee q$	$\sim(p \vee q)$	$\sim p$	$\sim q$	$\sim p \wedge \sim q$
T	T	T	F	F	F	F
T	F	T	F	F	T	F
F	T	T	F	T	F	F
F	F	F	T	T	T	T

■

In logical arguments it is often necessary to simplify a complicated sentence. In order for this simplification to result in a valid argument, it is essential that the replacement sentence be logically equivalent to the original sentence, for then the two sentences always have the same truth values. Thus, because of the logical equivalence shown in Example A.8, we can replace either of the sentences $\sim(p \vee q)$ or $\sim p \wedge \sim q$ by the other without affecting the validity of an argument.

We will close this section by stating a theorem containing several important logical equivalences that occur frequently in mathematical arguments. The proof of this theorem will be left to the exercises. Note the similarity between parts (a) through (h) of this theorem and parts (a) through (c) of Theorem 2.1 and parts (a) and (b) of Theorem 2.2.

THEOREM A.1   The following pairs of statements are logically equivalent.

(a) $p \wedge q$ and $q \wedge p$ (commutative law for conjunction)
(b) $p \vee q$ and $q \vee p$ (commutative law for disjunction)
(c) $(p \wedge q) \wedge r$ and $p \wedge (q \wedge r)$ (associative law for conjunction)
(d) $(p \vee q) \vee r$ and $p \vee (q \vee r)$ (associative law for disjunction)
(e) $p \vee (q \wedge r)$ and $(p \vee q) \wedge (p \vee r)$ (distributive law)
(f) $p \wedge (q \vee r)$ and $(p \wedge q) \vee (p \wedge r)$ (distributive law)
(g) $\sim(p \vee q)$ and $\sim p \wedge \sim q$ (De Morgan's law)
(h) $\sim(p \wedge q)$ and $\sim p \vee \sim q$ (De Morgan's law)
(i) $p \rightarrow q$ and $\sim q \rightarrow \sim p$ (law of the contrapositive)

EXERCISES A.2

In Exercises 1–10 construct a truth table for each symbolic sentence.

1. $(p \vee q) \wedge [\sim(p \wedge q)]$

2. $(\sim p \vee q) \wedge (\sim q \wedge p)$

3. $(p \vee q) \rightarrow (\sim p \wedge q)$

4. $(\sim p \wedge q) \rightarrow (\sim q \vee p)$

5. $(p \rightarrow q) \rightarrow (p \vee r)$

6. $p \rightarrow (\sim q \vee r)$

7. $(\sim q \wedge r) \leftrightarrow (\sim p \vee q)$

8. $\sim[p \wedge (q \vee r)]$

9. $[(p \vee q) \wedge r] \rightarrow [(p \wedge r) \vee q]$

10. $(r \wedge \sim q) \leftrightarrow (q \vee p)$

In Exercises 11–16 show that the given statements are tautologies.

11. $\sim p \vee p$

12. $(p \rightarrow q) \vee (\sim q \wedge p)$

13. $(\sim p \wedge q) \rightarrow \sim(q \rightarrow p)$

14. $\sim(\sim p \wedge q) \rightarrow (\sim q \vee p)$

15. $\sim[((p \rightarrow q) \wedge (\sim q \vee r)) \wedge (\sim r \wedge p)]$

16. $[(p \wedge q) \rightarrow r] \rightarrow [\sim r \rightarrow (\sim p \vee \sim q)]$

In Exercises 17–24 show that the given pairs of statements are logically equivalent.

17. p and $\sim(\sim p)$

18. p and $p \vee (p \wedge q)$

19. $\sim(p \rightarrow q)$ and $\sim q \wedge (p \vee q)$

20. $p \leftrightarrow q$ and $(\sim p \vee q) \wedge (\sim q \vee p)$

21. $p \rightarrow (q \rightarrow r)$ and $(p \wedge q) \rightarrow r$

22. $(p \rightarrow q) \rightarrow r$ and $(p \vee r) \wedge (q \rightarrow r)$

23. $(p \vee q) \rightarrow r$ and $(p \rightarrow r) \wedge (q \rightarrow r)$

24. $p \rightarrow (q \vee r)$ and $(p \rightarrow q) \vee (p \rightarrow r)$

25. Prove Theorem A.1 parts (a) and (b).

26. Prove Theorem A.1 parts (c) and (d).

27. Prove Theorem A.1 parts (e) and (f).

28. Prove Theorem A.1 part (h).

29. Prove Theorem A.1 part (i).

30. The sentence $[(p \rightarrow q) \wedge \sim q] \rightarrow \sim p$ is called **modus tollens.** Prove that modus tollens is a tautology.

31. The sentence $[p \wedge (p \rightarrow q)] \rightarrow q$ is called **modus ponens.** Prove that modus ponens is a tautology.

32. The sentence $[(p \vee q) \wedge \sim p] \rightarrow q$ is called the law of **disjunctive syllogism.** Prove that disjunctive syllogism is a tautology.

33. Define a new connective named ''exclusive or'' and denoted $\veebar$ by regarding $p \veebar q$ to be true if and only if exactly one of p or q is true.

   (a) Write a truth table for ''exclusive or.''

   (b) Show that $p \veebar q$ is logically equivalent to $\sim(p \leftrightarrow q)$.

34. The **Sheffer stroke** is a connective denoted | and defined by the truth table below.

p	q	$p \mid q$
T	T	F
T	F	T
F	T	T
F	F	T

The following parts prove that all of the basic connectives can be written using only the Sheffer stroke.

   (a) Show that $p \mid p$ is logically equivalent to $\sim p$.

   (b) Show that $(p \mid p) \mid (q \mid q)$ is logically equivalent to $p \vee q$.

   (c) Show that $(p \mid q) \mid (p \mid q)$ is logically equivalent to $p \wedge q$.

   (d) Show that $p \mid (q \mid q)$ is logically equivalent to $p \rightarrow q$.

A.3   Methods of Proof

Mathematics is probably the only human endeavor which places such a central emphasis on the use of logic and proof. Being able to think logically and to read proofs certainly increases mathematical understanding, but more importantly these skills enable us to apply mathematical ideas in new situations. In this section we will discuss basic methods of proof so that the reader will have a better understanding of the logical framework in which proofs are written.

   A **theorem** is a mathematical statement that is true. Theorems are essentially conditional statements, although the wording of a theorem may obscure this fact. For instance, Theorem 1.3 is worded

A set with n elements has exactly 2^n subsets.

In this wording the theorem does not seem to be a conditional statement; yet we can express this theorem as a conditional statement by writing

If S is a set with n elements, then S has exactly 2^n subsets.

When the theorem is expressed as a conditional statement, the premise and conclusion of the conditional statement are called the **hypothesis** and **conclusion** of the theorem.

By a **proof** of a theorem we mean a logical argument that establishes the theorem to be true. The most natural form of proof is the **direct proof.** Suppose that we wish to prove the theorem $p \to q$. Since $p \to q$ is true whenever p is false, we need only show that whenever p is true, so is q. Therefore, in a direct proof we assume that the hypothesis of the theorem, p, is true and demonstrate that the conclusion, q, is true. It then follows that $p \to q$ is true. The following example illustrates this technique.

EXAMPLE A.9

Suppose that we wish to prove the theorem: If n is an even integer, then n^2 is an even integer.

To prove this result by a direct proof, we assume the hypothesis and prove the conclusion. Accordingly, we assume that n is an even integer and will prove that n^2 is even. Since n is even, $n = 2m$ for some integer m. Then

$$n^2 = (2m)^2 = 4m^2 = 2(2m^2),$$

so that n^2 is divisible by 2. Hence, n^2 is an even integer.   ■

EXAMPLE A.10

Consider the theorem: If x is a real number and $x^2 - 1 = 0$, then $x = -1$ or $x = 1$.

Since $x^2 - 1 = 0$, factoring gives $(x + 1)(x - 1) = 0$. But if the product of two real numbers is 0, at least one of them must be 0. Consequently, $x + 1 = 0$ or $x - 1 = 0$. In the first case $x = -1$, and in the second $x = 1$. Thus, $x = -1$ or $x = 1$.   ■

The argument in Example A.10 uses the **law of syllogism,** which states

$$[(p \to q) \land (q \to r)] \to (p \to r).$$

For if we let p, q, r, and s be the statements

$$p: x^2 - 1 = 0$$
$$r: (x + 1)(x - 1) = 0$$
$$s: x + 1 = 0 \text{ or } x - 1 = 0$$
$$q: x = -1 \text{ or } x = 1,$$

then the argument in Example A.10 shows that

$$(p \rightarrow r) \wedge (r \rightarrow s) \wedge (s \rightarrow q).$$

Hence, by two applications of the law of syllogism we conclude that $p \rightarrow q$, that is, the theorem is proved.

Another type of proof is based on the law of the contrapositive, which states that the sentences $p \rightarrow q$ and $\sim q \rightarrow \sim p$ are logically equivalent. To prove the theorem $p \rightarrow q$ by this method, we give a direct proof of the statement $\sim q \rightarrow \sim p$ by assuming $\sim q$ and proving $\sim p$. The law of the contrapositive then allows us to conclude that $p \rightarrow q$ is also true.

EXAMPLE A.11

Suppose we wish to prove the theorem: If $x + y > 100$, then $x > 50$ or $y > 50$.

The theorem to be proved has the form $p \rightarrow q$, where p and q denote the statements below.

$$p\!: x + y > 100 \qquad \text{and} \qquad q\!: x > 50 \text{ or } y > 50$$

We will establish the contrapositive of the desired result, which is $\sim q \rightarrow \sim p$. Consequently, we will assume that $\sim q$ is true and show that $\sim p$ is true. Using Example A.8, we see that $\sim q$ and $\sim p$ are the sentences

$$\sim q\!: x \leq 50 \text{ and } y \leq 50 \qquad \text{and} \qquad \sim p\!: x + y \leq 100.$$

Suppose that $x \leq 50$ and $y \leq 50$. Then

$$x + y \leq 50 + 50 = 100.$$

Hence, $\sim p$ is proved, that is, $\sim q \rightarrow \sim p$. It now follows from the law of the contrapositive that $p \rightarrow q$ is true. ▪

EXAMPLE A.12

We will prove the theorem: If n is an integer and n^2 is odd, then n is odd.

The contrapositive of this theorem is: If an integer n is not odd, then n^2 is not odd. This sentence can also be expressed in the form: If n is an even integer, then n^2 is even. But this is precisely the result proved in Example A.9, and so its contrapositive, which is the theorem to be proved, is also true. ▪

A very different style of proof is a **proof by contradiction.** In this method of proof we prove the theorem $p \rightarrow q$ by assuming p and $\sim q$ are true and deducing a false statement r. Since $(p \wedge \sim q) \rightarrow r$ is true but r is false, we can conclude that the premise $p \wedge \sim q$ of this conditional sentence is false. But then its negation $\sim(p \wedge \sim q)$ is true, which is logically equivalent to the desired statement $p \rightarrow q$. (See Exercise 1.)

EXAMPLE A.13   We will prove the theorem: If n is the sum of the squares of two odd integers, then n is not a perfect square.

Proving this theorem by contradiction seems natural because the theorem expresses a negative idea (that n is *not* a perfect square). Thus, when we deny the conclusion, we obtain the positive statement that n is a perfect square.

Accordingly we will use a proof by contradiction. Therefore we assume the hypothesis and deny the conclusion, and so we assume both that n is the sum of the squares of two odd integers and that n is a perfect square. Since n is a perfect square, we have $n = m^2$ for some integer m. But also n is the sum of the squares of two odd integers. Thus, since an odd integer is one more than an even integer, we can express n in the form

$$n = (2r + 1)^2 + (2s + 1)^2$$

for some integers r and s. It follows that

$$n = (2r + 1)^2 + (2s + 1)^2 = (4r^2 + 4r + 1) + (4s^2 + 4s + 1)$$
$$= 4(r^2 + s^2 + r + s) + 2,$$

so that n is even. Thus, $m^2 = n$ is even. It is easy to see that if m^2 is even, then m must also be even. (Note that this result is the converse of the theorem proved in Example A.9, so we cannot deduce it from Example A.9. This result can be easily proved, however, as in Exercise 3.) We deduce, therefore, that m is even, and so $m = 2p$ for some integer p. Thus, $n = m^2 = (2p)^2 = 4p^2$ is divisible by 4. But we saw above that $n = 4(r^2 + s^2 + r + s) + 2$, which is not divisible by 4. Hence, we have derived a false statement, namely that n is both divisible by 4 and not divisible by 4. Thus, assuming the hypothesis and denying the conclusion has led to a false statement. It therefore follows that if the hypothesis is true, the conclusion must be true also. Consequently the theorem has been proved.   ■

EXAMPLE A.14   Show that there is no rational number r such that $r^2 = 2$. (Recall that a rational number is one that can be written as the quotient of two integers.)

The theorem to be proved can be written as the conditional sentence: If r is a rational number, then $r^2 \neq 2$. Again proving this theorem by contradiction seems natural because the theorem expresses a negative idea (that r^2 is *not* equal to 2). Thus, when we deny the conclusion, we obtain the positive statement that there is a rational number r such that $r^2 = 2$.

Accordingly we will use a proof by contradiction. Thus, we assume the hypothesis and deny the conclusion, and so we assume that there is a rational number r such that $r^2 = 2$. Because r is a rational number, it can be expressed in the form $\dfrac{m}{n}$, where m and n are integers. Moreover, we may choose m and n to have no common factors greater than 1, so that the fraction $\dfrac{m}{n}$ is in lowest terms.

Then we have $\left(\dfrac{m}{n}\right)^2 = 2$, from which it follows that $m^2 = 2n^2$. Hence, m^2 is an even integer. Thus, by Exercise 3, m must be even, that is, $m = 2p$ for some integer p. Substituting this value for m in the equation $m^2 = 2n^2$ yields $4p^2 = 2n^2$, so that $2p^2 = n^2$. Hence, n^2 is even, and so it follows as above that n is even. But then both m and n are even, that is, m and n have a common factor of 2. This fact contradicts our choice of m and n as having no common factors greater than 1, and so we deduce that the conclusion to our theorem must be true. Thus, the theorem is proved. ■

We have discussed three basic methods of proof in this section, the direct proof, proof of the contrapositive, and proof by contradiction. There are other types of proofs as well. One method of proof that is quite important in discrete mathematics is proof by induction, which is discussed in Section 2.5. Another type of proof is a proof by cases, in which the theorem to be proved is subdivided into parts, each of which is proved separately. The next example demonstrates this technique.

EXAMPLE A.15   Show that if n is an integer, then $n^3 - n$ is even.

Since every integer n is either even or odd, we will consider these two cases.

Case 1: n is even

Then $n = 2m$ for some integer m. Therefore,

$$n^3 - n = (2m)^3 - 2m = 8m^3 - 2m = 2(4m^3 - m),$$

which is even.

Case 2: n is odd

Then $n = 2m + 1$ for some integer m. Hence,

$$
\begin{aligned}
n^3 - n &= (2m + 1)^3 - (2m + 1) \\
&= (8m^3 + 12m^2 + 6m + 1) - (2m + 1) \\
&= 8m^3 + 12m^2 + 4m \\
&= 2(4m^3 + 6m^2 + 2m),
\end{aligned}
$$

which is even.

Because $n^3 - n$ is even in either case, we conclude that $n^3 - n$ is even for all integers n. ■

To close this section, we will briefly consider the problem of disproving a statement $p \rightarrow q$, that is, of showing that it is false. Because a conditional statement is false only when its premise is true and its conclusion is false, we must find an instance in which p is true and q is false. Such an instance is called a **counter-**

example to the statement. For example, consider the statement: If p is a prime positive integer, then p is odd. (Recall that a positive integer is called **prime** if it is larger than 1 and its only positive integer divisors are itself and 1.) To disprove this statement, we must find a counterexample, that is, a prime positive integer that is not odd. In this case 2 is the desired counterexample, since 2 is a prime positive integer (its only positive integer divisors are 1 and 2) and 2 is even.

EXERCISES A.3

1. Prove that $\sim(p \wedge \sim q)$ is logically equivalent to $p \rightarrow q$.

2. Prove that the law of syllogism is a tautology.

3. Prove that if m is an integer and m^2 is even, then m is even. (*Hint:* Prove the contrapositive.)

4. Prove as in Example A.14 that there is no rational number r such that $r^2 = 3$.

Prove the theorems in Exercises 5–12. Assume that all the symbols used in these exercises represent positive integers.

5. If a divides b, then ac divides bc for any c.

6. If ac divides bc, then a divides b.

7. If a divides b and b divides c, then a divides c.

8. If a divides b, then $a \leq b$.

9. If p and q are primes and p divides q, then $p = q$.

10. If a divides b and a divides $b + 2$, then $a = 1$ or $a = 2$.

11. If xy is even, then x is even or y is even.

12. For all positive integers n greater than 10, $12(n - 2) < n^2 - n$.

Prove or disprove the results in Exercises 13–22. Assume that all the numbers mentioned in these exercises are integers.

13. The sum of two odd integers is odd.

14. The product of two odd integers is odd.

15. If $ac = bc$, then $a = b$.

16. If 3 divides xy, then 3 divides x or 3 divides y.

17. If 6 divides xy, then 6 divides x or 6 divides y.

18. If 3 divides x and 3 divides y, then 3 divides $ax + by$.

19. If a and b are odd, then $a^2 + b^2$ is even.

20. If a and b are odd, then $a^2 + b^2$ is not divisible by 4.

21. For all integers n, n is odd if and only if 8 divides $n^2 - 1$.

22. The product of two integers is odd if and only if both of the integers are odd.

23. Prove or disprove: For every positive integer n, $n^2 + n + 41$ is prime.

24. Prove that in any set of three consecutive odd positive integers other than 3, 5, and 7, at least one number is not prime.

25. Prove that for each positive integer n, $n^2 - 2$ is not divisible by 3.

26. Prove that for each positive integer n, $n^4 - n^2$ is divisible by 6.

27. Prove that if p is a prime positive integer, then $\log_{10} p$ is not expressible as the quotient of two integers.

28. Prove that there are infinitely many primes.

Suggested Readings

1. Lucas, John. *An Introduction to Abstract Mathematics*. Belmont, CA: Wadsworth Publishing Company, 1986.
2. Kenelly, John W. *Informal Logic*. Boston: Allyn and Bacon, 1967.
3. Mendelson, Elliott. *Introduction to Mathematical Logic*. Princeton, N.J.: Van Nostrand, 1964.
4. Polya, G. *How to Solve It*. 2nd ed. Garden City, N.Y.: Doubleday and Company, 1957.
5. Solow, Daniel. *How to Read and Do Proofs*. New York: John Wiley and Sons, 1982.

BIBLIOGRAPHY

Aho, Alfred V., John E. Hopcroft, and Jeffrey D. Ullman. *Data Structures and Algorithms*. Reading, MA: Addison-Wesley, 1983.

Aho, Alfred V., John E. Hopcroft, and Jeffrey D. Ullman. *The Design and Analysis of Computer Algorithms*. Reading, MA: Addison-Wesley, 1974.

Anderson, Ian. *A First Course in Combinatorial Mathematics*. London: Oxford University Press, 1974.

Behzad, Mehdi, Gary Chartrand, and Linda Lesniak-Foster. *Graphs & Digraphs*. Boston: Prindle, Weber, & Schmidt, 1979.

Bogart, Kenneth P. *Introductory Combinatorics*. Marsfield, MA: Pitman, 1983.

Bondy, J. A., and U. S. R. Murty. *Graph Theory with Applications*. New York: North Holland, 1976.

Busacker, Robert G., and Thomas L. Saaty. *Finite Graphs and Networks: An Introduction with Applications*. New York: McGraw-Hill, 1965.

Chachra, Vinod, Prabhakar M. Ghare, and James M. Moore. *Applications of Graph Theory Algorithms*. New York: North Holland, 1979.

Chartrand, Gary. *Graphs as Mathematical Models*. Boston: Prindle, Weber & Schmidt, 1977.

Christofides, Nicos. *Graph Theory: An Algorithmic Approach*. New York: Academic Press, 1975.

Doerr, Alan, and Kenneth Levasseur. *Applied Discrete Structures for Computer Science*. Chicago: Science Research Associates, 1985.

Even, Shimon. *Graph Algorithms*. Rockville, MD: Computer Science Press, 1979.

Gersting, Judith. *Mathematical Structures*. New York: W. H. Freeman, 1982.

Grimaldi, Ralph P. *Discrete and Combinatorial Mathematics*. Reading, MA: Addison-Wesley, 1985.

Harary, Frank. *Graph Theory*. Reading, MA: Addison-Wesley, 1969.

Horowitz, Ellis, and Sartaj Sahni. *Fundamentals of Computer Algorithms*. Rockville, MD: Computer Science Press, 1984.

Horowitz, Ellis, and Sartaj Sahni. *Fundamentals of Data Structures in Pascal*. Rockville, MD: Computer Science Press, 1984.

Hu, T. C. *Combinatorial Algorithms*. Reading, MA: Addison-Wesley, 1982.

Johnsonbaugh, Richard. *Discrete Mathematics*. revised edition. New York: Macmillan, 1984.

Knuth, Donald E. *The Art of Computer Programming, vol. 1: Fundamental Algorithms*. 2nd edition. Reading, MA: Addison-Wesley, 1973.

Knuth, Donald E. *The Art of Computer Programming, vol. 2: Seminumerical Algorithms,* 2nd edition. Reading, MA: Addison-Wesley, 1981.

Knuth, Donald E. *The Art of Computer Programming, vol. 3: Searching and Sorting*. Reading, MA: Addison-Wesley, 1973.

Kolman, Bernard, and Robert C. Busby. *Discrete Mathematical Structures for Computer Science*. Englewood Cliffs, NJ: Prentice-Hall, 1984.

Lawler, Eugene L. *Combinatorial Optimization: Networks and Matroids*. New York: Holt, Rinehart, and Winston, 1976.

Liu, C. L. *Elements of Discrete Mathematics*. 2nd edition. New York: McGraw-Hill, 1985.

Liu, C. L. *Introduction to Combinatorial Mathematics*. New York: McGraw-Hill, 1968.

Minieka, Edward. *Optimization Algorithms for Networks and Graphs*. New York: Marcel Dekker, 1978.

Molluzzo, John C., and Fred Buckley. *A First Course in Discrete Mathematics*. Belmont, CA: Wadsworth, 1986.

Mott, Joe L., Abraham Kandel, and Theodore P. Baker. *Discrete Mathematics for Computer Scientists & Mathematicians*. 2nd edition. Englewood Cliffs, NJ: Reston, 1986.

Nijenhuis, Albert, and Herbert S. Wilf. *Combinatorial Algorithms*. 2nd edition. New York: Academic Press, 1978.

Ore, Oystein. *Graphs and Their Uses*. New York: L. W. Singer, 1963.

Pfleger, Shari Lawrence, and David W. Straight. *Introduction to Discrete Mathematics*. New York: John Wiley, 1985.

Polimeni, Albert D., and Joseph H. Straight. *Foundations of Discrete Mathematics*. Monterey, CA: Brooks/Cole, 1985.

Reingold, Edward, Jurg Nievergelt, and Narsingh Deo. *Combinatorial Algorithms*. Englewood Cliffs, NJ: Prentice-Hall, 1977.

Roberts, Fred S. *Applied Combinatorics*. Englewood Cliffs, NJ: Prentice-Hall, 1984.

Roberts, Fred S. *Discrete Mathematical Models*. Englewood Cliffs, NJ: Prentice-Hall, 1976.

Roberts, Fred S. *Graph Theory and Its Applications to Problems of Society*. Philadelphia: Society for Industrial and Applied Mathematics, 1978.

Roman, Steven. *An Introduction to Discrete Mathematics*. Philadelphia: Saunders, 1986.

Ross, Kenneth A., and Charles B. Wright. *Discrete Mathematics*. Englewood Cliffs, NJ: Prentice-Hall, 1985.

Ryser, Herbert John. *Combinatorial Mathematics*. Washington, DC: Mathematical Association of America, 1963.

Sedgewick, Robert. *Algorithms*. Reading, MA: Addison-Wesley, 1983.

Stanat, Donald F., and David F. McAllister. *Discrete Mathematics in Computer Science*. Englewood Cliffs, NJ: Prentice-Hall, 1977.

Tarjan, Robert Endre. *Data Structures and Network Algorithms*. Philadelphia: Society for Industrial and Applied Mathematics, 1983.

Tremblay, J. P., and R. Manohar. *Discrete Mathematical Structures with Applications to Computer Science*. New York: McGraw-Hill, 1975.

Tucker, Alan. *Applied Combinatorics*. 2nd edition. New York: John Wiley, 1984.

Wirth, Nicklaus. *Algorithms + Data Structures = Programs*. Englewood Cliffs, NJ: Prentice-Hall, 1976.

ANSWERS TO ODD-NUMBERED EXERCISES

Chapter 1

Exercises 1.1 (page 9)

1. 33; *A-B-D-F-G* or *A-C-E-F-G*   **3.** 43; *B-D-E-G*
5. 20.7; *A-D-H-K*   **7.** 2.1; *A-C-E-H-J*

9. 23; *A-D-F-G*

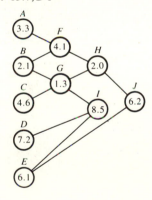

11. 24; *B-C-F-G*

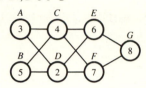

13. 15.7; *D-I*

15. 0.29; *E-F-C-D-I*

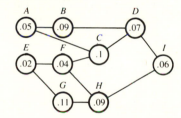

17. 27 minutes   **19.** 15 days

Exercises 1.2 (page 16)

1. 120   **3.** 6720   **5.** 28   **7.** 840   **9.** 604,800   **11.** 25.2
13. 720   **15.** 56   **17.** 362,880   **19.** 720   **21.** 288
23. 210   **25.** 60   **27.** 20,118,067,200   **29.** 60

Exercises 1.3 (page 22)

1. F   **3.** F   **5.** F   **7.** T   **9.** F   **11.** F   **13.** F
15. yes; 26   **17.** yes; 40   **19.** 1, 2, 4, 5, 6, 7, 8, 11, 12; 54
21. 128   **23.** 32   **25.** $n - m + 1$ **27.** 31   **29.** 12.7 days

Exercises 1.4 (page 31)

1. yes; 2   **3.** no   **5.** no   **7.** 3, 13; 5, 13
9. $-7, 3, 11, 3; -1, 0, 5, 3$ **11.** 001101   **13.** 101101

15.

k	j	a_1	a_2	a_3
1		1	0	1
2		1	0	1
2	1	1	1	1
2	2	0	1	1

17.

k	j	a_1	a_2	a_3	a_4
1		1	1	0	1
2		1	1	0	1
3		1	1	0	1
3	1	1	1	1	1
3	2	0	1	1	1
3	3	0	0	1	1

19. 58 minutes; 0.8 seconds   **21.** 385,517 years; 6.4 seconds   **23.** $3n + 1$

Chapter 2

Exercises 2.1   (page 39)
1. $\{1, 2, 3, 4, 5, 6, 7, 8, 9\}$; $\{3, 5\}$; $\{2, 7, 8\}$; $\{1, 4, 6, 9\}$; $\{2, 7, 8\}$
3. $\{1, 2, 3, 4, 7, 8, 9\}$; $\varnothing$; A; $\{3, 5, 6, 7\}$; $\{1, 2, 4, 5, 6, 8, 9\}$
5. $\{(1, 7), (1, 8), (2, 7), (2, 8), (3, 7), (3, 8), (4, 7), (4, 8)\}$ **7.** $\{(a, x), (a, y), (a, z), (e, x), (e, y), (e, z)\}$
9.
11.

13. $A = \{1\}, B = \{2\}, C = \{1, 2\}$ **15.** $A = \{1, 2\}, B = \{1, 3\}, C = \{2, 3\}$ **17.** $\varnothing$ **19.** $A - B$ **21.** $B - A$
23. $A - B$ **25.** mn **27.** $B \subseteq A$ **39.** $A = \{1\}, B = \{2\}, C = \{3\}, D = \{4\}$

Exercises 2.2   (page 45)
1. symmetric and transitive   **3.** reflexive, symmetric, and transitive
5. reflexive and symmetric   **7.** reflexive, symmetric, and transitive
9. reflexive, symmetric, and transitive   **11.** reflexive and transitive   **13.** even integers and odd integers
15. sets $\{n: p$ is the largest prime dividing $n\}$, p any prime   **17.** For any prime p, the set of integers greater than 1 that are divisible by p but not divisible by any prime greater than p.
19. $\{(1, 1), (5, 5), (1, 5), (5, 1), (2, 2), (4, 4), (2, 4), (4, 2), (3, 3)\}$ **23.** There may be no element related to x; that is, $x R y$ may not be true for any y.   **25.** 2^{n^2} **27.** $2^{n-1} - 1$ **29.** 15

Exercises 2.3   (page 51)
1. $q = 7; r = 4$ **3.** $q = 0; r = 25$ **5.** $q = 9; r = 0$
7. $q = 8; r = 9$ **9.** $p \equiv q$ **11.** $p \not\equiv q$ **13.** $p \not\equiv q$ **15.** $p \equiv q$
17. [2]   **19.** [4]   **21.** [6]   **23.** [1]   **25.** [2]   **27.** [8]   **29.** [4]   **31.** [4]   **33.** [2]   **35.** [11]
37. 8 P.M.   **39.** 9   **41.** $-10{,}224$ and $29{,}202$ **43.** No, $10 \in A$ but $10 \notin B$
45. $3 R 11$ and $6 R 10$, but both $9 R 21$ and $18 R 110$ are false.

Exercises 2.4   (page 62)
1. function with domain X **3.** not a function with domain X
5. function with domain X **7.** not a function with domain X **9.** not a function with domain X **11.** function with domain X **13.** 8   **15.** $\frac{1}{4}$ **17.** 2   **19.** -9 **21.** 3
23. 0   **25.** -4 **27.** -5 **29.** 5.21   **31.** -0.22 **33.** 0.62   **35.** 9.97   **37.** $8x + 11; 8x - 5$
39. $5(2^x) + 7; 2^{5x+7}$ **41.** $|x| \log_2 |x|; |x \log_2 x|$ **43.** $x^2 - 2x + 1; x^2 - 1$ **45.** one-to-one, not onto
47. one-to-one and onto   **49.** onto, not one-to-one   **51.** one-to-one, not onto [neither]   **53.** $\frac{x}{5}$ **55.** $-x$ **57.** x^3
59. does not exist   **61.** $Y = \{x \in X: x > 0\}; g^{-1}(x) = -1 + \log_2\left(\dfrac{x}{3}\right)$ **63.** n^m

Exercises 2.5   (page 71)

1. arithmetic progression with common difference 2   **3.** geometric progression with common ratio -3 **5.** geometric progression with common ratio $\frac{1}{2}$ **7.** neither an arithmetic nor a geometric progression   **9.** 43   **11.** 12   **13.** $-39{,}366$ **15.** 102   **17.** 110   **19.** $\frac{364}{81}$ **21.** 0.52217   **23.** 915   **27.** 1, 1, 2, 3, 5, 8, 13, 21, 34, 55   **29.** \$125   **31.** approximately \$39,093.47   **33.** No base for the induction was established.   **35.** The proof of the inductive step is faulty since $x - 1$ and $y - 1$ need not be positive integers. (If $x = 1$, then $x - 1 = 0$; and if $y = 1$, then $y - 1 = 0$.) Hence, the induction hypothesis cannot be applied to $x - 1$ and $y - 1$.   **37.** Any constant is a polynomial; and if S is a polynomial, k is a positive integer, and c is a constant, then $S + cx^k$ is a polynomial.   **53.** Mr. and Mrs. Lewis both shook n hands.

Exercises 2.6   (page 80)

1. 64   **3.** 128   **5.** 256   **7.** 21   **9.** 792   **11.** 20   **13.** 120   **15.** $\frac{52!}{13!\,39!}$ **33.** $\frac{n(n-3)}{2}$

Chapter 3

Exercises 3.1   (page 90)

1. $V = \{A, B, C, D\}; E = \{\{A, B\}, \{A, C\}, \{B, C\}, \{B, D\}, \{C, D\}\}$
3. $V = \{F, G, H\}; E = \{\{F, G\}\}$
5. $V = \{A, B, C, D\}; E = \{\{A, C\}, \{A, D\}, \{B, C\}, \{B, D\}\}$

7. ·   **9.**   **11.**   **13.** yes   **15.** no   **17.** no   **19.** yes

23.   **25.** $B, C, D, E, 4; A, C, F, 3$ **27.** $B, 1; A, 1$ **29.** $B, C, E, F, 4; A, C, D, 3$

31.   **35.** $3, 6, 10, \frac{n(n-1)}{2}$ **37.** 10

39.
$$\begin{bmatrix} 0 & 1 & 1 & 1 \\ 1 & 0 & 1 & 1 \\ 1 & 1 & 0 & 1 \\ 1 & 1 & 1 & 0 \end{bmatrix} \begin{array}{l} v_1: v_2, v_3, v_4 \\ v_2: v_1, v_3, v_4 \\ v_3: v_1, v_2, v_4 \\ v_4: v_1, v_2, v_3 \end{array}$$

41.
$$\begin{bmatrix} 0 & 1 & 1 & 1 \\ 1 & 0 & 1 & 0 \\ 1 & 1 & 0 & 1 \\ 1 & 0 & 1 & 0 \end{bmatrix} \begin{array}{l} v_1: v_2, v_3, v_4 \\ v_2: v_1, v_3 \\ v_3: v_1, v_2, v_4 \\ v_4: v_1, v_3 \end{array}$$

43.
$$\begin{bmatrix} 0 & 1 & 0 \\ 1 & 0 & 0 \\ 0 & 0 & 0 \end{bmatrix} \begin{array}{l} v_1: v_2 \\ v_2: v_1 \\ v_3: \text{none} \end{array}$$

45.   **47.**   **49.**   **51.**

53. yes   **55.** no

57.
$$A + B = \begin{bmatrix} 5 & 12 & 3 \\ -7 & 7 & -2 \\ 17 & 8 & 3 \end{bmatrix} = B + A, \ B + C = \begin{bmatrix} -2 & 1 & 2 \\ 5 & 4 & -4 \\ 16 & 7 & 7 \end{bmatrix}$$

$$A + (B + C) = \begin{bmatrix} 3 & 9 & 4 \\ -2 & 5 & -4 \\ 24 & 7 & 7 \end{bmatrix} = (A + B) + C, \ A + Z = A$$

59.

$$AB = \begin{bmatrix} 16 & 4 \\ 5 & 1 \\ 5 & 5 \end{bmatrix}, BC = \begin{bmatrix} 2 & -1 \\ 0 & 2 \\ 4 & 0 \end{bmatrix}$$

$$(AB)C = \begin{bmatrix} 20 & 4 \\ 6 & 1 \\ 10 & 5 \end{bmatrix} = A(BC), AS = \begin{bmatrix} 0 & 8 & 20 \\ -4 & 0 & 8 \\ 12 & 16 & 4 \end{bmatrix} = SA$$

61.

$$AB = \begin{bmatrix} 3 & 8 & 23 \\ 1 & 2 & 3 \end{bmatrix}, AC = \begin{bmatrix} 4 & 1 & 9 \\ -4 & 3 & -2 \end{bmatrix}$$

$$AB + AC = \begin{bmatrix} 7 & 9 & 32 \\ -3 & 5 & 1 \end{bmatrix} = A(B + C), B + C = \begin{bmatrix} 1 & 3 & 5 \\ 1 & 1 & 7 \\ -2 & 2 & -3 \end{bmatrix}$$

$$BC = \begin{bmatrix} -8 & 7 & 0 \\ -14 & 5 & -11 \\ 4 & 0 & 5 \end{bmatrix} \quad CB = \begin{bmatrix} -1 & -2 & -5 \\ 5 & 8 & 11 \\ -9 & -11 & -5 \end{bmatrix}$$

$$B^2 = \begin{bmatrix} 5 & 10 & 19 \\ 3 & 2 & -9 \\ 0 & 2 & 10 \end{bmatrix}$$

65. n

67. The vertices corresponding to primes p where $2p > n$

Exercises 3.2 (page 103)

1. a graph **3.** not a graph **5.** not a graph **7.** not a graph **9.** parallel edges: none; loops: a, c **11.** parallel edges: a, b; loops: d **13.** parallel edges: a, b, c; loops: f **15.** parallel edges: a, b, c, d; loops: none **17.** (i) c (1); a, c (2); a, c, b (3) (ii) c (1) (iii) c; c; c (iv) a (1); b (1) **19.** (i) e (1); b, c (2); a, f, b, a, d (5) (ii) e (1); a, d (2); b, c (2); a, f, c (3); b, f, d (3) (iii) e; b, c; a, d (4); a, e, c, f (4); b, f, d, e (4); a, d, e (3); c, d, f (3); b, c, e (3); a, b, f (3)

21. (a) **(b)** **(c)**

23. no **25.** yes **27.** no **29.** yes **31.** no **33.** no **35.** no **37.** no **39.** no **41.** yes, $a, b, e, k, v, s, t, u, w, r, h, g, j, q, p, n, m, i, f, d, c$ **43.** no **45.** yes, $a, b, d, h, j, i, g, e, f, c$ **47.** no **49.** no **51.** yes **53.** yes **55.** yes **57.** no

59. (a) **(b)** **(c)** **(d)**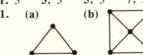

61. m and n even **65.** $n - 1$

Exercises 3.3 (page 117)

1. 3 **3.** 3 **5.** 3 **7.** 2 **9.** There are no edges. **11.** **(a)** **(b)**

13. Step 1. If any vertex is uncolored, color it red. Otherwise, stop.
Step 2. If any uncolored vertices are adjacent to red vertices, color them blue.
Step 3. If any uncolored vertices are adjacent to blue vertices, color them red and go to step 2. Otherwise, go to step 1.

15.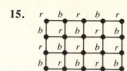

17.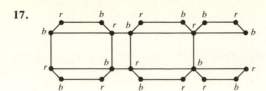

19. 5

21.

23.

25. n^n

27.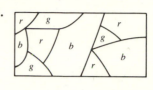

29.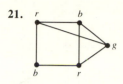

31. Three separate meeting times are needed with finance and agriculture meeting at the same time and likewise for budget and labor. **33.** 3 **35.** yes

37.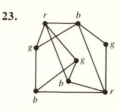

Exercises 3.4 (page 131)

1. $V = \{A, B, C, D\}; E = \{(A, B), (B, D), (C, A), (C, D)\}$

3. $V = \{A, B, C, D\}; E = \{(A, B), (B, A), (A, C), (C, D), (D, C)\}$

5.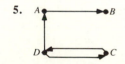

7. A• •B

C• •D

9.

11.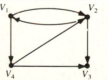

13.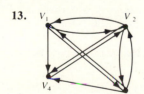

15.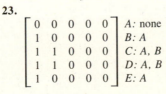

17. For vertex A: B, C; B, D; 2; 2.
For vertex B: A, C, D; A, C; 3; 2.

19. For vertex A: B, C, D, E; none; 4; 0.
For vertex B: C, D; A; 2; 1.

21.
$$\begin{bmatrix} 0 & 1 & 0 & 1 \\ 1 & 0 & 1 & 0 \\ 1 & 1 & 0 & 0 \\ 0 & 1 & 0 & 0 \end{bmatrix}$$
A: B, D
B: A, C
C: A, B
D: B

23.
$$\begin{bmatrix} 0 & 0 & 0 & 0 & 0 \\ 1 & 0 & 0 & 0 & 0 \\ 1 & 1 & 0 & 0 & 0 \\ 1 & 1 & 0 & 0 & 0 \\ 1 & 0 & 0 & 0 & 0 \end{bmatrix}$$
A: none
B: A
C: A, B
D: A, B
E: A

25.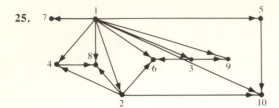

27. yes   **29.**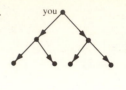

31.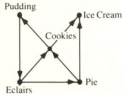

33. *(i)* *A, B* (1); *A, C, D, B* (3); *A, D, C, D, B* (4)
   (ii) *A, B* (1); *A, D, B* (2); *A, C, B* (2); *A, C, D, B* (3); *A, D, C, B* (3)
   (iii) *A, B; A, C, D, B; A, D, B*
   (iv) *A, B, A* (2); *C, D, C* (2); *A, C, B, A* (3); *A, D, B, A* (3); *A, C, D, B, A* (4); *A, D, C, B, A* (4)
35. *(i)* *A, B* (1); *A, B, B* (2); *A, B, B, B* (3)
   (ii) *A, B* (1)
   (iii) *A, B; A, B; A, B*
   (iv) *A, B, A* (2); *B, B* (1)
37. yes   **39.** yes   **41.**   **43.** $n - 1$ **45.** no   **47.** yes   **49.** yes

51. $(A, B), (B, E), (E, D), (D, C), (C, A), (B, D), (E, C), (A, E)$
53. $(A, B), (B, C), (C, F), (F, E), (E, D), (D, A), (A, E), (E, C), (B, E)$ **55.** Euler path: *b, a, c, f, g, e, d*
57. Euler path: *e, c, d, f, m, n, i, g, h, k, j, b, a*   **59.** none   **61.** 0011   **65.** *a, e, f; e, c, b; c, a, d*
67. *a, d, f; b, f, e; d, f, c; f, c, a; c, a, d*   **69.** yes, cookies, ice cream, eclairs, pie, pudding; one
71. vertex *B; B, A; B, A, C; B, D; B, E*   **73.** yes; Bears, Vikings, Packers, Bucs, Lions; one
77. Step 1. Select a vertex *V* and a directed edge leaving *V*.
   Step 2. If the other vertex on the last chosen directed edge is not *V*, then choose an unused directed edge leaving this other vertex. Repeat step 2.
   Step 3. If all of the directed edges have been used, then stop (a directed Euler circuit has been constructed). Otherwise choose an unused directed edge leaving a vertex already visited, and give this previously visited vertex a temporary name *A*.
   Step 4. If the other vertex on the last chosen directed edge is not *A*, then choose an unused directed edge leaving this other vertex. Repeat step 4.
   Step 5. Insert these newly chosen directed edges at the vertex *A*. Go to step 3.

Exercises 3.5   (page 146)

1. 5; *S, D, G, H, I, T*   **3.** 3; *S, C, E, T*   **5.** 6; *S, D, H, E, F, J, T*
7. 9; *S, E, F, L, G, B, C, H, M, T*   **9.** 7; *S, A, B, F, G, J, O, T*
11. 6   **13.** *C*, 3; *E*, 4; *H*, 5; *D*, 5; *F*, 6; *I*, 6; *A*, 8; *G*, 7; *B*, 9. The shortest paths are *S, C, E, F, G, A* and *S, C, E, F, G, B*.   **15.** *C*, 3; *E*, 2; *H*, 1; *F*, 3; *D*, 5; *G*, 6; *B*, 6; *A*, 8. The shortest paths are *S, C, D, G, A* and *S, H, B*.
17. *A*, 5; *B*, 10; *C*, 4; *D*, 3; *E*, 5; *F*, 2; *G*, 4. The shortest paths are *S, F, G, A* and *S, F, G, B*.
19. *C*, 5; *F*, 2; *A*, 7; *D*, 6; *G*, 5; *H*, 6; *E*, 14; *B*, 11; *I*, 8. The shortest paths are *S, F, G, H, A* and *S, F, G, H, I, B*.
21. *S, E, F, K, L, G, A, C, H, M, T*   **23.** *S, F, A, C, D, E, T*   **25.** *P, K; P, H; P, K, O; P, N; P, L; P, K, M*
27. 1, 0, 4, 0; 1, 0, 4, 0   **29.** 1, 1, 6, 10; 0, 1, 2, 6   **31.** 0, 2, 1, 4; 1, 0, 2, 2   **33.** 1, 0, 2, 4; 0, 1, 1, 4

Chapter 4

Exercises 4.1   (page 161)   **1.** yes   **3.** no   **5.** no   **7.** yes   **9.** yes   **11.** 16
13. Connect Lincoln to each other town, using only 6 lines.   **15.** 12

17.   **19.** $n - 1$ **21.** $\dfrac{n(n-1)}{2}$ **25.** 1, 2   **27.** yes

29.   **31.** no

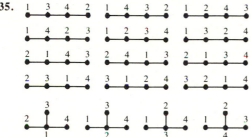

butane   isobutane

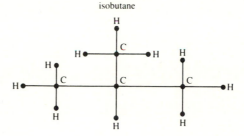

35.

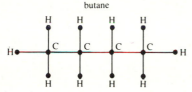

37. 3, 4; 4, 3; 2, 4; 4, 2; 2, 3; 3, 2; 1, 4; 4, 1; 1, 3; 3, 1; 1, 2; 2, 1; 1, 1; 2, 2; 3, 3; 4, 4   **39.** 2, 2, 1, 3, 3
41. 1, 5, 6, 5, 4, 5, 9
43.   **45.**

Exercises 4.2   (page 175)   **1.** yes   **3.** no   **5.** no   **7.** no
9. 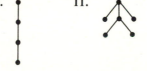   **11.**   **13.**

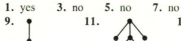

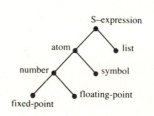

15. A vertex would have indegree greater than 1.
17. Step 1. Label R with a *.
   Step 2. If all vertices are labeled, stop. Otherwise find an unlabeled vertex U adjacent to a labeled vertex V. Direct the
         edge between U and V from V to U and label U with a *. Go to Step 2.

19. There is only one way.   **21.** *(i) A (ii) A, B, C, D, H, I (iii) J, K, L, E, F, G (iv) C (v) D, E, F (vi) H, I, J, K, L (vii) A, B, D*   **23.** *(i) E (ii) E, A, D, I, J (iii) B, K, G, F, H, C (iv) D (v)* none *(vi) G (vii) D, E*

25.   **27.**   **29.**

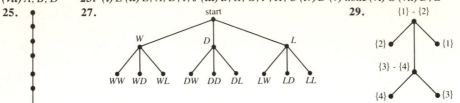

31.   **33.** 0   **35.** 4   **37.** 2   **39.**

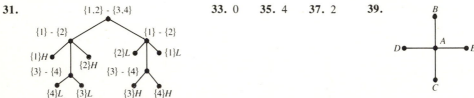

41.   **43.**   **45.**   **47.** *n*

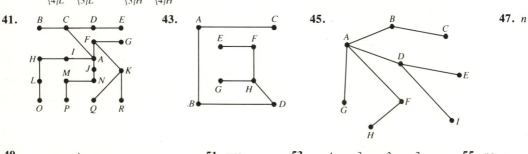

49.   **51.** yes   **53.**   **55.** no

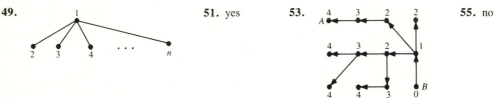

Exercises 4.3   (page 188)   **1.** *A, C, F, B, D, E, G, H*   **3.** *A, B, E, C, D, H, J, I, G, F*
5. *A, C, E, B, F, J, D, G, H, I*   **7.** *A, C, F, D, B, E, H, G*

9.   **11.**   **13.**   **15.**

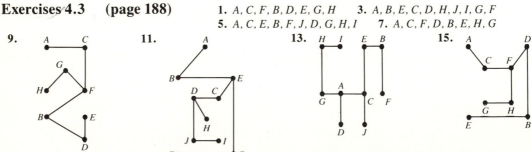

17. {*A, H*}, {*F, E*}, {*B, E*}, {*G, C*}, {*H, F*}   **19.** {*A, E*}, {*B, F*}, {*C, H*}, {*C, I*}   **21.** {*A, I*}, {*F, C*}

23. {A, D}, {A, E}, {C, G}   **25.** yes

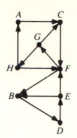

27. no   **29.** no   **31.** yes

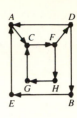

33. yes

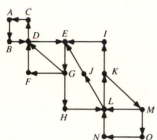

37. 2   **39.** $(n - 1)!$

43. The vertices are labeled by consecutive positive integers in the manner of the breadth-first search algorithm.

Exercises 4.4   (page 197)

1. a, c, g, f **3.** c, a, d, e, k, f, i, j **5.** g, f, c, a

7. k, e, f, i, j, d, c, a **9.** d, e, b, c **11.** m, j, g, h, e, n, a, b

13. {1, 5}, {5, 6}, {6, 4}, {4, 2}, {2, 7} {6, 3}   **15.**

17. b, c, d, e, k, f, i, j

19.

21. If the package cost 26 cents to mail, the greedy algorithm would use one 22–cent stamp and four 1–cent stamps. However, two 13–cent stamps would also do the job with fewer stamps.   **27.** $i, m, d, f, g, b, c, n, a$ **29.** $k, f, j, c, e, g, b, d, q, i, o$

31. b, k, e, f, i, c, j, d

Exercises 4.5   (page 211)

1.

3.

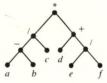

5.

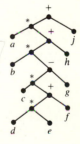

7. B   **9.** E   **11.** H   **13.** A, B, C **15.** A, B, D, F, C, E, G

17. $A, B, D, G, L, E, H, M, I, N, C, F, J, O, P, K, Q$ **19.** B, C, A **21.** F, D, B, G, E, C, A

23. $L, G, D, M, H, N, I, E, B, O, P, J, Q, K, F, C, A$ **25.** B, A, C **27.** D, F, B, A, E, G, C

29. $L, G, D, B, M, H, E, N, I, A, C, O, J, P, F, K, Q$ **31.** $+ * a b c$

33. $* / - a b c + d / e f$ **35.** $+ + * a + * b - * c + * d e f g h j$ **37.** $a b * c +$ **39.** $a b - c / d e f / + *$

41. $a b c d e * f + * g - * h + * j +$ **43.** 13   **45.** 13   **47.** -2 **49.** 39

51.

53.

55.

57.

59.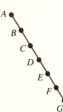

61. A
•

63.

67. List a vertex the last time it is passed as we follow the line counterclockwise around the binary tree.

69. As we follow the line counterclockwise around the binary tree, list a terminal vertex the first time it is passed and an internal vertex the first time it is passed if it has no left subtree, otherwise the second time it is passed.

Exercises 4.6 (page 227)

1. no **3.** no **5.** no **7.** $a = 1, b = 1, c = 1$

9.

11.

13.

15.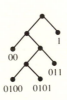

17.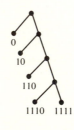

19. BATMAN **21.** TONTO **23.** GOGO **25.** THEHATS **27.** DOG **29.** QUIET

31.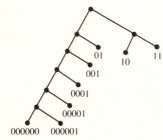

33.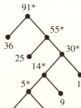

35. 277 **37.** 955 **39.** R: 0, I: 11, H: 101, V: 100

41. 1: 100, 2: 1111, 3: 1110, 4: 01001, 5: 0101, 6: 00, 7: 01000, 8: 011, 9: 110, 10: 101

49.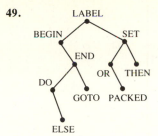

51.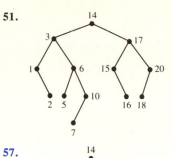

53.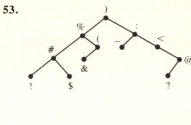

55.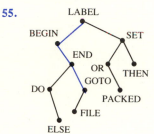

57.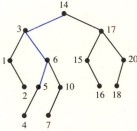

59.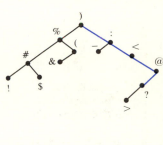

61. no

63.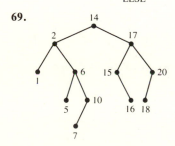

65.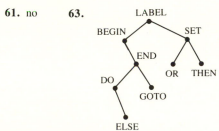

67.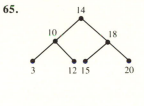

69.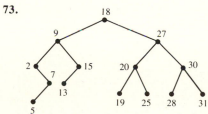

71.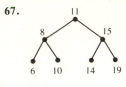

73.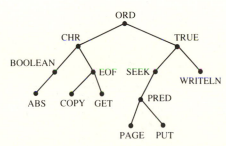

Chapter 5

Exercises 5.1 (page 238)

1. 2   **3.** 6   **5.** 0   **7.** yes   **9.** no   **11.** {3}
13. {1, 2, 3, 4, 5}   **15.** {1, 3, 4, 6}   **17.** $n!$ **19.** 0

21. Amy, Burt, Dan, and Edsel like only 3 flavors among them.
25. Timmack, Alfors, Tang, Ramirez, Washington, Jelinek, Rupp

Exercises 5.2 (page 245)

1. {1, 3, 6, 8, 9, 11, 13}   **3.** no   **5.** no   **7.** {{1, 2}, {3, 4}, {5, 6},
{7, 8}, {9, 10}, {12, 13}}, {{1, 4}, {3, 5}, {6, 7}}, {{1, 2}, {3, 4}}

9. {2, 4, 5, 7, 10, 12}, {1, 3, 6, 7}, {1, 2, 4}

11.
$$\begin{bmatrix} 1 & 0 & 0 & 1 & 0 & 0 \\ 1 & 0 & 1 & 0 & 0 & 0 \\ 0 & 1 & 1 & 0 & 0 & 1 \\ 1 & 0 & 1 & 1 & 0 & 0 \\ 0 & 0 & 0 & 1 & 1 & 0 \\ 0 & 0 & 1 & 0 & 0 & 0 \end{bmatrix}$$

13.
$$\begin{bmatrix} 1 & 1 & 1 & 0 & 0 & 0 \\ 1 & 1 & 0 & 1 & 0 & 0 \\ 0 & 1 & 1 & 0 & 1 & 0 \\ 1 & 0 & 0 & 1 & 0 & 1 \\ 0 & 0 & 0 & 1 & 1 & 1 \\ 0 & 0 & 1 & 0 & 1 & 1 \end{bmatrix}$$

15.
$$\begin{bmatrix} 1 & 1 & 0 & 0 & 0 & 0 \\ 0 & 0 & 1 & 0 & 0 & 0 \\ 1 & 1 & 0 & 1 & 0 & 0 \\ 0 & 0 & 0 & 0 & 1 & 0 \\ 0 & 1 & 0 & 0 & 0 & 1 \\ 0 & 0 & 1 & 0 & 0 & 0 \\ 0 & 0 & 0 & 0 & 1 & 0 \end{bmatrix}$$

17.
$$\begin{bmatrix} * & 0 & 0 & 0 & 0 & 0 \\ 0 & 0 & * & 0 & 0 & 0 \\ 0 & * & 0 & 0 & 0 & 0 \\ 0 & 0 & 0 & * & 0 & 0 \\ 0 & 0 & 0 & 0 & * & 0 \\ 0 & 0 & 0 & 0 & 0 & 0 \end{bmatrix}$$
$$\begin{bmatrix} 0 & 0 & 0 & * \\ * & 0 & 0 & 0 \\ 0 & * & 0 & 0 \\ 0 & 0 & 0 & 0 \\ 0 & 0 & 0 & 0 \end{bmatrix}$$ main diagonal

19. row named 5, columns named 2, 6, 8, 10; rows named 3, 5, column named 8; all rows

21.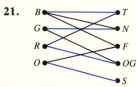

	T	N	F	OG	S
B	1*	1	1	0	0
G	0	1*	0	1	0
R	1	0	0	1*	0
O	0	0	1	0	1*

23.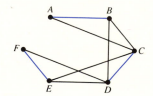

25.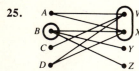

	W	X	Y	Z
A	1	1	0	0
B	0	1	1	1
C	1	0	0	0
D	1	1	0	0

Exercises 5.3 (page 255)

1.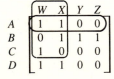
$$\begin{bmatrix} 0 & 1* & 0 & 1 \\ 1* & 0 & 0 & 1 \\ 1 & 1 & 0 & 0 \\ 1 & 0 & 0 & 0 \\ 2 & 1 & LS & LS \end{bmatrix} \begin{matrix} DS \\ DS \\ \\ \\ \end{matrix}$$

3. 3A, 2A, 2D   **5.** 1D, 2A, 3C, 4B   **7.** 1D, 2E, 3B, 4A   **9.** 1D, 2A, 3B, 5C   **11.** 1B, 3A, 4D
13. {1, B}, {2, C}, {3, A}   **15.** {1, A}, {2, D}, {3, E}, {4, B}, {5, C}   **17.** B, C, A, D   **19.** W, Z, Y, X
21. carrot, banana, egg, apple   **23.** Constantine to 1, Egmont to 2, Fungo to 3, Drury to 4, Arabella to 5

Exercises 5.4   (page 264)   **1.** row 2, columns 2 and 4   **3.** row 3, columns 1, 3, 4   **5.** $\{2, A, C\}$
7. $\{B, C, D, E\}$ **9.** impossible   **11.** $\{1, 3, 5, 6\}$ **13.** 7 hours
15. $\{1, 4, 5, 6, 7, 8, B\}$

Exercises 5.5   (page 271)   **1.** 13   **3.** 13   **5.** 18   **7.** 11   **9.** 16   **11.** 28
13. Addams to Chicago, Hart to Las Vegas, Young to New York, Herriman to Los Angeles   **15.** The Hungarian algorithm must be applied to a square matrix.

Chapter 6

Exercises 6.1   (page 283)   **1.** a network with source A and sink E **3.** not a network because arc (C, B) has negative capacity   **5.** a network with source D and sink B
7. not a flow because 5 flows into D and 6 flows out of D **9.** a flow with value 3   **11.** not a flow because 2 flows into D and 3 flows out of D **13.** not a cut because vertex C is not in S or T **15.** cut with capacity 40
17. cut with capacity 34   **19.**

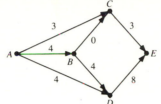

21.

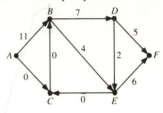

23.

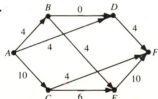

25. $S = \{A, B, C, D\}$ and $T = \{E\}$ **27.** $S = \{A, B, C\}$ and $T = \{D, E, F\}$

29. $S = \{A, C\}$ and $T = \{B, D, E, F\}$ **31.**

33. $F(X, Y) = 15$ and $F(Y, X) = 3$
35. For the flow in Exercise 10 take $X = \{D\}$, $Y_1 = \{A, B, C\}$, and $Y_2 = \{B, E, F\}$.

Exercises 6.2 (page 295)

1. 1   3. 2   5.

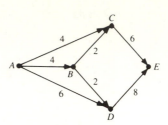

7.

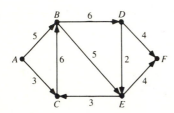

9. Increase the flow by 3 along the path A, C, E.
11. The given flow is maximal.
13. Increase the flow by 2 along the path A, D, B, E, F.
15. Increase the flow by 2 along the path A, B, D, C, F, E, G.
17. The given flow is maximal.

19.

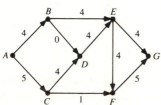

21.

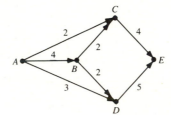

23.

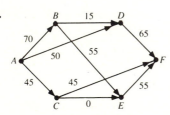

25.

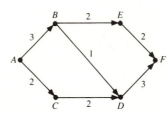

29.

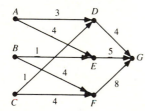

Exercises 6.3 (page 303)

1. 21   3. 28
5. $\{A, B, C, E\}$ and $\{D, F\}$ 7. $\{A, B, D\}$ and $\{C, E, F\}$
9. $\{A, B, C, D\}$ and $\{E\}$ 11. $\{A, C, F\}$ and $\{B, D, E, G\}$ 13.

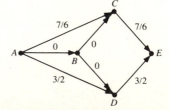

17. 2^{n-2} **19.** (s, A) and (F, t) **23.**

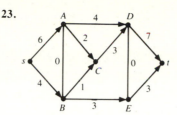

Exercises 6.4   (page 311)   **1.** bipartite; $V_1 = \{A, D, E\}$ and $V_2 = \{B, C, F\}$ **3.** not bipartite

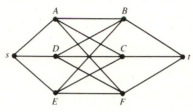

5. bipartite; $V_1 = \{A, D\}$ and $V_2 = \{B, C, E, F\}$ **7.** $\{(A, Y), (B, Z), (D, X)\}$ **9.** The given matching is maximal.

11. $\{(A, 1), (C, 3), (D, 2)\}$

13. $\{(a, E), (b, C), (c, B), (d, D), (e, A)\}$

15. A team cannot be assembled.

17. Only Professors Abel, Crittenden, and Forcade can teach courses 1, 3, 4, and 6.   Hence, it is impossible to schedule these four courses without assigning more than one course per professor.

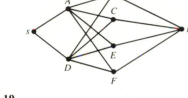

19.

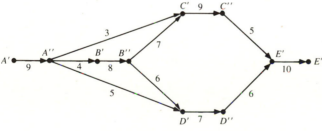

21.

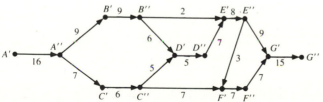

23.

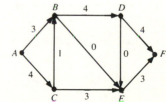

Chapter 7

Exercises 7.1 **(page 319)** **1.** 10 **3.** 56 **5.** 6
7. 84 **9.** 45 **11.** n

13. The $n = 6$ row is: 1, 6, 15, 20, 15, 6, 1. **15.** $(x + y)^6 = x^6 + 6x^5y + 15x^4y^2 + 20x^3y^3 + 15x^2y^4 + 6xy^5 + y^6$
17. $(3x - y)^4 = 81x^4 - 108x^3y + 54x^2y^2 - 12xy^3 + y^4$
19. 35 **21.** 252 **23.** 15

Exercises 7.2 **(page 325)** **1.** 13 **3.** 5 **5.** 14
7. 45 **9.** 32 **11.** 2^n

13. 168 **15. (a)** 216 **(b)** 120 **17. (a)** 720 **(b)** 144 **(c)** 36 **(d)** 48
19. 3219 **21.** 70 **23.** 48
25. 36,504 **27. (a)** 720 **(b)** 360 **(c)** 240 **(d)** 576 **29.** 112

Exercises 7.3 **(page 330)** **1.** 20 **3.** 10 **5.** 12
7. 15,120 **9.** 5040 **11.** n

13. 24 **15.** 360 **17.** 286 **19.** 210 **21.** 120 **23.** 84
25. 200 **27.** 10,584 **29. (a)** 495 **(b)** 5 **(c)** 72 **(d)** 54

Exercises 7.4 **(page 336)** **1.** 1260 **3.** 210 **5.** 45
7. 1820 **9.** 63,063,000 **11.** 35

13. 165 **15.** 10 **17.** 140 **19.** 3,864,861 **21.** 4200 **23.** 462
25. 1050 **27.** 165 **29.** 220

Exercises 7.5 **(page 341)** **1.** $\dfrac{5}{6}$ **3.** $\dfrac{1}{32}$ **5.** $\dfrac{1}{18}$

7. $\dfrac{5}{16}$ **9.** $\dfrac{4}{33}$ **11.** $\dfrac{1}{210}$

13. $\dfrac{63}{125}$ **15.** $\dfrac{1}{105}$ **17.** $\dfrac{1}{120}$ **19.** $\dfrac{5}{68}$ **21.** $\dfrac{175}{429}$ **23.** $\dfrac{3}{28}$ **25.** $\dfrac{140}{2187}$ **27.** $\dfrac{5}{12}$

Exercises 7.6 **(page 348)** **1.** 40 **3.** 160 **5.** 55
7. 27 **9.** $C(r, s)$ **11.** $k(k - 1)(k - 2)$

13. 61 **15.** 685,464 **17.** 35

19. $1 - \dfrac{5C(48, 13) - 10C(44, 13) + 10C(44, 13) - 5C(36, 13) + C(32, 13)}{C(52, 13)}$

23. $n!\left[\dfrac{1}{0!} - \dfrac{1}{1!} + \dfrac{1}{2!} - \dfrac{1}{3!} + \ldots + \dfrac{(-1)^n}{n!}\right]$

25. $S(n, 0) = 0, S(n, 1) = 1, S(n, n) = 1, S(n, n - 1) = C(n, 2)$, and $S(n, 2) = 2^{n-1} - 1$

Exercises 7.7 **(page 353)** **1.** $p << q$ **3.** $p << q$ **5.** $p >> q$
7. (2, 1, 4, 3, 6, 5) **9.** (2, 1, 5, 3, 4, 6) **11.** (5, 6, 4, 1, 2, 3)

13. none **15.** (5, 3, 1, 2, 4, 6) **17.** (6, 4, 1, 2, 3, 5)
19. (1, 2, 3, 4); (1, 2, 4, 3); (1, 3, 2, 4); (1, 3, 4, 2); (1, 4, 2, 3); (1, 4, 3, 2); (2, 1, 3, 4); (2, 1, 4, 3); (2, 3, 1, 4);
(2, 3, 4, 1); (2, 4, 1, 3); (2, 4, 3, 1); (3, 1, 2, 4); (3, 1, 4, 2); (3, 2, 1, 4); (3, 2, 4, 1); (3, 4, 1, 2); (3, 4, 2, 1); (4, 1, 2, 3);
(4, 1, 3, 2); (4, 2, 1, 3); (4, 2, 3, 1); (4, 3, 1, 2); (4, 3, 2, 1)

Chapter 8

Exercises 8.1 (page 362)

1. $s_2 = 10, s_3 = 39, s_4 = 133, s_5 = 424$ **3.** $s_2 = 9, s_3 = 65, s_4 = 457, s_5 = 3201$ **5.** $s_2 = 7, s_3 = 15, s_4 = 31, s_5 = 63$ **7.** $s_3 = 9, s_4 = 33$ **9.** $s_3 = 1, s_4 = 1$ **11.** $s_n = 2s_{n-1}$ for $n \geq 1$ and $s_0 = 1$ **13.** $a_n = 2.5a_{n-1}$ for $n \geq 2$ and $a_1 = 500; a_{12} \approx 11,920,929$ **15.** $v_n = 1.06v_{n-1} + \$100$ for $n \geq 2$ and $v_1 = \$2000$ **17.** $s_n = s_{n-1} + n$ for $n \geq 1$ and $s_0 = 1$ **19.** $c_n = c_{n-8} + 1$ for $n \geq 12, c_n = 0$ for $0 \leq n < 4, c_n = 1$ for $4 \leq n < 8$, and $c_n = 2$ for $8 \leq n < 12$ **21.** $D_{n+1} = n(D_{n-1} + D_n)$ for $n > 2, D_1 = 0$, and $D_2 = 1$ **23.** $r_n = 0.99r_{n-1}$ for $n \geq 1$ and the initial amount is $r_0; r_4 \approx 0.96r_0$

Exercises 8.2 (page 366)

1. $s_1 = 1, s_2 = 4, s_3 = 16, s_4 = 64, s_5 = 256$ **3.** $s_1 = -1, s_2 = 4, s_3 = -1, s_4 = 4, s_5 = -1$ **5.** $s_1 = 5, s_2 = 9, s_3 = 13, s_4 = 17, s_5 = 21$ **7.** $s_1 = -1, s_2 = 1, s_3 = -1, s_4 = 1, s_5 = -1$ **9.** $s_1 = 2, s_2 = 3, s_3 = 5, s_4 = 9, s_5 = 17$

11. $s_n = \frac{8}{3}(4^{n-1}) - \frac{5}{3}$ for $n \geq 1$ **13.** $s_n = -(-3)^{n-1}$ for $n \geq 1$ **15.** $s_n = 2^{n-1} + 2$ for $n \geq 1$

17. $s_n = \frac{4}{3}(7^{n-1}) - \frac{1}{3}$ for $n \geq 1$ **19.** $s_n = 500(2.5)^{n-1}$ for $n \geq 1$

21. $v_n = \frac{11,000}{3}(1.06^{n-1}) - \frac{5000}{3}$ **23.** $u_n = 1.01u_{n-1} - \$800$ for $n \geq 1$ and $u_1 = \$48,000$; $u_n = -\$32,000(1.01^{n-1}) + \$80,000; u_{12} \approx \$44,298.62, u_{60} \approx \$22,441.28$, and $u_{90} \approx \$2419.55$.

Exercises 8.3 (page 371)

1. $s_1 = 1, s_2 = 2, s_3 = 3, s_4 = 5, s_5 = 8$ **3.** $s_1 = 4, s_2 = 8, s_3 = 16, s_4 = 32, s_5 = 64$ **5.** $s_1 = 1, s_2 = 2, s_3 = -3, s_4 = 8, s_5 = -19$

7. $s_n = 2^{n+1}$ for $n \geq 1$ **9.** $s_n = 0.5(-1)^n + 0.1(5)^n$ for $n \geq 1$ **11.** $s_n = \frac{-2}{9}(-3)^n + \frac{1}{27}(9)^n$ **13.** 1, 1, 2, 3, 5, 8, 13, 21, 34, 55, 89, 144, 233, 377, 610, 987, 1597, 2584, 4181, and 6765 **15.** $\frac{F_2}{F_1} = 1, \frac{F_3}{F_2} = 2, \frac{F_4}{F_3} = 1.5, \frac{F_5}{F_4} = \frac{5}{3} \approx 1.667, \frac{F_6}{F_5} = 1.6, \frac{F_7}{F_6} = 1.625, \frac{1 + \sqrt{5}}{2} \approx 1.618$ **17.** The area of the original square is 64, and the area of the rectangle is 65. The difference in areas results from the fact that the pieces of the rectangle do not fit together to form a line segment along the diagonal. Rather, they form a small parallelogram of area 1.

19. $d_0 = 0, d_{k+m} = 1$, and $d_n = \frac{2}{3}d_{n+1} + \frac{1}{3}d_{n-1}$ for $1 \leq n \leq k + m - 1$

Exercises 8.4 (page 380)

1. $\lfloor 243 \rfloor = 243$ **3.** $\left\lfloor \frac{34}{7} \right\rfloor = 4$ **5.** $\lfloor 0.871 \rfloor = 0$

7. $\left\lfloor \frac{-(-34 + 2)}{2} \right\rfloor = \lfloor 16 \rfloor = 16$

9.

				J	I
13	56	87	42	3	1
13	56	42	87	2	
13	42	56	87	1	
13	42	56	87	3	2
13	42	56	87	2	
13	42	56	87	3	3
13	42	56	87		

11. There are no exchanges in the bubble sort for this list, as it is in order at the start.

13.

					J	I
3	4	3	(5)	(6)	4	1
3	4	(3)	(5)	6	3	
3	(4)	(3)	5	6	2	
(3)	(3)	4	5	6	1	
3	3	4	(5)	(6)	4	2
3	3	(4)	(5)	6	3	
3	(3)	(4)	5	6	2	
3	3	4	(5)	(6)	4	3
3	3	(4)	(5)	6	3	
3	3	4	(5)	(6)	4	4
3	3	4	5	6		

15.

					J	I
4	3	2	(5)	(2)	4	1
4	3	(2)	(2)	5	3	
4	(3)	(2)	2	5	2	
(4)	(2)	3	2	5	1	
2	4	3	(2)	(5)	4	2
2	4	(3)	(2)	5	3	
2	(4)	(2)	3	5	2	
2	2	4	(3)	(5)	4	3
2	2	(4)	(3)	5	3	
2	2	3	(4)	(5)	4	4
2	2	3	4	5		

17.

L	R	I	s_I	Is $s_I = t$?
1	100	$\left\lfloor \dfrac{101}{2} \right\rfloor = 50$	50	No, greater
51	100	$\left\lfloor \dfrac{151}{2} \right\rfloor = 75$	75	No, greater
76	100	$\left\lfloor \dfrac{176}{2} \right\rfloor = 88$	88	No, less
76	87	$\left\lfloor \dfrac{163}{2} \right\rfloor = 81$	81	No, greater
82	87	$\left\lfloor \dfrac{169}{2} \right\rfloor = 84$	84	No, less
82	83	$\left\lfloor \dfrac{165}{2} \right\rfloor = 82$	82	No, greater
83	83	83	83	Yes

19.

L	R	I	s_I	Is $s_I = t$?
1	300	$\left\lfloor \dfrac{301}{2} \right\rfloor = 150$	450	No, less
1	149	$\left\lfloor \dfrac{150}{2} \right\rfloor = 75$	225	No, greater
76	149	$\left\lfloor \dfrac{225}{2} \right\rfloor = 112$	336	No, greater
113	149	$\left\lfloor \dfrac{262}{2} \right\rfloor = 131$	393	No, greater
132	149	$\left\lfloor \dfrac{281}{2} \right\rfloor = 140$	420	No, less
132	139	$\left\lfloor \dfrac{271}{2} \right\rfloor = 135$	405	No, less
132	134	$\left\lfloor \dfrac{266}{2} \right\rfloor = 133$	399	No, greater
134	134	134	402	No, less
134	133			

Since $R < L$, the target t is not in the list.

21.

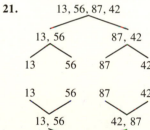

23.

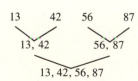

25.

34, 81, 46, 2, 53, 5, 4, 8, 26, 1, 0, 45, 26, 5, 48, 35

34, 81, 46, 2, 53, 5, 4, 8   26, 1, 0, 45, 26, 5, 48, 35

34, 81, 46, 2   53, 5, 4, 8   26, 1, 0, 45   26, 5, 48, 35

34, 81   46, 2   53, 5   4, 8   26, 1   0, 45   26, 5   48, 35

34   81   46   2   53   5   4   8   26   1   0   45   26   5   48   35

34   81   46   2   53   5   4   8   26   1   0   45   26   5   48   35

34, 81   2, 46   5, 53   4, 8   1, 26   0, 45   5, 26   35, 48

2, 34, 46, 81   4, 5, 8, 53   0, 1, 26, 45   5, 26, 35, 48

2, 4, 5, 8, 34, 46, 53, 81   0, 1, 5, 26, 26, 35, 45, 48

0, 1, 2, 4, 5, 5, 8, 26, 26, 34, 35, 45, 46, 48, 53, 81

31. There are at least two. At least one fewer comparison than usual is needed.   **33.** The second algorithm is more efficient for every positive integer.   **35.** $C_n = 2C_{n/2} + n - 1$ for $n = 2^k$ with $k \geq 0$ and $C_1 = 0$

Chapter 9

Exercises 9.1   (page 390)   **1.** $(x \wedge y) \vee x$ **3.** $((x' \vee y) \wedge x)'$ **5.** $(x'' \vee y') \wedge x'$
7. $(x' \wedge (y' \wedge x))'$

9.

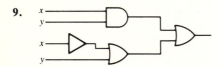

11.

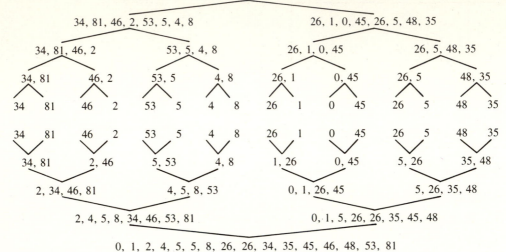

15. 0   **17.** 1

13.

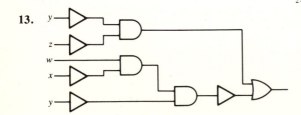

19.

x	y	Output
0	0	1
0	1	1
1	0	0
1	1	1

21.

x	y	z	Output
0	0	0	0
0	0	1	0
0	1	0	0
0	1	1	0
1	0	0	0
1	0	1	0
1	1	0	1
1	1	1	0

23.

x	y	Output
0	0	0
0	1	0
1	0	0
1	1	1

25.

x	y	Output
0	0	0
0	1	1
1	0	1
1	1	1

27.

x	y	z	Output
0	0	0	1
0	0	1	1
0	1	0	0
0	1	1	0
1	0	0	1
1	0	1	1
1	1	0	1
1	1	1	1

29. equivalent   **31.** equivalent   **33.** equivalent   **35.** not equivalent   **37.** equivalent   **39.** not equivalent
41. not equivalent
43.

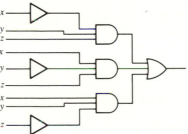

47. 1, undefined

Exercises 9.2   (page 399)

9. (c), (f), (g)   **11.** (c), (f), (g)   **13.** (j), (i), (e)   **15.** (j), (b)
17. (c), (a), (b), (b), (b), (b), (e), (e), (b), (b), (a), (c)   **19.** When $x = y = 0, z = 1$, the first is 0 and the second 1.   **21.** When $x = z = 0, y = 1$, the first is 0 and the second 1.
23. $(x' \wedge y) \vee (x \wedge y')$ **25.** $(x' \wedge y \wedge z) \vee (x \wedge y' \wedge z) \vee (x \wedge y \wedge z')$

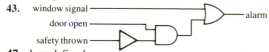

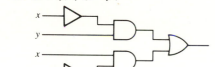

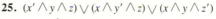

27. $(x' \wedge y \wedge z) \vee (x \wedge y' \wedge z) \vee (x \wedge y \wedge z)$

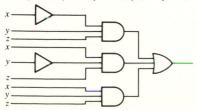

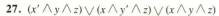

29. 12   **31.** 9   **33.** 14   **35.**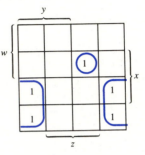

37. $(a \wedge b \wedge c') \vee ((b \vee c) \wedge d') \vee (a \wedge d)$ **39.** all

Exercises 9.3   (page 413)

1. $(x' \wedge y') \vee (x' \wedge y) \vee (x \wedge y)$
3. $(x' \wedge y' \wedge z') \vee (x' \wedge y' \wedge z) \vee (x' \wedge y \wedge z') \vee (x' \wedge y \wedge z)$
5. $(w' \wedge x' \wedge y' \wedge z') \vee (w' \wedge x' \wedge y \wedge z') \vee (w' \wedge x \wedge y' \wedge z') \vee (w' \wedge x \wedge y \wedge z') \vee (w \wedge x \wedge y' \wedge z)$
7. $x \vee y'$ **9.** $y \vee x' \vee (y' \wedge z')$ **11.** $(x \wedge z) \vee (w' \wedge x) \vee (w \wedge x' \wedge y') \vee (w' \wedge y \wedge z')$

13.   **15.** 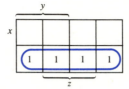   **17.**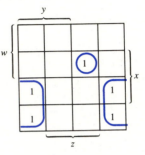

19. $x' \vee y$ **21.** x' **23.** $(w' \wedge z') \vee (w \wedge x \wedge y' \wedge z)$

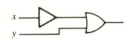  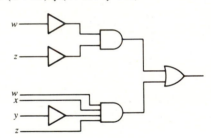

25. $(x' \wedge z) \vee (y' \wedge z)$ **27.** $(x' \wedge z) \vee (x \wedge y')$ **29.** $x \vee (y' \wedge z')$ **31.** $(w \wedge x) \vee (y' \wedge z')$
33. 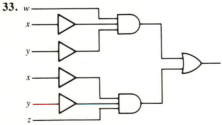   **35.** 32

Exercises 9.4   (page 423)

1. 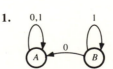   **3.**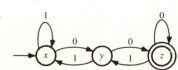

5.

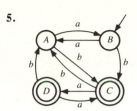

7.

	A	B	C
x	B	C	C
y	C	C	A

Initial state B,
accepting state A

9.

	1	2	3
a	2	3	1
b	3	1	2
c	1	2	3

Accepting state 2

11. y

13. A   **15.** yes   **17.** no   **19.**

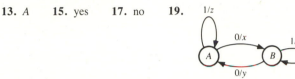

21.

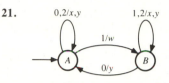

23.

	1	2	1	2
1	2	2	a	a
2	2	1	c	b
3	1	2	c	a

25.

	A	B	C	A	B	C
1	B	B	C	x	y	y
2	C	C	A	z	x	z
3	C	C	C	y	z	z

27. ywywwxx   **29.** yzzyyz   **31.** All strings containing a 1
33. All strings containing exactly n 1's, where $n \equiv 1 \pmod 3$
35.

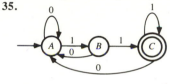

37.

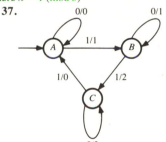

Appendix

Exercises A.1   (page 434)

1. false statement   **3.** true statement   **5.** not a statement
7. false statement   **9.** not a statement   **11.** true statement
13. $4 + 5 \neq 9$ **15.** Miss Piggy is married to Kermit the Frog.   **17.** Some birds cannot fly.
19. No man weighs 400 pounds.   **21.** No students do not pass calculus. (All students pass calculus.)
23. Someone does not enjoy cherry pie.   **25.** (a) One is an even integer and nine is a positive integer. (false)
(b) One is an even integer or nine is a positive integer. (true)
27. (a) The Atlantic is an ocean and the Nile is a river. (true) (b) The Atlantic is an ocean or the Nile is a river. (true)
29. (a) Birds have four legs and rabbits have wings. (false) (b) Birds have four legs or rabbits have wings. (false)
31. (a) Flutes are wind instruments and timpani are string instruments. (false) (b) Flutes are wind instruments or timpani
are string instruments. (true)   **33.** (a) If I go to the movies, then this is Friday. (b) If this isn't Friday, then I won't go to
the movies. (c) If I don't go to the movies, then this isn't Friday.   **35.** (a) If Hart runs for President, then he won't run
for the Senate. (b) If Hart runs for the Senate, then he won't run for President. (c) If Hart doesn't run for President, then he
is running for the Senate.

Exercises A.2 (page 438)

Note: only the last columns of truth tables are given.

1.

p	q	$(p \lor q) \land [\sim(p \land q)]$
T	T	F
T	F	T
F	T	T
F	F	F

3.

p	q	$(p \lor q) \to (\sim p \land q)$
T	T	F
T	F	F
F	T	T
F	F	T

5.

p	q	r	$p \to (\sim q \lor r)$
T	T	T	T
T	T	F	T
T	F	T	T
T	F	F	T
F	T	T	T
F	T	F	F
F	F	T	T
F	F	F	F

7.

p	q	r	$(\sim q \land r) \leftrightarrow (\sim p \lor q)$
T	T	T	F
T	T	F	F
T	F	T	F
T	F	F	T
F	T	T	F
F	T	F	F
F	F	T	T
F	F	F	F

9.

p	q	r	$[(p \lor q) \land r] \to [(p \land r) \lor q]$
T	T	T	T
T	T	F	T
T	F	T	T
T	F	F	T
F	T	T	T
F	T	F	T
F	F	T	T
F	F	F	T

17.

p	$\sim p$	$\sim(\sim p)$
T	F	T
F	T	F

19. If the truth table is arranged as in Exercise 1 above, then the column corresponding to the given statements is: $F, T, F,$ F. **21.** If the truth table is arranged as in Exercise 5 above, then the column corresponding to the given statements is: T, F, T, T, T, T, T, T. **23.** If the truth table is arranged as in Exercise 5 above, then the column corresponding to the given statements is: T, F, T, F, T, F, T, T.

25.

p	q	$p \land q$	$q \land p$	$p \lor q$	$q \lor p$
T	T	T	T	T	T
T	F	F	F	T	T
F	T	F	F	T	T
F	F	F	F	F	F

27. (e) If the truth table is arranged as in Exercise 5 above, then the column corresponding to both statements is: $T, T, T, T,$ T, F, F, F. **(f)** If the truth table is arranged as in Exercise 5 above, then the column corresponding to both statements is: T, T, T, F, F, F, F, F. **29.** If the truth table is arranged as in Exercise 1 above, then the column corresponding to both statements is T, F, T, T.

33. (a)

p	q	$p \veebar q$
T	T	F
T	F	T
F	T	T
F	F	F

(b) If the truth table is arranged as in part (a) above, then the column corresponding to both statements is F, T, T, F.

Exercises A.3   (page 444)

1.

p	q	$\sim q$	$p \wedge (\sim q)$	$\sim[p \wedge (\sim q)]$	$p \rightarrow q$
T	T	F	F	T	T
T	F	T	T	F	F
F	T	F	F	T	T
F	F	T	F	T	T

13. The statement is false. For example, $3 + 5 = 8$.   **15.** The statement is false. For example, if $a = 3$, $b = 2$, and $c = 0$, then $ac = bc$, but $a \neq b$.   **17.** The statement is false. For example, if $x = 4$ and $y = 9$, then 6 divides xy, but 6 does not divide either x or y.   **19.** The statement is true.   **21.** The statement is true.   **23.** The statement is false. For example, if $n = 41$, then $n^2 + n + 41 = 41^2 + 41 + 41 = 41(43)$.

Index